TRIUMPH
TR 2 & TR3
SERVICE INSTRUCTION MANUAL

TRIUMPH TR2
Incorporating Supplement for TR3 Model

Third Edition

Issued by
SERVICE DIVISION, STANDARD – TRIUMPH SALES LTD.
COVENTRY, ENGLAND

FOREWORD

This Manual has been prepared with a view to assisting Standard Distributors and Dealers, at Home and Overseas, to give an efficient repair and maintenance service to owners of this Model.

The book is divided into seventeen sections, which are separately indexed and indicated alphabetically. These sections deal with the main components, equipment, specialised tools and general data.

Dimensions and working clearances, together with other useful data, are summarised at the beginning of various sections with a view to facilitating reference by repairers.

The Manual covers the specification of this Model existing at the time of printing. Revised editions or supplements will be made available as developments are considered to justify such issues. In the meantime all our Agents are kept fully up-to-date on Service matters by the monthly issue of Service Information Sheets.

Although this Manual is primarily intended for the use and guidance of Standard Distributors and Dealers and other members of the Motor Trade, owners of this Model can purchase copies through their local Standard Distributor, but such orders will not be accepted direct by Standard-Triumph Sales Ltd.

CONTENTS

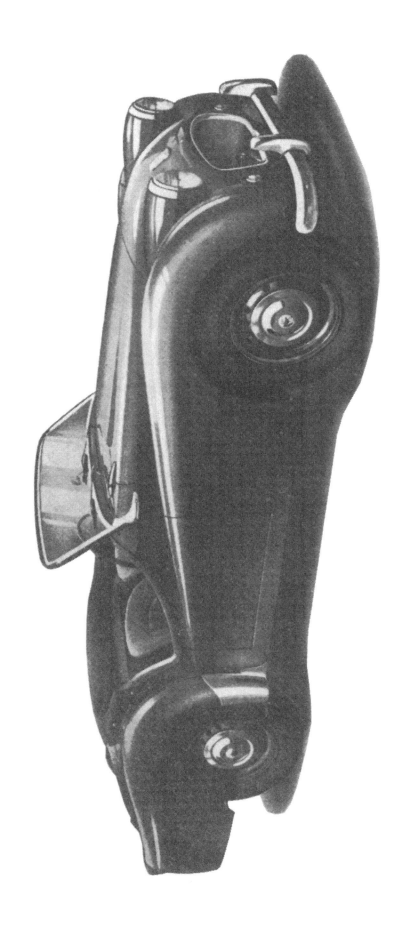

Service Instruction Manual

GENERAL DATA

SECTION A

GENERAL DATA

INDEX

ILLUSTRATIONS

GENERAL DATA

GENERAL DATA

Summaries of dimensions and tolerances, relative to various components are given at the commencement of the respective sections to which they refer. Whilst data given, in some instances, in this section appears elsewhere in the body of this manual, such information being frequently required, it is considered desirable that it should be summarised in this section for easy reference.

For the convenience of overseas readers, a table of metric equivalents is included in this section.

CHASSIS SPECIFICATION

Engine Details

Type	O.H.V. Push Rod Operated.
Bore of Cylinder	3.268″ (83 mm.)
Stroke of Crank	3.622″ (92 mm.)
Cubic Capacity (Swept Volume)	121.5 cu ins. (1,991 c.cs.)
Compression Ratio	8.5
Firing Order	1, 3, 4, 2
Compression Pressure (With three Sparking Plugs fitted and compression gauge in fourth cylinder engine warm, throttle set at tick over, using 20 SAE oil and operating the starter)	Average reading 120 lbs. per sq. in. (8.4 kgs. per sq. cm.)
Sparking Plug Make and Type	Champion No. L10S High speed work No. L11S.
Sparking Plug Reach	½″ (12.700 mm.)
Sparking Plug Gap	.032″ (.8 mm.)
Distributor	Lucas DM2 P.4
Distributor Break Gap	.015″ (.4 mm.)
Ignition Setting (Full Retard)	4° B.T.D.C. (Based on the use of fuel with a minimum Octane value of 80).
Vacuum Advance	Basic setting 4 divisions.
Inlet Rocker Clearance	Touring .010″ (.25 mm.) High Speed Motoring .013″ (.33 mm.)
Exhaust Rocker Clearance	Touring .012″ (.30 mm.) High Speed Motoring .013″ (.33 mm.)

The above measurements are based on a cold engine.

Crankshaft	Three journal molybdenum manganese steel stamping with integral balance weights.
Crankshaft Bearings	Vandervell bi-metal shell bearings.
Crankshaft Thrust	Four half semicircular white metal faced washers fitted in pairs either side of the centre bearing.
Connecting Rods	60-ton molybdenum manganese steel stamping with big end caps offset to camshaft side. Floating gudgeon pin secured by circlips.
Connecting Rod Bearings, Big End	Lead indium bronze bearings.
Small End	Clevite Bush.
Pistons	Aluminium alloy split skirt compensating type, graded F. G or H.
Piston Rings	All fitted above gudgeon pin.
Compression Rings	Cast iron, .062″ wide.
Scraper Ring	Cast iron, .156″ wide.
Camshaft	Special cast iron with four bearings and silent contour symmetrical cams. Driven by Duplex chain.

Camshaft Bearings ...... Front Bearing—cast iron sleeve ; 2nd, 3rd and 4th direct in crankcase. After Engine No. TS 9095E engines will be fitted with replaceable Vandervell shell bearings, See TR3 Supplement Engine Section "B".

Lubricating System ...... Wet Sump. Capacity 11 pints.

Oil Pump ...... Hobourn Eaton high capacity double eccentric rotor. Feed to main bearings, big end bearings and all camshaft bearings under pressure.

Oil Pressure ...... 70 lbs. sq. in. at 2,000 r.p.m. (4.9 kg. sq. cm.)

Oil Cleaner ...... Purolator by-pass flow system with replaceable cartridge.

Carburettors ...... Twin S.U. H.4. Standard needles FV. For high speed motoring G.C. needles.

Valve Timing ...... With valve rocker clearance set at .015″ (.38 mm.) Inlet Valve opens at 15° B.T.D.C. Exhaust Valve closes at 15° A.T.D.C. 15° is equivalent to .081″ piston travel or 1.5″ (3.81 cms.), measured round the flywheel adjacent to the starter teeth. Dims. on fan pulley = .72″.

Cooling System ...... Thermostatically controlled.

Pressurised Radiator Pressure release at 3¼—4¼ lbs.

Radiator Temperature Normal running should not exceed 185°F. (85°C.).

Capacity of Cooling System .... 13 pints (7.4 litres).
With Heater ...... 14 pints (8 litres).

Thermostat .... Commences to open at 150°F. (70°C.). Fully open at 197°F. (92°C.).

Frost Precautions ...... With "Smith's Bluecol" anti-freeze mixture. Other brands as recommended by their manufacturers.

Degrees of Frost (Fahrenheit)		
15°	25°	35°

Proportion		
10%	15%	20%

Amount of "Bluecol" (Pints)		
1.5	2.5	3

Piston Speed .... 2,850 ft./min. at 4,800 r.p.m. (This speed is equivalent to 100 m.p.h. in "Normal" top gear.)

Flywheel .... Cast Iron with induction hardened shrunk-on steel starter ring gear.

Transmission

Clutch .... Borg and Beck 9″ single dry plate. Hydraulically operated. Ball bearing clutch throw out.

Gearbox .... Four forward ratios and reverse. Synchromesh on 2nd, 3rd and top forward ratios. Silent helical gears. Oil filler combined with dipstick.

Ratios

	Overdrive Top	Top	3rd	2nd	1st	Rev.
Gearbox	.82	1.00	1.325	2.00	3.38	4.28
Overall	3.03	3.7	4.9	7.4	12.5	15.8

Rear Axle .... Hypoid Bevel Gears. Taper roller bearings on differential and for Hypoid Pinion Shaft. Ball bearings for road wheels. Shim adjustment for Pinion and Crown Wheel adjustment.

Rear Axle Ratio	......	3.7. (37T × 10T).
Wheels	....	Steel Disc Type with chrome nave plates (wire wheels optional extra.).
Suspension	....	Coil springs for independent front suspension with telescopic dampers Wide semi-elliptic springs at rear, controlled by piston type dampers.
Brakes	....	Lockheed Hydraulic 10" × 2¼" front, 9" × 1¾" rear. (After Commission No. TS.5481 10" × 2¼" front and rear.) Two leading shoe type used on front wheels, leading and trailing shoe type on rear wheels. Alloy cast iron brake drums. Foot operation hydraulic on all four wheels. Hand operation mechanical on rear wheels only.
Steering	....	High Gear Cam and Lever type unit. Optional for use on right or left hand drive. 17" (431 mm.) steering wheel with three spoke spring type.
Battery	....	12 volt, 51 amp. hour capacity, located under bonnet.
Performance Data		B.H.P. (Road Setting): 90 at 4,800 r.p.m. Maximum torque : 1,400 lb./ins. at 3,000 r.p.m., equivalent to 145 lbs./sq. ins. B.M.E.P. (See also Fig. 1).

Maximum Speeds
(Touring Trim)

Top Gear	110 m.p.h.	175 km.p.h.
3rd Gear	75	120
2nd Gear	45	75
1st Gear	25	40

Engine R.P.M. at	10 m.p.h.	10 km.p.h.
Top Gear	500	310
3rd Gear	660	410
2nd Gear	1,000	620
1st Gear	1,680	1,050
Rev. Gear	2,130	1,325

Acceleration Two Up

Gear	Speed	Time
Top	20—40 M.P.H. (32—64 Km.P.H.)	9 secs.
	30—50 M.P.H. (48—80 Km.P.H.)	9 secs.
Through Gears	0—50 M.P.H. (0—80 Km.P.H.)	8 secs.
	0—60 M.P.H. (0—96 Km.P.H.)	12 secs.

Fuel Consumption

Petrol	....	26—32 m.p.g. (10.87—8.83 litres per 100 km.).
Oil	....	3,000 m.p.g. (1,100 km. per litre.)

Car Dimensions

Wheelbase	....	7' 4"	224 cms.
Track—Front	......	3' 9"	114 cms.
Rear	......	3' 9½"	116 cms.
Front wheel alignment	......	" Toe in " ⅛".	
Ground clearance (under axle)	......	6"	15.2 cms.
Turning Circle (between Kerbs)	......	32' 0"	9.75 metres

Overall Dimensions

Length	....	12' 7"	384 cms.
Width	....	4' 7½"	141 cms.
Height (unladen)			
Hood erect	......	4' 2"	127 cms.
Top of Screen	......	3' 10"	117 cms.
Hood down and Screen removed	......	3' 4"	102 cms.
Luggage Space	......	See page 5 of this section.	

3

GENERAL DATA

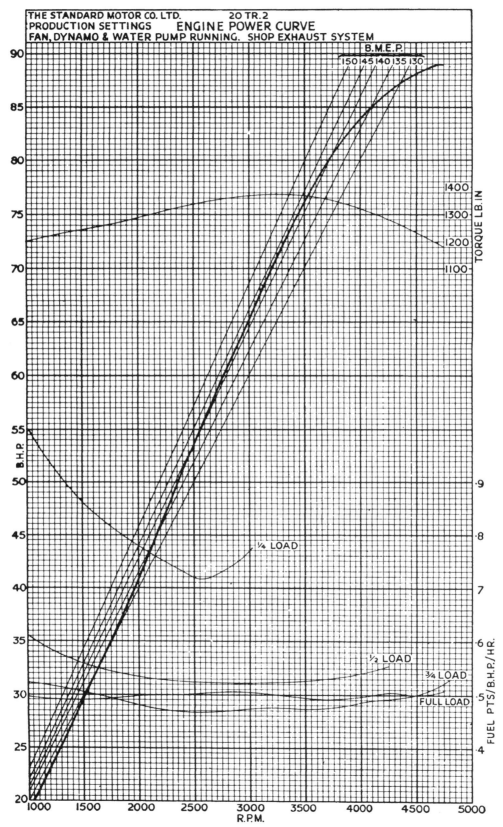

THE STANDARD MOTOR CO. LTD.   **20 TR.2**
PRODUCTION SETTINGS   **ENGINE POWER CURVE**
FAN, DYNAMO & WATER PUMP RUNNING. SHOP EXHAUST SYSTEM

Fig. 1   **Power Curve.**

GENERAL DATA

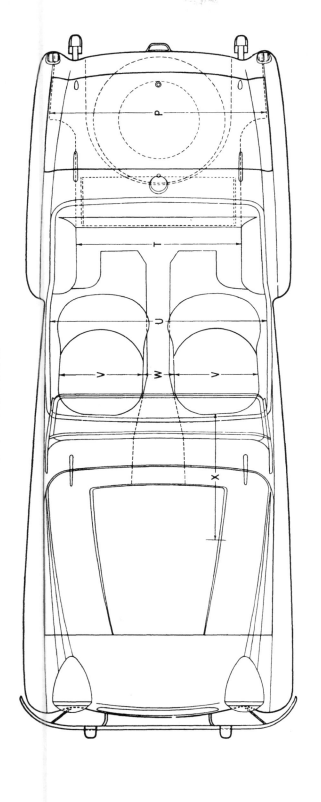

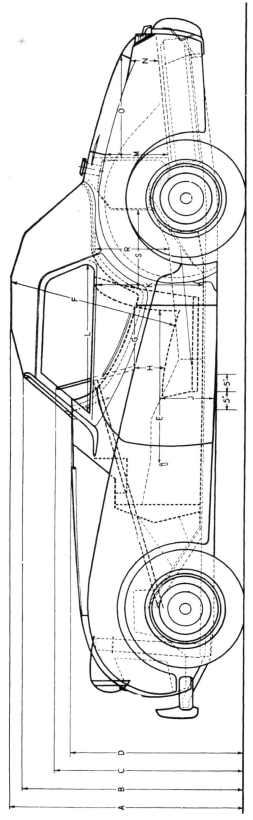

Fig. 2   Body Dimensions.

Body Dimensions (See Fig. 2).

A	Hood erect	50"	1,270 mm.
B	Top of Windscreen	46"	1,168 mm.
C	Top of Steering Wheel	40"	1,016 mm.
D	Road to Top of Scuttle	37"	940 mm.
E	Pedal to Squab	32½" to 42½"	825 to 1,079 mm.
F	Seat to Hood	36"	914 mm.
G	Squab to Steering Wheel	8" to 18"	203 to 457 mm.
H	Seat to Steering Wheel	6" app.	152 mm.
J	Seat to Floor	8½"	216 mm.
K	Squab Height	19"	482 mm.
L	Sidescreen Width	31⅛"	800 mm.

Boot

M	Height at Hinges	14½"	368 mm.
N	Height at Locks	7"	177 mm.
O	Length of Opening	Max. 26¾"	679 mm.
		Min. 18¾"	476 mm.
P	Width of Opening	Max. 45"	1,143 mm.
		Min. 41½"	1,054 mm.

Luggage Space Behind Seats

R	Depth of Space	Max. 23"	584 mm.
		Min. 13½"	342 mm.
S	Length of Space	Max. 20"	508 mm.
		Min. 15"	381 mm.
T	Width of Space	34½"	876 mm.
U	Width at Elbows	45"	1,143 mm.
V	Width of Seat	18"	457 mm.
W	Space between Seats	5½"	139 mm.
X	Passenger Leg Room	Max. 34"	863 mm.
		Min. 24"	609 mm.

Car Weight

Complete Car with
Tools, Fuel and
Water

| 18 cwts. | 3 qrs. | 7 lbs. | (955 kg.) |

Shipping Weight

| 17 cwts. | 2 qrs. | 21 lbs. | (902 kg.) |

Tyre Sizes and Pressure

Tyre Size .... 5.50″—15″.
Tyre Pressures
Front ...... 22 lbs./sq. in. 1.55 kgsq./.cm.
Rear ...... 24 lbs./sq. in. 1.7 kg./sq. cm.
Where cars are to be used for racing or special
high testing it is desirable that the Dunlop
Rubber Company be consulted for special tyres.

Water Capacity

| Cooling System | ...... | 13 pints | 7.4 litres |
| With Heater Fitted | ...... | 14 pints | 8 litres |

Oil Capacity

Engine—From Dry		11 pints	6.25 litres
Drain and			
Refill	......	10 pints	5.7 litres
Gearbox	....	1½ pints	.85 litres
„ with Overdrive		3½ pints	2.0 litres
From Dry			
Rear Axle	....	1½ pints	.85 litres

Petrol

Petrol Tank capacity   12½ galls. 57 litres

Body Specification

Two seater open sports, all weather equipment.
Detachable windscreen of Triplex safety glass.
Provision for fitting aero screens.
Steel body rust-proofed.
Front wings, rear wings and complete front
    panel are bolted on detachable type.
Door hinged at front.

SPIRE SPEED NUTS

1. GENERAL NOTES

These speed nuts are being used in increasing numbers on our products at the
present time in the place of nuts and lock
washers, as, in many instances, they simplify manufacturing processes and speed up
assembly work.

Although no particular skill is required in
their application, an elementary know-
ledge of the correct way to fit them is
necessary. It is not intended to refer to
each type of speed nut in detail and, in any
case, the types at present in use are likely to
be increased as production proceeds and the
desirability of their employment becomes
apparent.

2. DESCRIPTION

Spire speed nuts provide a compensating
thread lock. As the screw is tightened, the
two arched prongs move inwards to engage
and lock against the flanks of the screw
thread. The prongs compensate for tolerance variations in the screw. A spring
locking action is provided by compression
of the arch in both prongs and base as the
screw is tightened. The combined forces
of the threaded lock and that provided by
the spring prevent loosening due to vibration.

3. TIGHTENING TORQUES

Unlike normal threaded nuts, spire speed
nuts do not require a great deal of torque
when tightening the screw. The retention
of the screw by the nut depends on spring
tension alone. When tightening a screw
into a speed nut, only sufficient torque
should be used to produce the thread and
spring lock shown in Fig. 3. Excessive

Fig. 3   **Showing an Untightened Spire Nut on the
left of the illustration and on the other
side a fully tightened one.**

torque will only distort the ends of the
prongs and affect their spring tension and
may even break them.
Spire speed nuts can be used indefinitely
providing they have not been damaged by
over-tightening.

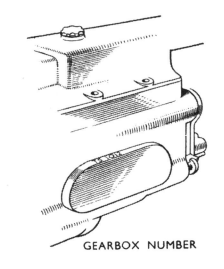

GEARBOX NUMBER

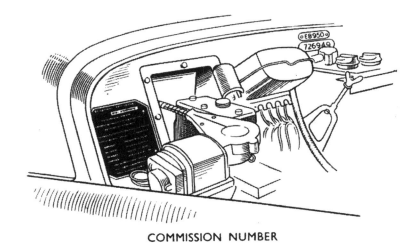

COMMISSION NUMBER

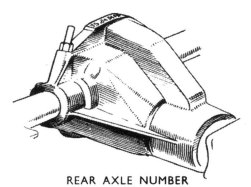

REAR AXLE NUMBER

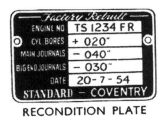

RECONDITION PLATE

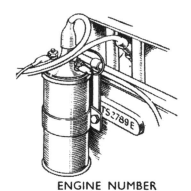

ENGINE NUMBER

Fig. 4   **Commission Numbers.**

4. COMMISSION NUMBER   (Chassis Number)

This number is found on a plate attached to the bulkhead under the bonnet at the right-hand side (see Fig. 4).   It has the prefix letters " TS."

NOTE : **It is important that this number is quoted when writing to the Company concerning the car and particularly when ordering spare parts.**

5. BODY NUMBER

This number is stamped on an oval plate affixed in the centre of the bulkhead under the bonnet (see Fig. 4).   It is a number with six numerals.

6. ENGINE NUMBER

This number is stamped on a boss situated on the cylinder block casting below No. 3 plug (see Fig. 4).   It has a prefix " TS " and a suffix letter " E."

Factory Rebuilt Engines

All factory rebuilt engines have the previous number erased and the new number stamped on a plate which is attached to the same boss (see Fig. 4).

This plate also gives information as to the size of the crank pins and journals, also the date on which the unit was rebuilt. This number has a prefix " TS " and a suffix " FR."

7. GEARBOX NUMBER

This number is stamped on the left-hand side of the box on the upper wall of the cast oval (see Fig. 4).   This number has the prefix " TS."

8. REAR AXLE NUMBER

This number is stamped on the upper rim of the flange to which the rear cover plate is attached (see Fig. 4).   This number has the prefix " TS."

RECOMMENDED LUBRICANTS

BRITISH ISLES

COMPONENT	SHELL	ESSO	DUCKHAM'S	VACUUM	WAKEFIELD	B.P. ENERGOL
ENGINE Summer	Shell X-100 30	Essolube 30	Duckham's NOL " Thirty "	Mobiloil A	Castrol XL	Energol S.A.E. 30
Winter	Shell X-100 20/20W	Essolube 20	Duckham's NOL " Twenty "	Mobiloil Arctic	Castrolite	Energol S.A.E. 20
Upper Cylinder Lubricant	Shell Donax U	Essomix	Duckham's Adcoids	Mobil Upperlube	Castrollo	Energol U.C.L.
GEARBOX	Shell X-100 30	Essolube 30	Duckham's NOL " Thirty "	Mobiloil A	Castrol XL	Energol S.A.E. 30
REAR AXLE **STEERING GEARBOX**	Shell Spirax 90 E.P.	Esso Expee Compound 90	Duckham's Hypoid 90	Mobilube G.X. 90	Castrol Hypoy	Energol EP S.A.E. 90
PROPELLER SHAFT JOINTS	Shell Spirax 140 E.P.	Esso Expee Compound 140	Duckham's NOL EPT 140	Mobilube G.X. 140	Castrol Hi-Press	Energol E.P. S.A.E. 140
FRONT WHEEL HUBS	Shell Retinax A	Esso High Temperature Grease	Duckham's LB10	Mobilgrease No. 5	Castrolease W.B.	Energrease C3
REAR WHEEL HUBS and ENGINE WATER PUMP (*Hand Gun*)		Esso Grease	Duckham's H.B.B.	Mobil Hub Grease	Castrolease Heavy	
CHASSIS. Grease Nipples (*Hand or Pressure Gun*)			Duckham's Laminoid Soft	Mobilgrease No. 4	Castrolease CL	
Oil Points (*Oil Can*) **Body and Chassis**	Shell X-100 20/20W	Essolube 20	Duckham's NOL " Twenty."	Mobil Handy Oil	Castrolite	Energol S.A.E. 20
REAR ROAD SPRINGS	Shell Donax P	Esso Penetrating Oil	Duckham's Laminoid Liquid	Mobil Spring Oil	Castrol Penetrating Oil	Energol Penetrating Oil
	ALTERNATIVELY USE REAR AXLE OR ENGINE OIL					
HANDBRAKE CABLES	Shell Retinax A	Esso Graphite Grease	Duckham's Keenol KG 16	Mobil Graphited Grease	Castrolease Brake Cable Grease	Energrease C3G
BRAKE RESERVOIR	GENUINE LOCKHEED HYDRAULIC BRAKE FLUID					

9

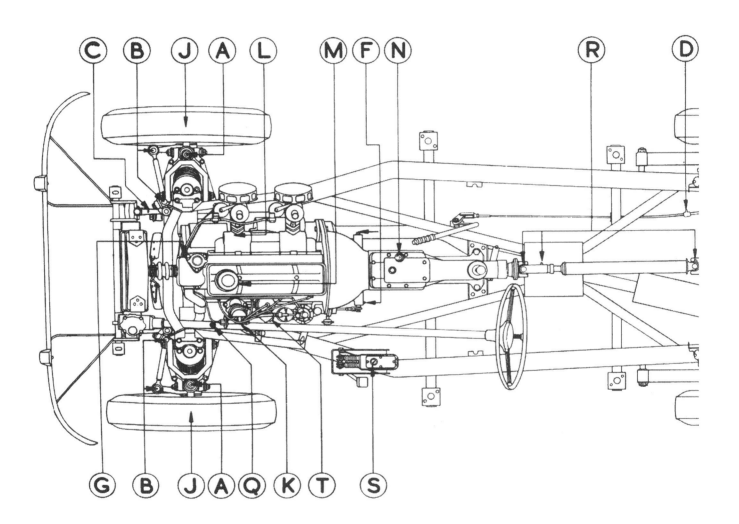

Fig. 5   Lubrication Chart.

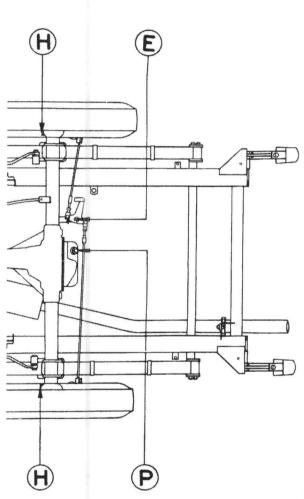

Ref.	ITEMS			DETAILS		Mileage Interval (Thousands of Miles)
A	Steering Swivels .... (4 nipples)			THREE OR FOUR STROKES	GREASE GUN	1
B	Steering	Outer Tie Rod Ball Joints .... (4 nipples)				1
C		Slave Drop Arm Pivot (1 nipple)				1
	Lower Wishbone Outer Bushes (4 nipples)					1
D	Handbrake	Cable .... (1 nipple)		FIVE STROKES		5
E		Compensator ...... (2 nipples)				5
F	Clutch Shaft Bearings .... (2 nipples)					5
G	Engine Water Pump ...... (1 nipple)					5
H	Hubs	Rear .... (2 nipples)				5
J		Front .... (2 nipples) Fitted up to Commission No. TS. 5348 only				5
K	Ignition Distributor			OIL AS RECOMMENDED	OIL CAN	5
	Handbrake Lever					5
	Carburettor Dashpots and Control Linkages					5
	Door Locks, Hinges, Bonnet Safety Catch, Boot and Spare Wheel Locks					5
L	Dynamo					10
M	Engine Sump	250 MILES		TOP UP OIL LEVEL		
				DRAIN & REFILL WITH NEW OIL		2½
	Oil Filler Cap			WASH		5
N	Gearbox			TOP UP OIL LEVEL		5
				DRAIN & REFILL WITH NEW OIL		10
P	Rear Axle					5
Q	Steering Gearbox			TOP UP OIL LEVEL		5
R	Propeller Shaft	Splines ...... (1 nipple)		THREE OR FOUR STROKES WITH OIL GUN		5
		Universal Joints (2 nipples)				5
	Road Springs			CLEAN AND OIL		5
	Air Cleaners			OIL AS RECOMMENDED		5
S	Hydraulic Brake and Clutch Reservoir			TOP UP FLUID LEVEL		5
T	Oil Cleaner			RENEW CARTRIDGE		10

RECOMMENDED LUBRICANTS
OVERSEAS COUNTRIES

	COMPONENT	DUCKHAM'S	VACUUM	WAKEFIELD	B.P. ENERGOL	SHELL	ESSO	S.A.E.
ENGINE	Air Temp. °F. Over 70°	Duckham's NOL "Forty"	Mobiloil "AF"	Castrol XXL	Energol Motor Oil S.A.E. 40	Shell X-100 40	Essolube 40	40
	40° to 70°	Duckham's NOL "Thirty"	Mobiloil "A"	Castrol XL	Energol Motor Oil S.A.E. 30	Shell X-100 30	Essolube 30	30
	10° to 40°	Duckham's NOL "Twenty"	Mobiloil Arctic	Castrolite	Energol Motor Oil S.A.E. 20W	Shell X-100 20/20W	Essolube 20	20
	—10° to 10°	Duckham's NOL "Ten"	Mobiloil 10W	Castrol Z	Energol Motor Oil S.A.E. 10W	Shell X-100 10W	Essolube 10	10
	Below—10°	Duckham's NOL "Five"	Mobiloil 5W	Castrol ZZ	Energol Motor Oil S.A.E. 5W	Shell X-100 5W	Esso Extra Motor Oil "Zero"	5
	Upper Cylinder Lubricant	Duckham's Adcoids	Mobil Upperlube	Castrollo	Energol U.C.L.	Shell Donax U	Esso Upper Motor Lubricant	—
GEARBOX	Over 70°	Duckham's NOL "Fifty"	Mobiloil BB	Castrol XXL	Energol Motor Oil S.A.E. 50	Shell X-100 50	Essolube 50	50
	Over 10° to 70°	Duckham's NOL "Thirty"	Mobiloil A	Castrol XL	Energol Motor Oil S.A.E. 30	Shell X-100 30	Essolube 30	30
	Below 10°	Duckham's NOL "Twenty"	Mobiloil Arctic	Castrolite	Energol Motor Oil S.A.E. 20W	Shell X-100 20/20W	Essolube 20	20
STEERING GEARBOX	Over 10°	Duckham's Hypoid 90	Mobilube GX 90	Castrol Hypoy	Energol EP S.A.E. 90	Shell Spirax 90EP	Esso XP Compound 90	EP 90
REAR AXLE	Below 10°	Duckham's Hypoid 80	Mobilube GX 80	Castrol Hypoy 80	Energol EP S.A.E. 80	Shell Spirax 80EP	Esso XP Compound 80	EP 80
PROPELLOR SHAFT JOINTS		Duckham's NOL EPT 140	Mobilube GX 140	Castrol Hi-Press	Energol EP S.A.E. 140	Shell Spirax 140EP	Esso XP Compound 140	EP 140
FRONT WHEEL HUBS		Duckham's LB10	Mobilgrease M.P.	Castrolease W.B.	Energrease C3	Shell Retinax A	Esso Bearing Grease	—
REAR WHEEL HUBS and ENGINE WATER PUMP (Hand Gun)		Duckham's H.B.B.		Castrolease Heavy			Esso Chassis Grease	—
CHASSIS Grease Nipples (Hand or Pressure Gun)		Duckham's Laminoid Soft		Castrolease CL				
	Oil Points (Oil Can) Body & Chassis	Duckham's NOL "Twenty"	Mobiloil Arctic	Castrolite	Energol Motor Oil S.A.E. 20W	Shell X-100 20/20W	Esso Handy Oil	20
REAR ROAD SPRINGS		Duckham's Laminoid Liquid	Mobilgrease M.P.	Castrol Penetrating Oil	Energol Penetrating Oil	Shell Donax P	Esso Penetrating Oil	—
		ALTERNATIVELY USE REAR AXLE OR ENGINE OIL						
HANDBRAKE CABLES		Duckham's Keenol KG16	Mobilgrease M.P.	Castrolease Brake Cable Grease	Energrease C3G	Shell Retinax A	Esso Spring Grease	—
BRAKE RESERVOIR		GENUINE LOCKHEED HYDRAULIC BRAKE FLUID						

NUT TIGHTENING TORQUES

Operation	Description	Detail No.	Specified Torque Range lb./ft.	Remarks
ENGINE				
CYLINDER HEAD	1/2″ UNF and UNC Stud	106960 106959	100—105	Tighten nuts with engine cold.
CONNECTING ROD CAPS	7/16″ UNF Bolt	105312	55—60	
MAIN BEARING CAPS	1/2″ × 13 NC Setscrew	57121	85—90	
FLYWHEEL ATTACHMENT TO CRANKSHAFT	3/8″ × 24 NF Setscrew	102065	42—46	
TIMING CHAIN WHEEL TO CAMSHAFT	5/16″ × 18 NC Setscrew	56370	24—26	
MANIFOLD ATTACHMENT	3/8″ NC Stud	58688 102475 107055	22—24	
OIL PUMP ATTACHMENTS	5/16″ × 24 UNF Stud	HN.2008	12—14	
REAR OIL SEAL ATTACHMENT	1/4″ × 20 UNC Setscrew	UN.0755	8—10	
CLUTCH ATTACHMENT	5/16″ × 18 UNC Setscrew	HU.0856	20	
ATTACHMENT OF END PLATES	5/16″ × 18 UNC Bolt	HU.0856	14—16	Tapped into Aluminium
ATTACHMENT OF OIL FILTERS	5/16″ × 18 × 24 UNC Bolts Cap Nut Bolt	{HB.0874 {HB.0882 DN.3408 HB.0856	18—20	
TIMING COVER	5/16″ × 18 and 24 NC Setscrew	HU.0805 HU.0857	14—16	
SUMP ATTACHMENT	5/16″ × 18 NC Setscrew	100749	16—18	
PULLEY TO WATER PUMP SPINDLE	5/16″ × 24 UNF Simmonds Nyloc Nut	TN.3208	16—18	
DYNAMO BRACKET TO BLOCK	5/16″ × 18 UNC Setscrew	HU.0856	16—18	
DYNAMO TO BRACKET AND PEDESTAL	5/16″ × 24 UNF Setscrew and Bolt	59115 HU.0808	16—18	
ROCKER PEDESTAL	3/8″ NF and NC Stud	108205	24—26	
OIL GALLERY PLUGS	7/16″ × 14 UNC 3/8″ × 16 UNC	102785 HU.0954	32—36 24—26	Tighten on to copper washer.
ATTACHMENT OF STARTER MOTOR	3/8″ × 24 NF Bolt	NB.0915	26—28	
WATER PUMP ATTACHMENT	3/8″ × 16 UNC Bolt 3/8″ × 16 UNC Bolt	HB.0971 HB.0968	26—28 26—28	
PETROL PUMP ATTACHMENT	5/16″ NF and NC Stud	31ST 131C056	12—14	
THERMOSTAT ASSEMBLY TO CYLINDER HEAD	5/16″ × 18 UNC Bolt 5/16″ × 18 UNC Bolt	HB.0878 HB.0866	16—18 16—18	
INLET TO EXHAUST MANIFOLD	5/16″ × 24 UNF Stud	100419	12—14	
DYNAMO TO PEDESTAL FRONT	5/16″ × 24 UNF Bolt	59115	16—18	

NUT TIGHTENING TORQUES (continued)				
Operation	Description	Detail No.	Specified Torque Range lb./ft.	Remarks
GEARBOX				
FRONT COVER TO GEARBOX	5/16" × 18 NC Setscrew	55771	14—16	
EXTENSION TO GEARBOX	5/16" × 18 UNC Bolt	HB.0866 HB.0858	14—16	
TOP COVER TO GEARBOX	5/16" × 18 Bolts and Setscrews	HU.0851 HB.0871 HB.0873	14—16	
ATTACHMENT OF ENGINE TO GEARBOX	5/16" × 18 NC and NF Bolt and Stud	HB.0858 125C056	14—16	
REAR MOUNTING TO GEARBOX EXTENSION	1/2" × 20 UNF Bolt	HB.1112	50—55	
FRONT SUSPENSION				
BACK PLATE AND TIE ROD LEVERS TO VERTICAL LINKS	1/4" × 24 UNF Setscrews and Bolts	HB.0925 HB.0922 HU.0905	24—26	
WHEEL STUDS AND NUTS	7/16" NF	100869	45—55	
BALL PIN TO VERTICAL LINK	1/2" × 20 UNF Nut—Slotted	2211 LN	55—65	To suit pin hole.
TOP WISHBONE TO FULCRUM PIN	7/16" × 20 UNF Nut—Slotted	2210 LN	26—40	To suit pin hole.
SPRING PAN TO WISHBONE	1/4" × 24 UNF Stud 1/4" × 24 UNF Bolt	107350 107351	26—28	
TIE ROD TO IDLER LEVER AND DROP-ARM	1/4" × 24 UNF Simmonds Nyloc Nut	TN.3209	26—28	
TOP INNER FULCRUM PIN TO CHASSIS	1/4" × 24 UNF Bolt 1/4" × 24 UNF Setscrew	HB.0913 HU.0908	26—28	
LOWER FULCRUM BRACKET TO CHASSIS	3/16" × 24 UNF Bolt	HB.0805	16—18	
LOWER WISHBONE TO FULCRUM PIN	7/16" × 20 UNF Nyloc Nut	TN.3210	26—28	
FRONT HUB TO STUB AXLE	1/2" × 20 UNF Nut—Slotted	LN.2211	Tighten up and unscrew one flat.	
REAR AXLE				
BEARING CAPS TO HOUSING	1/4" × 24 UNF Setscrew	100878	34—36	
HYPOID PINION FLANGE	1/4" × 18 UNF	100892	85—100	To suit split pin holes.
CROWN WHEEL TO DIFFERENTIAL CASE	7/16" × 24 UNF 3/8" × 24 UNF	107880 109735	22—24 35—40	Fitted from Commission No. TS.2181.

NUT TIGHTENING TORQUES (continued)

Operation	Description	Detail No.	Specified Torque Range lb./ft.	Remarks
REAR COVER ATTACHMENT	⅜″ × 24 UNF Setscrew	HU.0805	16—18	
BACKING PLATE ATTACHMENT	⅜″ × 24 UNF Setscrew	HU.0908	26—28	
HUB TO AXLE SHAFT	⅝″ × 18 UNF Nut—Slotted	100892 112635	110—125 125—145	From axle No. TS.8039
REAR SUSPENSION				
SPRING FRONT END TO FRAME	½″ × 20 UNF Bolt	106251	28—30	
SPRING SHACKLE (NUT TO PIN)	⅜″ × 24 UNF Nut Shackle Pin	HN.2009 104953	26—28	
ROAD SPRING TO REAR AXLE	Clip Nyloc Nut ⅜″ × 24 UNF	107688 YN.2909	28—30	
SHOCK ABSORBER TO FRAME BRACKET	⅜″ × 24 UNF Setscrew ⅜″ × 24 UNF Nyloc Nut	HU.0908 TN.3209	26—28	

GENERAL DATA

FRACTIONAL AND METRICAL EQUIVALENTS

Inches Frac.	Dec.	mm.	Inches Frac.	Dec.	mm.	Inches Frac.	Dec.	mm.
	.0039	.100	19/64"	.2968	7.540		.6500	16.510
1/128"	.00781	.200		.3000	7.620	21/32"	.6562	16.668
	.0118	.300	5/16"	.3125	7.937		.6693	17
1/64"	.0516	.3968		.3150	8	43/64"	.6719	17.065
	.0157	.400	21/64"	.3281	8.334	11/16"	.6875	17.462
	.0197	.500	11/32"	.3437	8.731		.7000	17.780
	.0236	.600		.3500	8.890	45/64"	.7031	17.859
	.0276	.700		.3543	9		.7087	18
1/32"	.0312	.794	23/64"	.3594	9.128	23/32"	.7187	18.256
	.0315	.800	3/8"	.3750	9.525	47/64"	.7344	18.652
	.0354	.900	25/64"	.3906	9.921		.7480	19
	.0394	1		.3937	10	3/4"	.7500	19.050
3/64"	.0469	1.191		.4000	10.160	49/64"	.7656	19.446
	.0500	1.270	13/32"	.4062	10.319	25/32"	.7812	19.843
1/16"	.0625	1.587	27/64"	.4219	10.716		.7874	20
5/64"	.0781	1.984		.4331	11	51/64"	.7969	20.240
	.0787	2	7/16"	.4375	11.112		.8000	20.230
3/32"	.0937	2.381		.4500	11.430	13/16"	.8125	20.637
	.1000	2.540	29/64"	.4531	11.509		.8268	21
7/64"	.1094	2.778	15/32"	.4687	11.906	53/64"	.8281	21.034
	.1181	3		.4724	12	27/32"	.8437	21.431
1/8"	.1250	3.175	31/64"	.4844	12.303		.8500	21.590
9/64"	.1406	3.572	1/2"	.5000	12.700	55/64"	.8594	21.827
	.1500	3.810		.5118	13		.8661	22
5/32"	.1562	3.969	33/64"	.5156	13.096	7/8"	.8750	22.225
	.1575	4	17/32"	.5312	13.493	57/64"	.8906	22.621
11/64"	.1719	4.365	35/64"	.5469	13.890		.9000	22.859
3/16"	.1875	4.762		.5500	13.970		.9055	23
	.1968	5		.5512	14	29/32"	.9063	23.018
	.2000	5.080	9/16"	.5625	14.287	59/64"	.9219	23.415
13/64"	.2031	5.159	37/64"	.5781	14.684	15/16"	.9375	23.812
7/32"	.2187	5.558		.5906	15		.9449	24
15/64"	.2344	5.953	19/32"	.5937	15.081		.9500	24.129
	.2362	6		.6000	15.240	61/64"	.9531	24.209
1/4"	.2500	6.350	39/64"	.6094	15.478	31/32"	.9687	24.606
17/64"	.2656	6.745	5/8"	.6250	15.875	63/64"	.9844	25
	.2756	7		.6299	16	1"	1.000	25.400
9/32"	.2812	7.144	41/64"	.6406	16.271			

STANDARD MEASURE AND METRIC EQUIVALENTS

English to Metric (linear)

1 inch	= 2.54 centimetres
1 foot	= 30.4799 centimetres
1 yard	= 0.914399 metre
1 mile	= 1.6093 kilometre
10 miles	= 16.093 kilometres

Metric to English (linear)

1 centimetre	= 0.3937 inch
1 metre	= 39.3702 inches
	= 1.0936 yard
1 kilometre	= 0.62137 mile

English to Metric (square measure)

1 square inch	= 6.4516 square centimetres
1 square foot	= 9.203 square decimetres
1 square yard	= .836126 square metre

Metric to English (square measure)

1 square centimetre	= .15500 square inch
1 square metre	= 1550.01 square inches
	= 10.7639 square feet
	= 1.196 square yard

16

GENERAL DATA

English to Metric (cubic measure)

1 cubic inch	=	16.387 cubic centimetres
1 cubic foot	=	28.317 litres
1 gallon (0.1605 cu. ft.)	=	4.546 litres

Metric to English (cubic measure)

1 litre (1,000 cu. cms.)	=	0.22 gallons, or 1.7598 pints
1 cubic centimetre	=	0.61 cubic inches

English to Metric (weight)

1 pound (Avoirdupois)	=	0.45359 kilogrammes
1 cwt. (112 pounds)	=	50.8 kilogrammes
1 ton (2,240 pounds)	=	1,016 kilogrammes

Metric to English (weight)

1 kilogramme	=	2.20462 pounds
100 kilogrammes	=	1.968 cwt.
1,000 kilogrammes	=	0.9842 tons

Service Instruction Manual

ENGINE

ENGINE

INDEX

ENGINE

ILLUSTRATIONS

PART AND DESCRIPTION	DIMENSIONS NEW	CLEARANCE NEW	REMARKS
Crankshaft			
Journal diameter	2.4795" 2.4790"	.0010" to .0025"	
Bearing Internal Diameter	2.4815" 2.4805"		
Bearing Housing Internal Diameter	2.6255" 2.6250"		
Undersize bearings are available in the following sizes :—.010", —.020", —.030", —.040".			
Crankshaft End Float			
Intermediate Journal Length	1.7507" 1.7498"	.0048" to .0117"	Clearance of .004" to .006" is specified and obtained by selective assembly of Thrust Washers.
Intermediate Bearing Cap Width. (Plus thickness of two Thrust Washers.)	1.7450" 1.7390"		
Main Bearing Cap Width	1.5050" 1.4950"		
Big End			
Crank Pin Diameter	2.0866" 2.0860"	.0016" to .0035"	
Bearing Internal Diameter	2.0895" 2.0882"		
Internal Diameter of Bearing Housing	2.2335" 2.2327"		
Bearing Width	.9670" .9650"		
Undersize bearings are available in the following sizes :—.010", —.020", —.030", —.040".			
Big End Float			
Crankpin Width	1.1915" 1.1865"	.007" to .014"	
Con. Rod Width	1.1795" 1.1775"		
Ovality and Taper			
Journals and Crankpins	Should not exceed .002"		

1

ENGINE—Dimensions and Tolerances

PART AND DESCRIPTION	DIMENSIONS NEW	CLEARANCE NEW	REMARKS
Small End			
Bore for Bush	1.0000″ .9950″		Press Fit in Con. Rod.
Bush External Diameter	1.0005″ .995″		
Internal Diameter of Bush	.8752″ .8748″	.0002″ at 68°F.	
Gudgeon Pin Diameter	.87510″ .87485″		
Piston Rings			
Compression Ring Width	.062″ .061″	.0015″ to .0035″	
Groove Width	.0645″ .0635″		
Scraper Ring Width	.156″ .155″	.001″ to .003″	
Groove Width	.158″ .157″		
Ring Gap in Cylinder Sleeves		.003″ to .010″	

Piston Rings are obtainable in the following oversizes : +.010″, +.020″, +.030″, +.040″.

Pistons and Cylinder Sleeves	F	G	H
Bore Diameter	3.2676″ 3.2673″	3.2680″ 3.2677″	3.2684″ 3.2681″
Top Diameter of Piston Skirt	3.2626″ 3.2622″	3.2630″ 3.2626″	3.2634″ 3.2630″
Bottom Diameter of Piston Skirt	3.2641″ 3.2637″	3.2645″ 3.2641″	3.2649″ 3.2645″

Top—Skirt Clearance	.0054″ .0047″	Applicable to " F," " G " & " H " Pistons
Bottom—Skirt Clearance	.0039″ .0032″	

ENGINE—Dimensions and Tolerances

PART AND DESCRIPTION	DIMENSIONS NEW	CLEARANCE NEW	REMARKS
Height of Cylinder Sleeves above face of Cylinder Block		.003″ to .0055″	
Pistons are available in the following oversizes: +.020″, + 030″, +040″.			
Camshaft			
Front Journal Diameter	1.8720″ 1.8710″	.0028″ to .0047″	
Front Journal Bearing Bore	1.8757″ 1.8748″		
External Diameter of Front Bearing	2.2498″ 2.2493″	Push Fit in Cylinder Block.	
Bore in Block for Front Bearing	2.2507″ 2.2498″		
Diameter of 2nd, 3rd and Rear Camshaft Journal	1.7157″ 1.7152″	.0026″ to .0046″	
Bore in Cylinder Block for 2nd, 3rd and Rear Journals	1.7198″ 1.7183″		
End Float		.003″ to .0075″	
Valves and Valve Guides			
Inlet Stem Diameter	.3110″ .3100″	.001″ to .003″	
Inlet Guide Diameter	.3130″ .3120″		
Exhaust Stem Diameter	.3715″ .3705″	.003″ to .005″	
Exhaust Guide Diameter	.3755″ .3745″		
Included Angle of Valve Faces	90°		
Inlet Valve Head Diameter	1.5620″ 1.5580″		
Width of Inlet Valve Seating	.0469″ approx.		

PART AND DESCRIPTION	DIMENSIONS NEW	CLEARANCE NEW	REMARKS
Exhaust Valve Head Diameter	1.3030″ 1.2990″		
Width of Exhaust Valve Seating	.0469″ approx.		

Oil Pump

Outer Rotor Outside Diameter	1.5975″ 1.5965″	.0055″ to .0075″	
Housing Internal Diameter	1.6040″ 1.6030″		
Depth of Rotor	1.4995″ 1.4985″	.0005″ to .0025″	
Housing Depth	1.5010″ 1.5000″		
Bush in Cylinder Block	.5010″ .5005″	.0015″ to .0030″	
Distributor Driving Shaft	.4990″ .4980″		
End Float of Distributor and Tachometer Gear Assembly		.003″ to .007″	

Inner Rotor

Major Diameter	1.1720″ 1.1710″		
Minor Diameter	.7310″ .7290″		
Rotor Depth	1.4995″ 1.4985″	.0005″ to .0025″	
Housing Depth	1.5010″ 1.5000″		
Clearance on Rotors Min.		.0005″ to .0025″	
Max.		.001″ to .004″	

4

PART AND DESCRIPTION	DIMENSIONS NEW	CLEARANCE NEW	REMARKS

Valve Springs

Outer Springs
Inlet and Exhaust

Fitted Length	1.560″		
Fitted Load	38 lbs.		
Free Length, approx.	1.980″		

Inner Spring
Inlet

Fitted Length	1.500″		
Fitted Load	33 lbs.		

Exhaust

Fitted Length	1.450″		
Fitted Load	36.5 lbs.		

Inlet and Exhaust

Free Length, approx.	2.080″		

Auxiliary Inner Spring
Exhaust Valve Only

Fitted Length	1.140″		
Fitted Load	10 lbs.		
Free Length, approx.	1.540″		

Valve Insert Dimensions

	Combustion head		Insert			
	Bore	Depth	O/D	I/D	Depth	Detail
Inlet	1.717″ 1.716″	.253″ .250″	1.723″ 1.722″	1.471″ 1.466″	.253″ .250″	111913
Exhaust	1.439″ 1.438″	.253″ .250″	1.445″ 1.444″	1.193″ 1.188″	.253″ .250″	102941

The seating of both valves is .044″ × 89°

NOTE : To convert lbs. to Kgs. divide by 2.204.

,, ,, ins. to Millimetres multiply by 25.4.

5

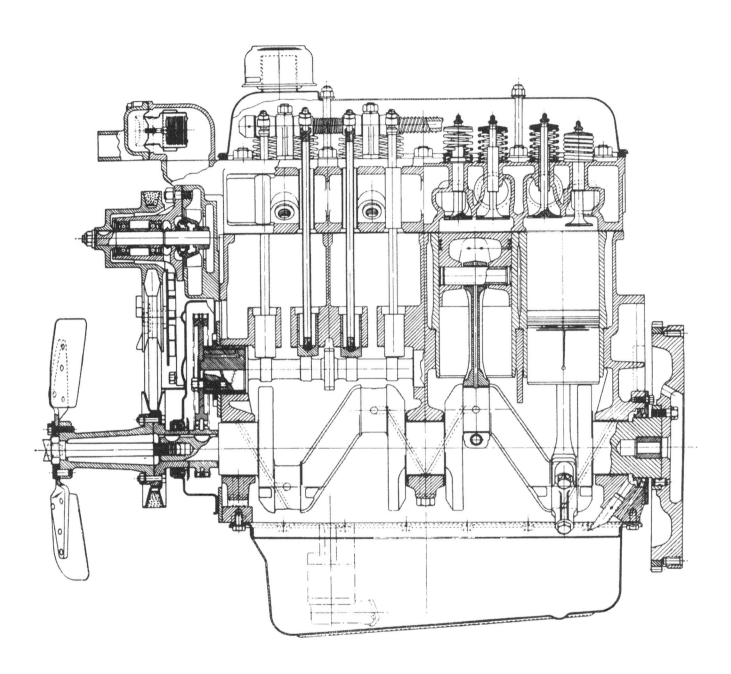

Fig. 1   Longitudinal view of Engine.   For illustration purposes the sump oil filter has been omitted.

6

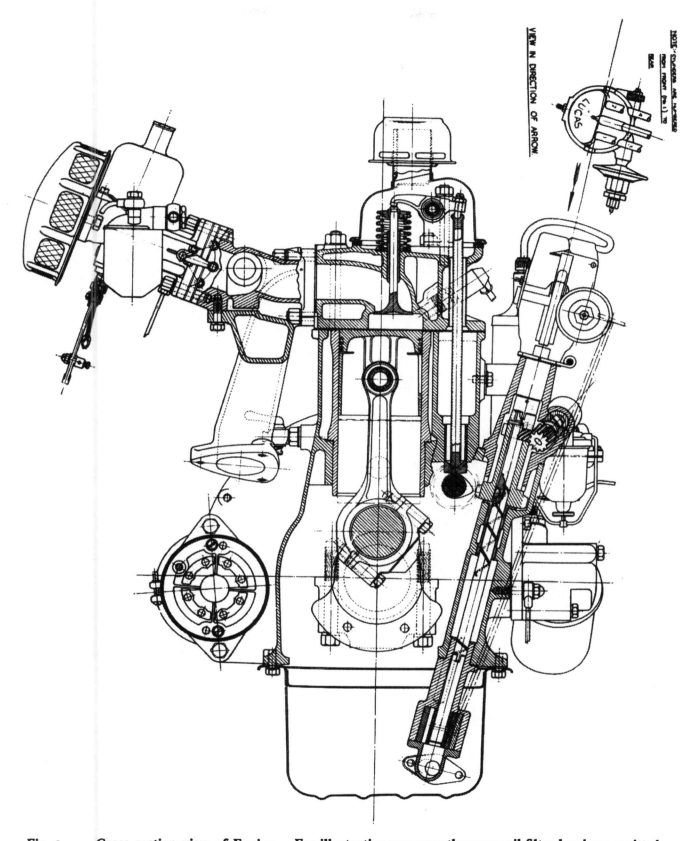

NOTE: CYLINDERS ARE NUMBERED FROM FRONT (No. 1) TO REAR.

VIEW IN DIRECTION OF ARROW.

Fig. 2   **Cross section view of Engine.   For illustration purposes the sump oil filter has been omitted.**

ENGINE

1. GENERAL DESCRIPTION
(Figs. 1 and 2)

(a) **The Engine** has four cylinders and the overhead valves are push rod operated, the 83 mm. bore and 92 mm. stroke give a capacity of 1,991 cubic centimetres. The compression is 8.5 to 1.

A low compression kit (see page 27) is available and reduces the compression ratio to 7.5 to 1.

(b) **The Cylinder Block** is an integral casting in cast iron, the abutments for the cylinder sleeves, the three rear camshaft bearings and the crankshaft bearing housings are machined in a single unit. The main bearing housings are line bore machined; the bearing caps are not interchangeable and are stamped together with the casting to assist identification.

After Engine No. 9095E four Vandervel bi-metal bearings were fitted to accommodate the camshaft. A recognition feature of engines so fitted with these bearings will be that three setscrews retaining the three rearmost bearings will clearly be seen on the left-hand side of the cylinder block. See TR3 Supplement Engine Section "B".

CYLINDER BORE & PISTON MARK	CYLINDER BORE DIAMETERS
'F'	3·2676" TO 3·2673"
'G'	3·2680" TO 3·2677"
'H'	3·2684" TO 3·2681"

Fig. 3   Cylinder Sleeves and Dimensions.

(c) **The Cylinder Sleeves** (Fig. 3) are of the wet type, being centrifugally cast in nickel chrome iron and provided with flanged upper faces, having two pairs of flats at 90° to one another.

These two pairs of flats provide alternative fitting positions to deal with piston slap which normally occurs due to wear along the axis of thrust.

The sleeves are machined all over and ground on their upper faces. The lower portion of each liner is provided externally with a reduced diameter, surmounted by a flanged face for spigoting into machined recesses in the cylinder block and a water seal provided by a plastic covered steel joint.

Fig. 4   A Figure of Eight Joint.

The Figure of Eight joint (Fig. 4) is made of steel and is plastic coated to provide the necessary sealing properties. Care must be exercised when handling or storing these joints and they should always be examined for chipping or peeling of the plastic coat before use.

If doubt exists as to the condition of the plastic coat the joint should be discarded. Only in the cases of extreme emergency should they be used and then with a liberal application of a sealing compound.

The sleeves are spigot mounted and held in position by the combustion head, the initial position of the sleeve allowing this to stand proud of the cylinder block .003" minimum to .0055" maximum (Fig. 5). The bores are graded F, G or H, and the appropriate symbol is engraved on the upper face of each sleeve. (See page 2.)

8

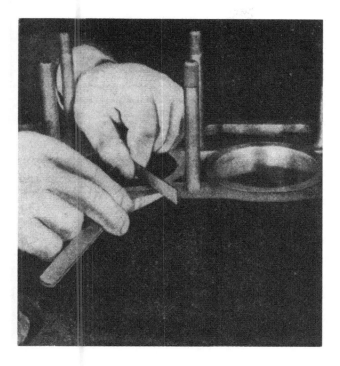

angle to the centre line of the connecting rod. The caps are dowelled to the connecting rods and located by these dowels. This form of cap provides a more convenient position for tightening and loosening bolts, and also has the added virtue of allowing the bearing caps to be removed progressively from below without the danger of their dropping into the repair pit immediately the bolts have been withdrawn. This connecting rod design permits the piston and connecting rod assembly to pass upward through the sleeve bores and also has an important advantage in reducing the stresses in the connecting rod bolts. The bolts themselves are secured by a locking plate made from 20-gauge material.

With the bearing cap removed, it is possible to examine and replace the bearings without removing the piston assembly from the engine.

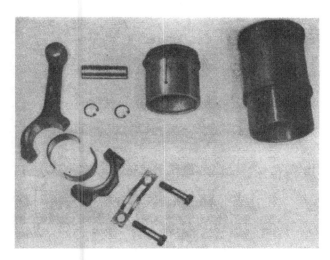

Fig. 6 **The Piston and Connecting Rod Assembly in exploded form.**

(d) **The Connecting Rods** (Fig. 6) are molybdenum manganese steel stampings being provided with phosphor bronze small end bushes and precision type big end bearings. The rod is drilled from the big end bearing end to the small end bearing to provide for the passage of oil under pressure from the main supply. The big end bearing cap is of a special design, the cap securing bolts being inclined at an

Fig. 7 **The Piston and Connecting Rod Assembly. Note position of cap in relation to split in piston skirt.**

9

35

(e) Aeroflex Compensating Pistons (Fig 6) are employed, which are made from a special aluminium alloy and each provided with two compression rings and one oil scraper ring.

The pistons are graded F, G or H (dimensions on page 2) and this symbol is stamped on the crowns. The piston skirt has a $\frac{1}{32}$" slot on the non-pressure side and is fitted to the connecting rod so that this slot is away from the point of maximum thrust, Fig. 7 (facing the camshaft side of the engine).

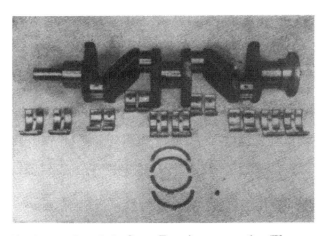

Fig. 8   Crankshaft, Bearings and Thrust Washers.

(f) The Crankshaft (Fig. 8) is forged from molybdenum manganese steel, being provided with balance weights which are an integral part of the crankshaft throws, adjacent to the three main bearings.

This shaft is accommodated in three precision type white metal steel back bearings, which are housed in the cylinder block, being secured in position by bearing caps and two bolts and spring washers per journal. Crankshaft thrust is taken by steel white metal covered washers which are fitted in two halves on either side of the centre main bearing housing, being located circumferentially by means of projections on the lower half of each pair of washers.

In the case of extreme necessity and knowing that the crankshaft is in good condition, it is possible to change the main bearings without first removing the engine from the chassis. It is essential however that extreme care

be taken when replacing the front and rear oil seals. This operation is described on page 32 and 33.

(g) The Valves are overhead, push rod operated. The push rods themselves are tubular being fitted with a ball at one end and a cup at the other, both being spot welded into position.

All valves are made from a chrome nickel silicon valve steel stamping, the inlet valve having a larger head and a smaller stem than the exhaust valve. The stems have a hardened tip. The exhaust valves fitted to engines after Engine No. TS. 481 E were made from a high nickel chromium tungsten valve steel stamping, and the stem was stellite tipped.

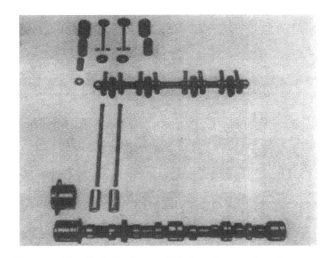

Fig. 9   Exploded view of Valve Operating Gear.

Inlet valves are provided with two springs. Three springs are used on the exhaust valves only (Fig. 9). Valve springs are located by a valve collar and held in position by split taper collars. The close coil of the valve springs must always be fitted to the cylinder head.

(h) The Camshaft (Fig 9) is of special iron alloy having chilled cam faces and is provided with four journals. The front journal is accommodated in a flanged cast iron bearing, whilst the other journals are mounted direct in the cylinder block.

In the near future it is proposed to fit four Vandervel bi-metal bearings to accommodate the camshaft. A recognition feature of engines so fitted with these bearings will be that three set-screws retaining the three rearmost bearings will clearly be seen on the left-hand side of the cylinder block. The front bearing is pressed into the front bearing sleeve.

The camshaft operates directly on flat based hollow cylindrical chilled cast iron tappets which in turn engage hardened spherical-ended push rods, the upper extremities of which are hardened and cup-shaped, accommodating hardened ball ended screws, which are mounted on the outer ends of the respective rockers. Camshaft end thrust is taken by the flanged front bearing, against the timing wheel and a shoulder on the shaft itself. End float is measured by a feeler gauge between the camshaft chain wheel and the front bearing housing or by a dial indicator. To reduce the end float a replacement bearing of increased length must be fitted. To increase the end float it will be necessary to rub the bearing down on a sheet of emery cloth placed on a surface plate to reduce its length.

The rockers are of case hardened steel and provided with phosphor bronze bushes which are lubricated under pressure from the main oil supply. The eight rockers themselves are carried on a hollow rocker shaft which is in turn mounted on four pedestal brackets, the oil being fed along the rocker shaft to the various rockers.

(i) **The Cooling System.** (see Section "C") is thermostatically controlled and pressurised; an impeller pump is utilised to assist the circulation of the cooling fluid.

A four-bladed $12\frac{1}{2}$" fan is mounted on rubber bushes and is attached to the crankshaft. The fan pulley is drilled in its outer periphery and aligning this hole with a pointer welded to the timing chain cover sets Nos. 1 and 4 pistons at T.D.C. (see Fig. 37).

The radiator is attached to the body at the upper corners and secured to the chassis at its sides.

(j) **The Fuel System** (see Section "P") incorporates a petrol shut off cock in the pipe line from the tank to the pump, this is situated on the left-hand chassis member adjacent to the engine. Petrol is supplied by an A.C. Type UE Pump to the twin S.U. Type H4 carburettors. Each carburettor has its individual A.C. air cleaner. The vacuum pipe to the distributor is taken from the front carburettor.

(k) **The Hobourn-Eaton Double Rotor Oil Pump** (Fig. 14) is of the submerged type and is self priming; oil is drawn from the engine sump through a gauze filter. The oil is fed to the oil gallery and to the Purolator oil filter.

(l) **Coil Ignition** is employed and the distributor (Lucas DM.2 Type V.167) has a vacuum and centrifugal automatic advance incorporated. It is suppressed for radio and television.

(m) **The Engine Mountings** are of the flexible type, the front bearer being assembled on the rubber blocks on either side of the chassis frame, the gearbox itself being supported on a rubber pad secured to a cross member of the chassis frame.

(n) **The Flywheel** is manufactured from cast iron and is fitted with a shrunken starter ring of heat treated steel. It is located on the crankshaft by a dowel and secured by four bolts with lock plates. The flywheel is marked by an arrow which, when aligned with a scribe line on the cylinder block, sets Nos. 1 and 4 pistons at T.D.C.

When fitting the flywheel to the crankshaft ensure that both components are free from burrs. After fitting, the run-out should be checked by a D.T.I. to ensure the run-out does not exceed .003". Failure to observe this point may lead to clutch disorders and vibration.

There are two dowel holes in the flywheel 90° removed from one another; this will enable the flywheel to be turned 90° should the teeth of the starter ring gear become increasingly worn and a replacement not be readily available. It must be remembered that the timing mark must be obliterated and a second stamped on the flywheel.

(o) **To Fit Replacement Starter Ring Gear.** When it is necessary to fit a replacement ring gear, certain precautions should be taken to ensure its future life. The installation can be

11

carried out whilst the flywheel is still cold. The ring should be immersed in boiling water or its temperature raised by some other means; a temperature higher than boiling water is not recommended for the heat properties of the ring may be destroyed. The ring must be fitted with the leading edges of the teeth toward the starter motor. Should a press not be available, fitting of the ring gear can be carried out using four "G" clamps and tapping the ring into position with a brass rod (Fig. 10).

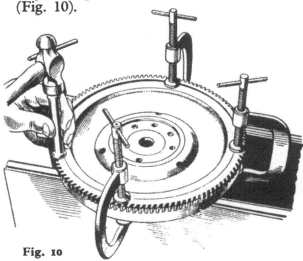

Fig. 10

Indicating the use of " G " Clamps when fitting a Replacement Starter Ring.

(p) **Crankcase Ventilation** (Fig. 11) is effected by permitting air to be drawn out of the engine. To enable this ejection a large bore pipe in the form of an inverted "U" is fitted into the left-hand side of the cylinder block by means of an adapter welded to its end. The exposed end is cut away at an angle to provide a wider opening facing away from the slipstream.

The passage of air (the slipstream) created by the cooling fan or the movement of the car causes a depression at the angle opening of the inverted "U" pipe and air is drawn out of the cylinder block.

Fresh air is taken in through the rocker cover oil filler cap, circulating round the valve springs and rockers before passing down the push rod tubes into the cylinder block to replace air which is being drawn out. It is essential therefore that the filler cap is kept as clean as possible to allow free passage of air. This cap, which has a gauze

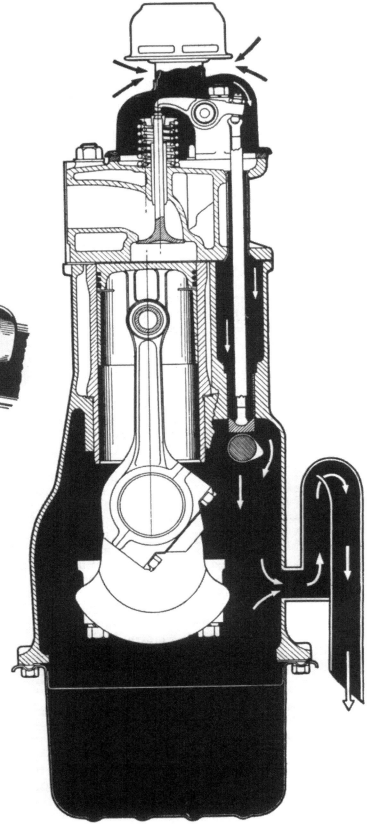

Fig. 11 **A diagrammatic view of Crankcase Ventilation.**

12

filter incorporated in it, should be washed in petrol and drained on each occasion when the engine oil is changed.

2. ENGINE LUBRICATION
(Figs. 12 and 13)
Description

Lubrication of the engine is by a Hobourn-Eaton pump. The pump is driven by a shaft which is mounted in a bush pressed into the cylinder block, and is provided with a helical gear which engages with a similar gear on the camshaft.

Oil is drawn into the pump through a primary gauze filter and passes through a channel in the pump casting to an annular space around the oil pump shaft. The annular space round the drive shaft is closed by the bush, and the oil thus forced through a hole in the cylinder block into the head of the external oil filter where some of this oil passes directly into the oil gallery which extends the length of the cylinder block ; the remainder of the oil passes into the bowl of the oil filter under the pressure of the oil pump. When the oil pressure exceeds 70 to 80 lbs. per sq. inch it opens a spring loaded ball valve and passes into the sump. The oil on its way to the base of the filter is forced through the filtering media and passes up an annular space around the bowl holding bolt through a restrictor into the sump.

The oil passes from the gallery to the three main bearings, through drillings in the crankshaft to the big end bearing ; then through further drillings in the connecting rods to the small end bushes and gudgeon pins. Splash lubrication is further assisted by a drilling into the oil passage between the small end and big end just below the piston skirt on each connecting rod.

By drillings from the channels leading to the main bearing oil is conveyed to the front, second and rear camshaft bearings. In the case of the third camshaft bearing this is fed direct from the oil gallery through a metering hole. A by-pass from the rear camshaft bearing conveys oil upwards through a drilling in the combustion head and rearmost rocker pedestal to the rocker shaft. Oil passes along the hollow shaft and through radial holes to the rockers, leaving each rocker by a hole drilled vertically to each tappet ball pin.

The oil is prevented from escaping by the rocker cover and after lubricating the valve springs and ball pins, returns downwards through the push rod tubes lubricating the push rod tappets before entering the sump.

Oil from the front camshaft bearing lubricates the timing chain where four slots cut at 90° to each other on the face of the flange adjacent to the camshaft timing wheel allow oil to escape on to the timing wheel. The oil is thrown out by centrifugal force on to the underside of the flanged portion of the wheel on which the teeth are cut.

Six holes are drilled obliquely, alternatey, from the back and the front of the wheell at equal intervals from the underside of the flange into the space between the two toothed rings. These holes allow the oil to be thrown on to the underside of the timing chain, ensuring its lubrication.

3. OIL PUMP

The oil pump is of the double rotor type as shown in Fig. 14.

The smaller centre rotor is driven by a short shaft on which it is pressed and pegged in position. The two rotors are contained in a housing at the base of the oil pump casting, which is provided with a cover plate having a ground face, allowing only sufficient clearance on the two rotors to provide for lubrication. The centres of the rotors are offset.

The rotor shaft has at its upper extremity a recess which engages a tongue on the lower end of the drive shaft. The driving shaft is mounted in a phosphor bronze bush which is pressed into the cylinder block, and at its upper end a helical gear is secured by means of a Woodruffe key. The helical gear on this shaft engages with a similar gear which is an integral part of the camshaft.

The centre rotor, by its engagement with the outer rotor, drives the latter at a slightly lower speed owing to the difference in sizes.

Owing to the relative movement of the outer rotor around the inner rotor; and the close fit of the cover plate, oil is forced round between the lobes of the rotor and forced out of a hole in the top of the rotor casing and upward through a drilled passage to the annular space around the

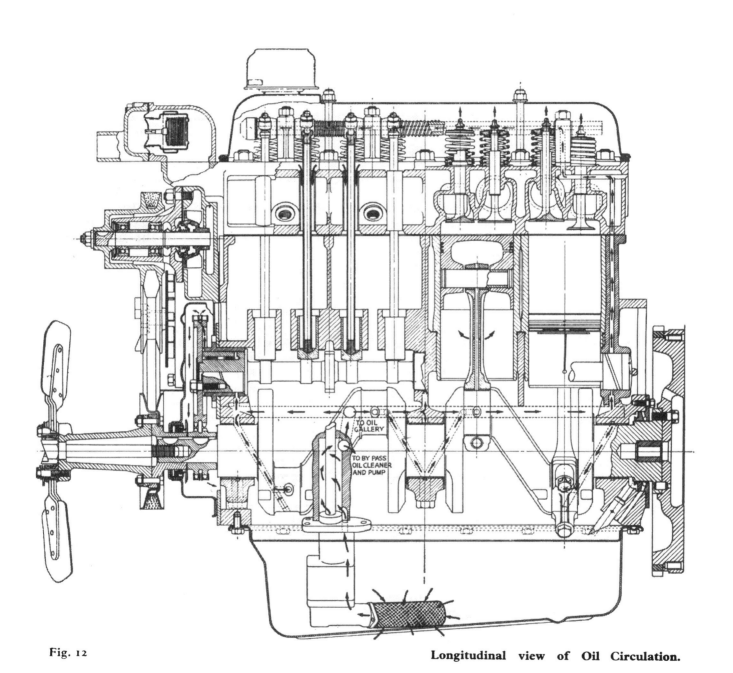

Fig. 12   **Longitudinal view of Oil Circulation.**

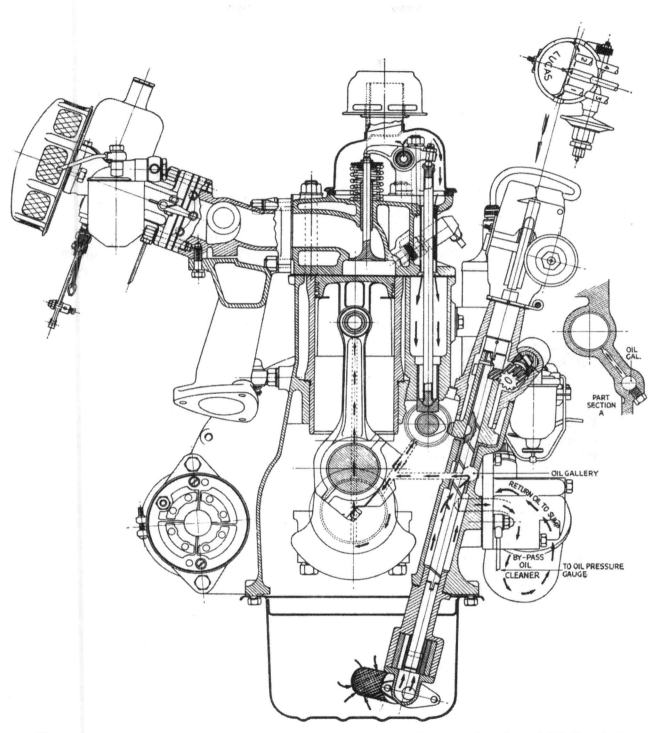

Fig. 13   **Cross section view of Oil Circulation.**

Labels on figure: LUCAS, OIL GAL., PART SECTION A, OIL GALLERY, RETURN OIL TO SUMP, BY-PASS OIL CLEANER, TO OIL PRESSURE GAUGE

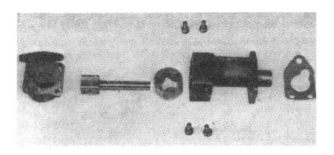

Fig. 14   **Exploded view of Oil Pump.**

distributor drive shaft. From this annular space oil is circulated round the engine as described in "Engine Lubrication."

(a) To Remove Oil Pump from the Engine

(i) Drain the oil from the sump (preferably when the engine is warm) and jack up the car.

(ii) Remove the sump securing bolts and, lowering it at the front, first manoeuvre the sump and tray past the oil pump gauze filter.

(iii) Remove the three pump securing bolts and remove the pump and filter as a unit.

(b) To Dismantle Oil Pump

Remove the two bolts securing the primary filter to the flange on the oil pump elbow. Take note of the position of the filter in relation to the elbow for re-assembly, *i.e.*, the tube projecting inwards should be as near as possible to the bottom of the sump, thus ensuring there is a clearance between the filter and the sump bottom.

To complete the dismantling it is now only necessary to remove the four setscrews. The inner rotor and shaft and the outer rotor can now be removed and the dismantling is complete.

(c) Servicing Oil Pump

As this pump provides a generous surplus of oil to that which is necessary for the engine lubrication, and owing to the design of the unit, very little wear is likely to occur in service, and little maintenance should be necessary to the unit during the life of the engine.

In actual practice, excepting the re-

mote possibility of failures due to defective materials, no adjustments are likely to be required until approximately 200,000 miles have been covered, and then it is only likely to be limited to the elimination of end float in the rotors, and can be satisfactorily dealt with by lapping the joint faces of the pump body and cover. The clearance new between the rotors and cover plate should be from .0005"—.0025" and where a serious drop in oil delivery from the pump is associated with development of excessive end float, steps should be taken to lap the cover plate and body.

(d) Engagement of Oil Pump (Fig. 15) and Distributor Driving Gear

This drive is taken from the helical gear on the camshaft through a similar gear unit mounted on the oil pump driving shaft.

The shaft has a tongue at the lowermost end which engages the oil pump mounted in the sump.

The helical gear unit is secured to the shaft by a Woodruffe key. The upper gear of this unit drives the tachometer and the boss-like extension is fitted with a mills pin to prevent the gear and shaft from rising. The head has an offset recess into which the distributor shaft will seat.

When correctly engaged the slot in the distributor driving boss, with No. 1 cylinder at T.D.C. on the compression stroke, should assume a position approximately "five minutes to five" with the offset towards the rear of the engine (Fig. 16). In this position the slot will point directly towards the exhaust valve rod sealing tube for No. 1 cylinder, the distributor rotor will face No. 1 sparking plug, and the keyway in the helical gear will be aligned with the oil dipstick when fitted.

See also "13 Ignition and Distributor Timing." Page 24.

3. CRANKSHAFT AND MAIN BEARINGS (Fig. 8)

The crankshaft is of molybdenum manganese forging with ground journals and crankpins.

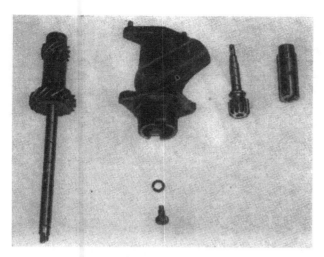

Fig. 15   Exploded view of Distributor and Tachometer Drive Details.

The main bearings are of the precision type, bi-metal steel backed. No hand fitting is required and in no circumstances should the bearing caps be filed with a view to taking up wear. The filing of bearing caps will make them unserviceable for future use when new bearings are ultimately used.

Where excessive bearing wear has occurred the only satisfactory cure is to replace worn

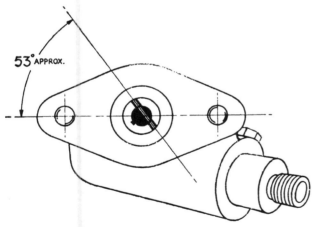

Fig. 16   Position of Slot in Distributor Boss when No. 1 cylinder is at T.D.C. on compression stroke.

bearings ensuring first, however, that the crankshaft journals and pins are in good order and that there is no question of a regrind being necessary. Where a crankshaft journal is worn, scored or tapered in excess of .002″ regrinding is necessary.

When a regrind is found to be necessary a decision will have to be made as to the suitable undersize bearings which will meet the particular case. The reduced diameter of journal to suit the various undersize bearings may be calculated by subtracting —.020″, —.030″ or —.040″, the sizes of bearings available from the original dimensions on page 1.

(a) Main Bearing Clearance

The crankshaft journal diameter and the internal dimension of the bearings is given on page 1. The clearance new for the main bearings is .001″—.0025″, if the worn clearance exceeds .006″ or if the journals have become scored, the crankshaft will have to be reground and undersized bearings fitted. The crankshaft should be measured with a micrometer gauge and if the reading is less than 2.477″ (for a crankshaft that has not previously been ground) the shaft is due for reconditioning.

With regard to the main bearings, when the worn internal dimensions exceed 2.483″ (for the standard size bearings) replacements should be fitted undersized to suit the amount which has to be removed from the undersizes available, *viz* : —.010″, —.020″, —.030″ and —.040″.

(b) Crankshaft End Float

The float specified for the crankshaft is .004″—.006″ when new, which should be measured as shown in Fig. 17. Where, after the fitting of new thrust washers, end float is below .004″ the steel face of the thrust washers should be rubbed down on a piece of emery cloth placed on a surface plate as shown in Fig. 18. Do not reduce the white metal bearing surface.

The illustration shows the end float being measured by the feeler gauge method. An alternative method is the use of a Dial Test Indicator which will give a more positive reading if the dial is at " zero " when the crankshaft is at the limit of its float.

After a considerable mileage, wear may occur on the face of the crankshaft abutting the thrust washers. It may be necessary to fit oversize thrust washers,

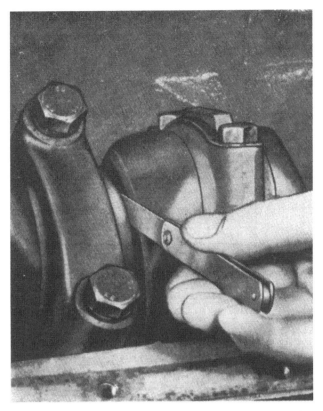

Fig. 17   Measuring Crankshaft End Float. This operation can be carried out with a Dial Test Indicator.

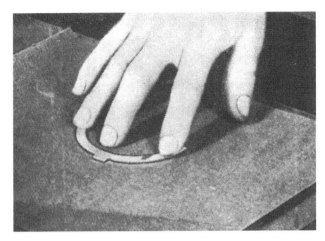

Fig. 18   Reducing the thickness of a Thrust Washer. This must only be carried out on the Steel Side.

and although this may rarely happen, oversize thrust washers +.005″ may be obtained by a special order on the Spares Department under their normal detail number specifying that the oversize thrust washers are required.

5. CONNECTING ROD BEARINGS (Fig. 8)

The connecting rod, a molybdenum manganese steel stamping, is provided with a phosphor bronze small end bush and the precision type lead indium bronze steel backed bearing at the big end. Like the main bearings, no hand fitting is necessary and in no circumstances should the bearing caps be filed to take up wear.

Where excessive journal wear has occurred the only satisfactory cure is to replace the bearings ensuring first, however, that the crankshaft journals and pins are in good order. Where a journal or pin is worn,

scored or tapered in excess of .0020″ regrinding is necessary.

When a regrind is found to be necessary a decision will have to be made as to the most suitable undersize bearings which will meet the particular case. The reduced diameter of the pin to suit the various undersize bearings may be calculated by subtracting —.0100″, —.0200″, —.0300″ or —.0400″ from the original size as listed on page 1. The small end bushes, dimensions given on page 2, should be pressed into the rods and subsequently reamed to $\frac{7}{8}″$±.0005″. A gudgeon pin selected to give a clearance of .0002″ at 68°F. This clearance will be indicated by a light finger push fit, with the piston warmed by immersion in hot water. The connecting rod centres are 6.250″± .002″ and there is no offsetting of the rod in relation to the bearing housings. The connecting rod cap is located in relation to the rod by means of dowel bush, as shown in Fig. 6.

Before installing a connecting rod it should be checked for alignment after first removing the bearing shells. The rod should be checked for bend, in which the piston will not be perpendicular to the crankpin, or if the gudgeon pin is not on the same plane as the crank pin the rod is twisted, see Fig. 19. Appropriate action should be taken to deal with the various causes of misalignment with a suitable bending bar. The connecting rod aligning fixture shown in Fig. 19 is obtained from Messrs. V. L. Churchill and Company Limited.

6. PISTON ASSEMBLY AND CYLINDER SLEEVES

The piston and cylinder bore dimensions are given on page 2. As indicated in this list of tolerances and limits, three sizes of pistons are used in conjunction with suitable bore dimensions. The three sizes of pistons and cylinder sleeves are indicated by the stamping of F, G or H on the crown of each piston and the upper flange of each cylinder sleeve as shown in Fig. 20.

Piston ring dimensions and clearances are also given on page 2. Where the worn clearance between the piston skirt and the cylinder sleeve bore exceed .007″ at the top and .005″ at the bottom reboring or replacement becomes necessary if a satisfactory repair job is to be executed.

The connecting rod should be fitted to the piston assembly with its bearing cap towards the split portion of the piston skirt

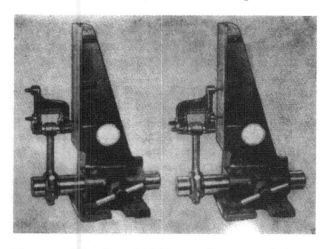

Fig. 19   The Churchill Fixture No. 335. Left-hand examining for " twist."   Right-hand examining for " bend."

and then should be assembled into the cylinder sleeves with the gudgeon pin in diametrical relation to pairs of opposite flats on the upper flanged faces of the cylinder sleeves. When assembling the sleeve and piston into the cylinder block, position the bearing cap of the connecting rod towards the camshaft side of the engine, or away from the point of maximum thrust. When cases of light wear occur and cause piston knock, an improvement can be effected by withdrawing the sleeve and rotating this 90° and so employ the alternate pair of flats as shown in Fig. 20.

The importance of using cylinder sleeve retainers to prevent relative movement of these parts is stressed.

Fig. 20   The Identification Letters stamped on the Piston Crown and the Cylinder Sleeves. Note also the flats on the outer periphery.

When the sleeves are installed in the block the flanged face should stand proud of the cylinder block by .003″ minimum—.0055″ maximum, and checked as shown in Fig. 5.

7. FIGURE OF EIGHT JOINTS (Fig. 4)

These joints are between the lower flanged face of the cylinder sleeves and the machined recesses in the cylinder block. They are metal and the plastic coating ensures that they afford a good water tight joint. Failure to do so will mean that water will leak from the cylinder block water jacket into the sump.

It is essential that these joints are handled and stored with great care to prevent damage to the plastic coat.

These joints are fitted one to each pair of cylinder sleeves. Before fitting, the sleeves and block should be thoroughly cleaned with a wire brush to ensure all scale and foreign matter is removed, and a light coating of " Wellseal " jointing compound applied to both sleeves and block. Extreme cleanliness is essential.

Sinking of the cylinder sleeves is prevented by the use of these metal joints. The sleeves should stand .003″ to .0055″ above the face of the cylinder block and a routine check should be made whenever the combustion head is removed. Should the cylinder sleeve(s) be below the specified limits new figure of eight joints should be fitted.

8. CAMSHAFT AND TIMING GEARS

The camshaft is of cast iron, having chilled faces for the cams and journals. With the camshaft a cast iron flanged front bearing is used, the other three journals making direct contact with the cylinder block.

In the near future it is proposed to fit four Vandervel bi-metal bearings to accommodate the camshaft A recognition of an engine so fitted with these bearings will be that three setscrews retaining the three rearmost bearings will be clearly visible on the left-hand side of the cylinder block. The front bearing is pressed into the front bearing sleeve.

The camshaft is driven by a double roller silent chain which engages with a sprocket on the crankshaft and one spigotted on the

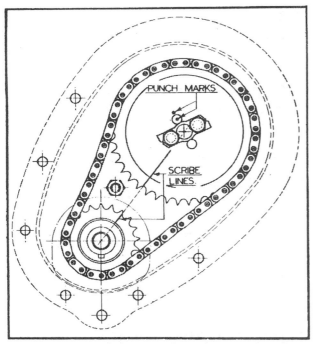

Fig. 21 Showing Wheel Markings for Valve Timing. Note the Keyway in the Crankshaft Sprocket pointing downward.

end of the camshaft and secured by two bolts.

Four holes are provided in the camshaft timing gear, which are equally spaced but

offset from a tooth centre. When the chain wheel is fitted at 90° to its initial position, which location we will identify as position "A", a $\frac{1}{2}$ tooth of adjustment is obtained. If on the other hand the wheel is turned " back to front " from position "A" a $\frac{1}{4}$ tooth of adjustment is obtained, whilst a 90° movement in the reversed position will give $\frac{3}{4}$ of a tooth variation from that given by position " A."

When the timing has been correctly set the faces of the two gears are marked with a scribed line drawn radially in such a manner that if the lines were produced outwards on the respective gears they would pass through the centres of the two gears.

In addition, to avoid any possibility of the camshaft position being incorrect, a centre punch mark is made on the end of the camshaft through an unoccupied bolt hole and on the face of the timing gear adjacent to the setscrew hole ; Fig. 21 shows the marking of the timing wheels.

The helical gear for the distributor and tachometer drive and the cam for operating the fuel pump are integral parts of the camshaft.

End float of the camshaft is taken between the flange on the front camshaft bearings and the rear face of the timing wheel.

This end float can be increased by reducing the length of the front bearing sleeve by rubbing the rearmost end on a sheet of emery cloth placed on a surface plate, to decrease the end float it will be necessary to replace the front bearing.

After grinding operations on the camshaft have been completed it is degreased, bonderized and whilst still warm immersed in a solution of " Dag " (colloidal graphite). This process considerably improves the bearing surfaces and gives additional wearing properties.

9. TO REMOVE CAMSHAFT

The camshaft may be removed from the engine while the unit is still in the chassis and the following procedure is used.
(a) Remove the front cowl and radiator as described in " Removal of Engine," page 28.
(b) Remove the cylinder head as described in " Decarbonising " and " Valve Grinding," page 25. Immediately after removal of the cylinder head,

sleeve retainers (Churchill Tool No. S.138) should be applied as shown in Fig. 22.

In the event of sleeve movement, new figure of eight washers should be fitted. Remove push rods and tappets.

(c) Disconnect tachometer drive. Remove distributor assembly complete with pedestal by removing the two securing nuts at the crankcase. Do not slacken clamp bolt. Remove distributor and oil pump helical driving gear.

(d) Check that the petrol has been turned off, remove petrol pipe and pump. (See "Fuel" Section P.)

(e) Loosen off dynamo and remove fan and fan assembly by withdrawing four bolts and the extension bolt.

(f) Remove the timing cover by withdrawing the seven setscrews, four bolts and one nut. Note the timing markings on the gear wheels and camshaft; this will assist in the re-assembly (see Fig. 21).

(g) Release the locking plate and withdraw the two setscrews. The timing chain can be lifted off the chain wheel and both components moved clear.

(h) The front camshaft bearing is next removed by withdrawing the two setscrews and locking washers. The bearing can be lifted away.

(i) The camshaft can now be drawn forward out of the cylinder block.

10. REFITTING CAMSHAFT

Re-assembly is the reverse procedure to the

Fig. 22 **Showing one of the two Cylinder Sleeve Retainers required to prevent movement.**

removal. It is considered desirable to describe certain operations as follows :—

(a) When resetting the valve timing, the engine should be set with Nos. 1 and 4 pistons at T.D.C. In this position the crankshaft timing wheel keyway is pointing vertically downwards, as shown in Fig. 21.

Rest the camshaft chainwheel on the camshaft spigot and turn the chainwheel about the camshaft until the identification punch mark on the end of the camshaft can be seen through the punch marked hole in the chainwheel. Secure the chainwheel to the camshaft leaving the two bolts finger tight.

Turn the camshaft chainwheel until the scribe line thereon aligns with the scribe line on the crankshaft sprocket. **Without moving** the camshaft remove the camshaft chainwheel and when removed fit the timing chain to this wheel and the one on the crankshaft in such a manner that the scribe lines remain aligned. Reposition the camshaft chainwheel and check by simulating pressure of the chain tensioner that the timing marks have retained their positions and re-adjust if necessary. Tighten bolts to correct torque loading and turn over tabs of locking plates.

(b) When refitting the oil pump and distributor driving helical gear, ensure that No. 1 piston is at T.D.C. on the compression stroke. In this position the correct engagement of the helical gear should allow the Woodruffe key to be positioned towards the front of the engine, pointing approximately towards the dipstick (Fig. 16). It may be found that the oil pump, shaft will not engage with the pump for the tongue and slot of these components are out of line. The engine will need to be turned over slowly until the shaft engages with the pump. Continue to turn the engine until the offset slot in the distributor drive boss attains the position as illustrated in Fig. 16. Disengage the helical gear and remove it from the housing. Turn the engine over until No. 1 piston attains the T.D.C. position on the compression stroke and replace the helical gear when the shaft will engage with the oil pump.

(c) Having refitted the cylinder head and rocker shaft it is advisable to apply oil to the ground surfaces where the rockers contact the valves, as these points do not immediately receive a supply of oil.

11. TO SET VALVE CLEARANCES

All adjustments should be made when the engine is cold.

(a) Remove the rocker cover from the engine.

(b) Turn the engine over by hand until the valves of any cylinder are on the point of rock. Note the number of this cylinder.

(c) Continue turning the engine for another complete revolution, this will ensure that the tappets of this cylinder are at the base of the cam (Fig. 23).

(d) Holding the ball pin in the rocker arm with a screwdriver, loosen the lock nut.

(e) Pressing down on the screwdriver to eliminate any slackness in the valve gear.

(f) Turn the screwdriver until a feeler gauge of .010″ for inlet valve or .012″ for exhaust valve will pass between the toe of the rocker and the tip of the valve stem. The ball pin or screwdriver is turned anti-clockwise to increase the gap and clockwise to decrease the gap.

(g) Holding the screwdriver steady, tighten the lock nut. Still applying pressure to the heel of the rocker check the gap and adjust if necessary.

(h) Repeat with the second valve of that cylinder.

(i) Having noted the number of this cylinder continue with the remaining three in the firing order 1, 3, 4, 2, by turning the engine half a revolution before making adjustments.

(j) Replace the rocker cover pressing, ensuring first that the cork seal is in sound condition and second, when placing the cover in position, that the right-hand side does not foul the combustion head securing nuts. Failure to observe either of these points may result in a serious loss of oil.

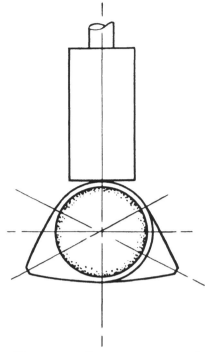

Fig. 23   Tappet on base or concentric position of cam.

12. TO SET VALVE TIMING IN THE ABSENCE OF TIMING WHEEL MARKINGS

It is assumed that, for the purpose of this instruction, the cylinder head and valve gear are in position and the crankshaft sprocket is keyed to the crankshaft but the camshaft chainwheel has yet to be fitted. The following procedure is recommended:

(a) Set valve rocker clearances for Nos. 1 and 4 cylinders to .015″ which is the valve timing clearances.

(b) Turn crankshaft until Nos. 1 and 4 pistons are at T.D.C.
This position may be found by placing the keyway in the crankshaft vertically downwards.

(c) Rotate the camshaft until the exhaust valve and inlet valve of No. 4 cylinder are at the point of balance in which the tappets will be in the position shown in Fig. 24. In this position the exhaust valve will just be about to close and the inlet just commencing to open. From the timing diagram, Fig. 25, it will be observed that the inlet valve opens at 15° B.T.D.C. and the exhaust valve closes at 15° A.T.D.C. 15° before or after T.D.C. is equivalent to .081″

(2.06 mm.) piston travel or 1.5"
(3.81 cm.) measured round the flywheel adjacent to the starter teeth.

(d) Offer up the camshaft chainwheel to
the camshaft itself but without moving
this shaft and adjust its engagement
with the chain until a pair of holes in
the chainwheel exactly match a pair in
the shaft. It may be necessary to turn
this wheel back to front to match
these holes.

(e) Having attained the correct position of
the chainwheel relative to the shaft,
encircle the wheel with the timing
chain.

(f) Without moving either crankshaft or
camshaft, position the loop of the chain
round the crankshaft sprocket in such
a manner that the holes in the chainwheel match those in the camshaft.

(g) The camshaft chainwheel is now secured to the camshaft by two bolts and
locking plates, the bolts are not locked
until a final check has been made.

(h) A final check can be made when the
engine is on a bench by marking the
rear of the cylinder block opposite
the T.D.C. mark on the flywheel
with Nos. 1 and 4 cylinders at T.D.C.

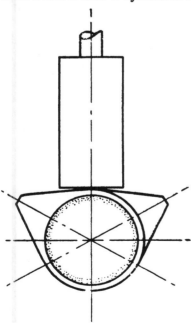

Fig. 24   **The Valve Tappet is at Point of Balance.**

The flywheel is then moved a
¼ turn anti-clockwise (viewed from the
front of the engine) and then turned

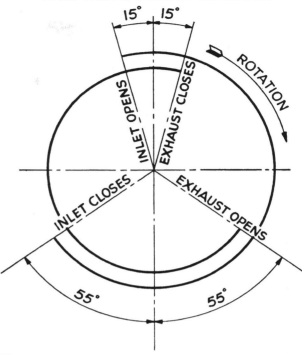

T.R.2. VALVE TIMING DIAGRAM

15° 15°

ROTATION

INLET OPENS

EXHAUST CLOSES

INLET CLOSES

EXHAUST OPENS

55°   55°

Fig. 25   **The TR2 Valve Timing Diagram.**

slowly in a clockwise direction. As the
flywheel is turned clockwise, insert a
.010" feeler gauge between the valve
stem and the rocker of No. 4 cylinder
inlet valve until a slight resistance is
felt, that is when the valve begins to
open. At this stage the movement of
the flywheel should be stopped; with a
pencil mark the flywheel opposite the
mark previously made on the cylinder
block.

Remove the feeler gauge from the inlet
valve.

Turn the flywheel clockwise until the
feeler gauge can be inserted between
the valve stem and the rocker of No. 4
cylinder exhaust valve, after which the
flywheel is turned to T.D.C. Proceed
to turn the flywheel slowly clockwise
and at the same time putting a slight
pull on the feeler gauge. The turning
of the flywheel should be stopped at a
point where the feeler gauge can be
removed and this indicates that the
exhaust valve has closed. A second
mark of the pencil is now made on the
flywheel opposite the mark on the cylinder block. With a rule measure the
distance from the T.D.C. mark on the
flywheel to each of the pencil marks.

If the timing is correct the two dimensions will be identical. Having finally proved the valve timing, the chainwheel locking tabs may be turned up.

(i) The timing gears are now marked with a scribe line as shown in Fig. 21.

(j) Fit the timing chain tensioner and secure with plain washer and split pin. Replace timing cover.

(k) The rocker clearances are now set to their working clearances of .012″ exhaust valve and .010″ for inlet valves (see page 22). When the car is used for high speed work the valve clearances for all valves is .013″.

13. IGNITION AND DISTRIBUTOR TIMING

See also "Engagement of Oil Pump and Distributor Driving Gear". (Page 16.)

It is important that the "Distributor and Tachometer Gear Assembly" is fitted with an end float of .003″ to .007″.

This can be measured in the following manner:—

(a) Measure and note the thickness of a $\frac{1}{2}$″ washer and assemble it with the distributor-tachometer driven gear to the oil pump driving shaft.

(b) Install this assembly in the cylinder block with the washer between the gear and the shaft bearing in the cylinder block. Ensure that the shaft is engaged in the oil pump.

(c) Over the gear assembly fit the distributor adapter.

(d) Utilising feeler gauges, ascertain the distance between the distributor adapter and its mating face on the cylinder block.

(e) When this measurement is compared with the thickness dimension of the washer the difference will represent the amount of "end float" or "interference".

Example

Thickness of washer .060″
Distance between faces .055″

The distance, being less than the washer, gives the gear assembly an "end float" of .005″.

Conversely

Thickness of washer .060″
Distance between faces .065″

The distance being greater than the washer, gives the gear assembly an "interference" of .005″. It will be necessary to fit shims or packings under the distributor adapter to obtain the correct end float.

Assuming the first instance to be the case, it will be necessary to add one packing of .002″ thickness to bring the end float to top limit. For the second instance it will be first necessary to "zero" the interference, i.e., .005″ and add sufficient packings to obtain the correct end float. The packing necessary in this case is .011″ for a middle limit end float.

(f) Remove the gear assembly, shaft and washer from the cylinder block.

(g) Turn the engine until the piston of No. 1 cylinder is at T.D.C. on compression stroke, in this position both valves will be closed.

(h) Fit the Woodruffe key to the oil pump driving shaft and insert the shaft in the block to engage the oil pump with its tongue. Rotate the shaft until the key is at right angles to the camshaft and points away from the engine.

(i) Position and lower the distributor-tachometer driven gear on the drive shaft until the keyway and the key engage. Continue a downward motion turning the gear clockwise to effect engagement with the driving gear on the camshaft. Caution must be exercised to prevent dislodging the Woodruffe key.

(j) When correctly engaged the offset slot in the gear assembly will be aligned with No. 1 pushrod sealing tube and the offset towards the rear of the engine. Similar to Fig. 16.

(k) Assemble the distributor adapter together with the necessary packings to obtain the correct end float. Secure with nuts and locking washers.

(l) Fit the distributor body with the rotor arm pointing to No. 1 push rod tube.

(m) Adjust the points to .015″ and with the contact points just commencing to separate the vernier adjuster on the third marking of its scale, secure the body to the adapter bracket with the nut and lock washer with a plain washer, under the lock washer.

(n) Advance the vernier a further 1 division, which is equivalent to advancing the ignition 4° on the flywheel B.T.D.C.

(o) Fit the distributor cover, connect the plug leads to the correct plugs (Fig. 26). The plugs having had their gaps set to .032″. Fit the H.T. and L.T. leads to the ignition coil.

14. TO DECARBONISE

We recommend the removal of the cylinder head for decarbonising after the first 5,000 miles. Attention after this running period has the advantage of allowing the initial casting stresses to resolve themselves and permits the consequent valve seat distortion to be counteracted by valve grinding. Failure to carry out this initial valve grinding is a frequent cause of excessive petrol consumption of new cars. Subsequent attention will not normally be required until further considerable amount of running has been done—normally after about 15,000 miles.

The above mentioned figures only take into consideration a car which is used under normal conditions. If the car is being used for competition and high speed work valve grinding is done as and when necessary.

The procedure recommended for decarbonising is as follows :—

(a) Disconnect the battery lead and plug leads from plugs.

(b) Drain the cooling system.

(c) Disconnect the fuel pipe clip, the top water and by-pass hoses and remove the thermo gauge bulb from the thermostat housing, then remove the latter from the cylinder head by withdrawing the two bolts.

(d) Remove the two rocker cover securing nuts and lift off the rocker cover.

(e) Remove the rocker shaft assembly by loosening off the four pedestal nuts progressively, allowing the assembly to rise as a unit.

(f) Remove the heater hose from the water shut-off cock at the rear of the cylinder head. (Where heater is fitted.)

(g) Disconnect the throttle and choke controls, the suction pipe and fuel feed pipe from the carburettors. Whilst there is no need to remove the carburettors this can be effected at the carburettor and manifold joints. (See "Fuel" Section P.)

(h) Remove the ten cylinder head nuts and lift the head from the block. **Do not**

Fig. 26   Plug Lead Attachment Sequence.

attempt to break the seal of the cylinder head by turning the engine as this will disturb the cylinder sleeves.

(i) Immediately the combustion head has been removed, place cylinder liner retainers in position (Fig. 22) and check the projection of the cylinder sleeves above the face of the cylinder block (Fig. 5). The flange of the cylinder sleeves should stand proud by .003″ minimum to .0055″ maximum. If the cylinder sleeves have sunk below .003″ new figure of eight joints will need to be fitted. (See page 19.)

Inspect also for cylinder sleeve movement and if any is suspected the cylinder sleeves and pistons will have to be removed and new figure of eight joints fitted.

Remove the push rods.

15. VALVE GRINDING

Lay the cylinder head on a bench so that the valve heads are supported, this will ensure that when pressure is exerted on the valve spring cap this spring will compress and the cotters easily removed.

The valves are numbered from the front of the engine and their positions perpetuated. The carbon should be cleaned off with a blunt instrument and finally cleaned with a petrol moistened rag.

Grind the valves into their appropriate seating, where valve faces are badly pitted they should either be renewed or replaced. No attempt should be made to grind a badly pitted valve into its seating or this will be unduly reduced.

When the necessity of recutting a valve seat

arises, it is important that the valve guides are concentric with the seats themselves. Where a valve guide is badly worn it should be replaced before the seat is recut.

While refacing valves, only remove sufficient metal to clean up the face, otherwise if too much is removed the edge will tend to curl up in service.

Where valve seats are badly worn or pitted they should be recut with an 89° cutter utilising a pilot of the same diameter as the valve stem. Should the valve seating become embedded in the cylinder head as shown in Fig. 27, it will first become necessary to employ a 15° cutter, to provide a clearance for the incoming or outgoing gases, following this with a cutter of $44\frac{1}{2}°$. This work should be carried out after the cylinder head has been cleaned.

The valve and guide data is given on pages 3 and 4.

16. REMOVAL OF CARBON

Remove the spark plugs, clean, set and test ready for replacement. If for any reason such as badly burnt or broken electrodes, and damaged insulation the plug should be replaced. For normal motoring Champion L10S ¼" reach; for high speed motoring L11S ¼" reach is recommended and the gap is to be set at .032". The normal life of a spark plug is 10,000 miles.

Clean the carbon from the cylinder head, finally wipe the chambers clean. Scrape the valve ports clean, exercising great care not to damage the valve seats. When the head is clean of carbon blow out with a compressed air line and wipe with a rag moistened with petrol. Ensure that the contact face is perfectly clean and flat.

Before cleaning the carbon from the tops of the pistons, smear a little grease around the top of the two bores and raise the piston almost to the top. Fill the other two bores and tappet chambers with non-fluffy cloth; this will safeguard against any carbon chips entering the lower extremities of the engine. It is suggested that the piston crowns are cleaned, utilising a stick of lead solder, which will not scratch the piston crown, in such a manner that the carbon deposit on the vertical wall of the piston and that deposit formed in each cylinder bore above the maximum travel point of the top piston ring is not disturbed. This carbon helps to insulate the piston rings from the

heat generated during combustion and provides a secondary oil seal.

The use of emery cloth or other abrasive for polishing is not recommended as particles of such abrasive may enter the bores and engine after re-assembly, causing serious damage.

Having cleaned two pistons, brush and blow away the carbon chippings, taking care not to allow any to drop into the cylinder block. Lower the clean pistons in their bores and wipe away the grease, remove the cloth stuffing from the other two piston bores and grease the tops. After greasing the tops of the cylinder bore raise these pistons and fill the remaining two bores with the rag. Repeat the cleaning operation. On completion of the piston cleaning, wipe and blow away the carbon chips and clear the block face, particularly around the cylinder sleeves and the tops of these

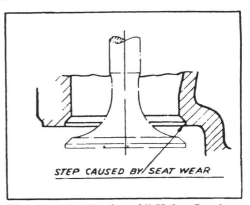

STEP CAUSED BY SEAT WEAR

Fig. 27   A " Pocketed " Valve Seating.

sleeves. Clean the grease from the cylinder bores and remove the cloth stuffing from the bores and tappet chambers.

The valve springs should be examined for damage and their length compared with new springs. If any doubt exists as to the condition they should be replaced. The exhaust valve is fitted with an auxiliary inner spring, making three springs in all. It should be noted that the close-coiled end of these springs is fitted nearest the cylinder head.

Ensure that the cylinder block and head faces are perfectly flat and clean, it should only be necessary then to apply a coating of grease to the cylinder gasket. Should it be decided to use a sealing compound, one of the non-setting type must be used for on future occasions when the head is removed, the cylinder sleeves may be disturbed because of their adherence to the gasket.

When refitting the cylinder head nuts, tighten them gradually in the sequence shown in Fig. 28 in order to produce an even pressure on the gasket and prevent undue strain in the cylinder block casting. It will be necessary to recheck the nut tightness when cold to 100—105 lbs. ft. Before tightening down the rocker shaft pedestals, screw back each adjusting screw and ensure that the ball ends of these screws engage correctly in the push rods. Failure to attend to this procedure can result in damage to the push rods. Smother the rocker gear with oil, particularly where the rockers bear on the valves.

Before replacing the rocker cover ensure that the cork joint is undamaged and shellaced to the cover, otherwise oil may leak through the joint.

After the first 500 miles the cylinder head nuts should be checked for tightness with the engine hot.

17. LOW COMPRESSION KIT— PART No. 502227

This kit was introduced for those owners who experienced difficulty in obtaining fuels of a high octane value.

The kit comprises of :—

8 Push Rods (longer than those normally fitted).
1 Combustion Head Gasket.
1 Low Compression Plate.
1 " Corgasyl " Combustion Head Gasket.

(a) Prepare the engine unit as for decarbonisation (see page 25.) No attempt should be made to break the combustion head seal by turning the crankshaft—this action will only disturb the cylinder liners on their lowermost seatings and water leakage will result. When the head has been removed fit liner retainers (Fig. 22) and check that the liners stand proud of the cylinder block .003″ to .0055″ (see page 19.)

(b) Apply a light coating of " Wellseal " jointing compound to both sides of the low compression plate and gaskets.

(c) Fit the copper cylinder head gasket (smooth face downwards), followed by the low compression plate and steel

" Corgasyl " gasket; this may be fitted either side up.

(d) Fit the longer push rods and lower the combustion head into position. Omitting the plain washers, tighten the combustion head nuts (Fig. 28) to the correct torque (100 to 105 lbs. ft.).

(e) Screw back the ball pins in the rockers and then fit shaft assembly to the com-

Fig. 28   Cylinder Head nut tightening sequence

bustion head and tighten nuts to 24—26 lbs. ft.
Adjust valves for clearance. (See page 22.)

(f) Reconnect fuel pipe, carburettor/distributor suction pipe, throttle and choke cables to carburettors.

(g) Replace rocker cover, ensuring first that the seal is in good order, also the thermostat housing, thermo gauge bulb.

(h) Refit heater hose (if heater is fitted), by-pass hose, top radiator hose. Replenish cooling system with coolant.

(i) Reconnect fuel pipe at pump. Connect battery lead.

NOTE—After the first 500 miles the cylinder nuts should be checked for tightness with the engine hot.

18. THE " PUROLATOR MICRONIC " OIL FILTER—TYPE 17F. 5102 (Fig. 29)

The Purolator Micronic filter consists of a plastic impregnated paper element which removes the finest particles of abrasive which invariably find their way into the engine. A filter of this type will stop not only the smaller microm sized particles of abrasive, but ensures a supply of clean oil to the engine at all times. The only attention which the filter needs is to see that the element is changed at periods not exceeding 8,000 miles. It is essential that this operation is carried out at specified periods

27

to ensure maximum filtration. To renew the element proceed as follows :—

(**a**) Clean the outside of the filter casing.

(**b**) Unscrew the centre bolt and remove the filter casing and element.

NOTE—The paper element, its perforated outer cover and element tube forms a complete element assembly.

Ensure that the top seal is retained in position in the groove in the filter head.

(**c**) Withdraw the element and clean the inside of the casing.

(**d**) Insert a new element into the filter casing.

(**e**) Fit the filter and new element to the filter head ensuring that the spigot formed on the head enters the centre tube of the element squarely. Tighten the centre bolt sufficiently to ensure an oil-tight joint. (14-16 lbs. feet.)

(**f**) Run the engine for a few minutes and inspect the filter for leaks. If leakage is noted between the filter casing and the head, the centre bolt must be unscrewed and the casing and element withdrawn. A new top seal should then be fitted. If leakage occurs at the bottom of the filter, withdraw the casing and element, remove the circlip from the centre bolt and withdraw the bolt from the casing ; collect the element support, bolt seal, washer and spring. Ease the remaining seal out of the bottom of the casing and fit a new seal in its place. Insert the centre bolt and fit the spring, the washer, a new bolt seal and the element support on to the part, fit circlip into its groove in the bolt. Place the element inside the casing and offer up the assembly to the filter head, screw the centre bolt home. A certain quantity of oil will be lost due to the removal of the filter casing, and the sump should be topped up after assembly of the filter.

The filter casing should not be disturbed until element renewal is required. To do this invites the hazard that the accumulated dirt on the outside of the filter may be allowed to contaminate the inside ; thus being

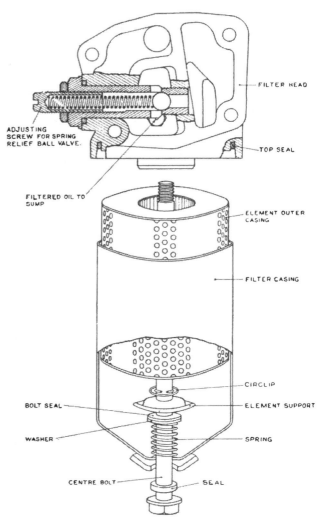

Fig. 29   **The Purolator Oil Filter.**

carried into the bearings when the engine is re-started.

Do no attempt to reset the pressure relief valve which is incorporated in the filter head. This is the main engine pressure relief valve and is set at the works to a predetermined figure.

19. REMOVAL OF ENGINE AND GEAR-BOX AS A UNIT

(**a**) Disconnect the battery. Turn off petrol at shut-off cock.

(**b**) The bonnet is removed by removing four hinge nuts, two at each hinge.

(**c**) Drain off the cooling fluid by opening the taps, one at the base of the radiator and the second situated below the inlet and exhaust manifold in the cylinder block.

(d) Drain off the oil from the engine and gearbox.

(e) Disconnect the head and side light cables at their snap connectors. Remove the bolts from the top brackets and the bolt in the centre of the cowling, this holds the bonnet lock connecting cable, release cable control at one side. Remove the twelve setscrews (six per side) situated under the wheel arches. Remove the starting handle bracket and the steady rods from under the cowling and finally the nut and bolt from the steady plate.

(f) To remove radiator disconnect top and bottom hoses, release the tie rods at the top and the bolts one either side at the base of the unit.

(g) Disconnect the lever linkages at the foremost carburettor; disconnect the inner and outer cables of the choke control and the fuel feed pipes at their banjo unions. Remove the carburettors from the manifold by undoing the four nuts—two at each flange.

(h) Remove the horns from their brackets by first removing the four fixing bolts (two to each horn). There is no need to disconnect the horns from their cables. Disconnect dynamo leads and remove dynamo from its bracket and remove fan belt.

(i) Remove front chassis cross tube by removal of three nuts and bolts at each flange.

(j) Remove the three nuts and washers at the exhaust flange and break the joint.

(k) Disconnect the flexible fuel pipe at the petrol tap, the oil pressure gauge pipe, starter motor cable, L.T. lead at the coil, the tachometer drive at distributor pedestal and withdraw the water temperature gauge bulb.

(l) To remove the seats, first remove the cushions and unscrew the sixteen nuts (eight to each seat) thus releasing the frame from the runners; it can then be lifted out.

(m) Free the rubber gear lever grommet by the removal of three self-tapping screws from the gearbox cover pressing and remove the latter by unscrewing the thirteen setscrews, hidden by the trim and floor covering.

(n) Remove the gear lever with grommet by loosening the locknut and unscrewing the lever.

(o) Remove the speedometer drive, the overdrive cable at the snap connector and the starter motor by removing two nuts and bolts.

(p) Drain the clutch hydraulic system. Disconnect the bundy tubing at the flexible hose at the left-hand side chassis member whilst holding the hexagon on the hose. Still holding the hexagon remove the hose securing nut and shakeproof washer; the flexible hose can now be withdrawn from its bracket.

(q) Uncouple the propeller shaft by removing the four nuts and bolts securing the two flanges. Remove the two nuts holding the gearbox to the chassis frame.

(r) Remove the four nuts and bolts (two each side) securing the engine mountings to the chassis.

(s) Fit slings to engine and lift out in a " nose up " position, as shown in Fig. 31.

Fig. 30   The front of Car prepared for Engine and Gearbox Removal.

29

20. DISMANTLING ENGINE

It is sound policy to clean the exterior of the engine and gearbox before commencing to dismantle.

(i) Detach gearbox by removing the nine nuts and bolts from the clutch bell housing.

(ii) Remove the clutch from the flywheel by withdrawing the six securing bolts.

(iii) Remove the flywheel by unlocking the tab washers and withdrawing the four bolts.

(iv) To remove the fuel pump, first disconnect the pipe to the carburettors and then remove the nuts and lock washers from the studs. It will be noticed that the rearmost stud accommodates the oil pressure gauge pipe clip.

(v) Remove rocker cover, together with oil filler cap.

(vi) Remove suction pipe from distributor and sparking plug leads, H.T. and L.T. leads at the ignition coil. Avoid loosening clamping bolt and remove distributor from pedestal, secured by two nuts with locking and plain washers. Lift out distributor and tachometer driving shaft assembly.

(vii) Remove the ignition coil.

(viii) From the front of thermostat housing remove the nut holding the clip for the

Fig. 31 The Engine and Gearbox being removed from the Chassis. Note the " nose up " attitude.

fuel and suction pipes; these two pipes are strapped together and can be lifted away. Remove the by-pass hose from the thermostat housing to the water pump housing after undoing the two hose clamps. Withdraw thermostat housing as a unit following the removal of the two bolts and lock washers securing it to the combustion head.

(ix) Remove water pump impeller after withdrawing one bolt and two nuts.

(x) Remove the water pump housing which is held by two bolts and spring washers.

(xi) Proceed to remove oil filter assembly by first removing the cap nut holding the oil pressure pipe banjo to the filter. This pipe can now be detached. The remaining three bolts can then be removed and the filter assembly taken away.

(xii) Remove dynamo bracket and pedestal.

(xiii) Remove fan assembly by withdrawing four bolts, followed by the extension bolt; the hub and hub extension can now be withdrawn from the crankshaft.

(xiv) Remove timing cover and packing, remove chain tensioner after withdrawal of split pin and washer. Observe the markings on the camshaft chainwheel and crankshaft sprocket which should correspond to Fig. 21 when No. 1 piston is at T.D.C. of compression stroke.

(xv) Release the tabs of the locking plate and withdraw the two bolts to release camshaft chainwheel, the chain can now be freed from the crankshaft sprocket. Camshaft chainwheel and chain can now be lifted away and the crankshaft sprocket and Woodruffe key removed from the crankshaft, followed by the shims.

(xvi) Lift rocker shaft assembly by removal of the four pedestal nuts.

(xvii) Remove the inlet and exhaust manifolds by removing eight nuts and six clamps.

(xviii) Remove combustion head by removal

of ten nuts and washers and lift out the push rods and tappets.

(xix) The camshaft can be withdrawn by first removing two bolts securing the front bearing, then the bearing and finally the camshaft.

(xx) Remove the nineteen sump securing bolts and remove the sump. Care should be taken not to damage the oil pump filter.

(xxi) Remove oil pump from inside cylinder block by unscrewing the three nuts and washers.

(xxii) Remove the front engine plate from the block by removing the five attachment bolts, and discard the packing.

(xxiii) Remove the bearing caps, bottom halves of the shell bearings and thrust washers by releasing the tabs of the locking plates and withdrawing the bolts. Remove also the big end bearing caps and bottom halves of the shell bearings by releasing the locking plates and withdrawing the bolts.

(xxiv) Lift out the crankshaft and collect the upper halves of the shell bearings.

(xxv) Collect the upper halves of the big end shell bearings and withdraw the connecting rods and pistons from cylinder block. The cylinder sleeves may be tapped out gently from below.

21. RE-ASSEMBLY OF ENGINE

When the engine is completely dismantled the following procedure is suggested for re-assembly.

The cylinder block and combustion head should be examined for leakage at the various core plugs. If these do show signs of leakage they must be renewed, their seatings thoroughly cleaned and new plugs fitted with jointing compound.

The main and big end journals of the crankshaft should be checked for wear against the dimensions listed on page 1.

Wear in excess of .0025" on the crank pins and the journals should be met by re-grinding, but where the bearing alone is seriously worn (in excess of .003") its replacement should suffice.

The bores of the sleeves should be measured and if more than .010" in excess of the dimensions quoted on page 2 they should be renewed. It should be noted that maximum wear occurs at the top of the bore.

The camshaft and camshaft bore should also be dimensionally examined. Journal wear in excess of .003" will necessitate a replacement shaft, whilst wear in the cylinder block bores of more than .0035" will entail a replacement block.

It is intended in the very near future to introduce replaceable camshaft bearings for all journals. At the time of going to press full details are not available and this matter will be dealt with in an issue of "Service Information."

The combustion head should be examined and due attention paid to valve guides, valve seats, valve springs and the valves themselves. Valve guides should be replaced if they are more than .003" oversize their original dimensions quoted on page 3.

Valve seats should be ground in, or if "pocketed" (Fig. 27), new seats should be shrunk in.

Valve springs should be thoroughly examined for cracks and dimensions compared with those quoted on page 5.

Valves should be examined to ensure that their stems are prefectly straight and the faces recut.

The block and the head should be thoroughly cleaned or blown out by compressed air to ensure that all foreign matter has been removed. Bolts, setscrews and nuts are to be tightened to the torque loadings given in General Data Section.

All joint washers, gaskets, locking washers, lock plates and split pins must be renewed

(i) Check that the two halves of the rear oil seal bear the same number (Fig. 32).

These are machined as a mated pair and failure to observe this instruction may result in oil leakage. Shellac the top half of the oil seal and attach it loosely to the cylinder block by its four bolts and lock washers. Shellac and similarly fit the lower half of the oil seal to the rear bearing cap. Ensuring that the crankshaft mandrel is clean (Fig. 33), lay it in the rear bearing housing (without the shell bearings).

Fit the bearing cap and tighten down sufficiently to nip the mandrel. Tighten the eight bolts to secure the oil seal to the cylinder block and bearing cap (torque loading of 8—10 lbs. ft). Remove bearing cap from block.

(ii) Fit the upper half of the main bearings to the cylinder block; thoroughly clean and lubricate; place the crankshaft in position.

(iii) Fit the lower halves of the main bearings to the bearing caps, and lubricate.

(iv) Thread the two top halves of the thrust washers at the side of the centre main bearing between the crankshaft and the cylinder block.

It is essential that the white metal side is toward the crankshaft.

Fit the thrust washers, one either side, to the centre bearing cap (Fig. 34) and lightly secure with the two bolts and lock washers to cylinder block. Fit the two remaining caps to the cylinder block with two bolts and two lock washers each.

(v) Commencing from the front of the engine, tighten the bearings cap bolts to the correct torque (see "General Data"). On tightening the rear bearing

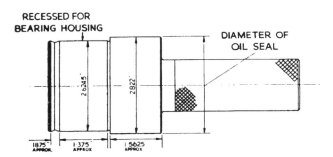

RECESSED FOR
BEARING HOUSING

DIAMETER OF
OIL SEAL

Fig. 33   **Crankshaft Mandrel for centring the Rear Oil Seal.**

be tightened down so that the oil seal division is flush.

(vi) Check the crankshaft end float by the use of the feeler gauges or by using the dial indicator gauge as shown in Fig. 17. Should the end float determined be greater than .006″, thicker thrust washers may be fitted; when the float is less than .004″, thinner washers are needed or the existing ones should be rubbed down on emery paper (Fig. 18).

(vii) Fit the front main bearing sealing block and tighten down the two cheese-headed bolts using a substantial screwdriver. Check that the face of the block is flush with the face of the cylinder block.

Plug the two cavities, one either end of the sealing block, with the sealing pad coated with shellac.

Fig. 32   **Rear Oil Seal for Crankshaft. These are a mated pair and should not be separated.**

cap, tap the oil seal lightly so that the joint between the two halves is flush.

In the absence of a crankshaft mandrel the oil seal attachment bolts will still be loose at this juncture. They should now be tightened to a torque loading of 8—10 lbs. ft. The bearing cap must

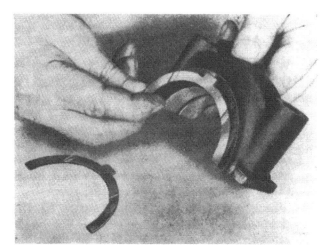

Fig. 34   **Fitting the Lower Thrust Washer to the Centre Bearing Cap.**

(viii) After dipping the felt packing strip into shellac force it into the recesses

either side of the rear main bearing cap with the aid of a $\frac{3}{16}''$ square brass drift (Fig. 35). Two lengths about 9" long are necessary. Completely fill the groove and cut the felt off $\frac{1}{64}''$ proud of the cylinder block face. It is suggested that the felt strip is cut into approximately $\frac{3}{4}''$ lengths for easy insertion.

(ix) Check the connecting rods for alignment in the Churchill Tool No. 335 or a similar tool. Press the Clevite bush into the small end of the connecting rod and ream out whilst in position using the Churchill Tool No. 6200A and reamer; dimensions are to be found on page 2. Assemble the piston to the connecting rods so that the split of the skirt faces the cap side of the rod (Fig. 7). It is suggested that the pistons be first submerged in hot water for a few moments and the gudgeon pin should then be a light push fit. Secure the gudgeon pin with circlips, one either side. Dry the piston and rod assemblies thoroughly.

(x) Fit the piston rings to the pistons, the two compression rings are uppermost with one oil scraper ring below. Lubricate freely. Move the rings so that their gaps are 180° removed from one another; failure to observe this point may lead to increased oil consumption. Wire brush the exterior of the cylinder liners to ensure that they are free from scale and all loose dirt on their machined surfaces. With the assistance of a piston ring compressor fit the piston assemblies to the cylinder sleeves bearing the same letter as the piston.

(xi) Arrange the piston and connecting rod assemblies now in their cylinder sleeves, so that the numbers stamped on the rods and caps run consecutively, i.e., 1, 2, 3, 4. Turn these assemblies upside down in pairs, 1 and 2, 3 and 4, with the flat of the liner adjacent to one another. The bearing caps are now all uppermost and must be turned face one way. Remove the bearing caps and fit the shell bearings to rods and caps. Fit one figure of eight packing, using a light coating of "Wellseal" jointing compound on the flanged faces of each

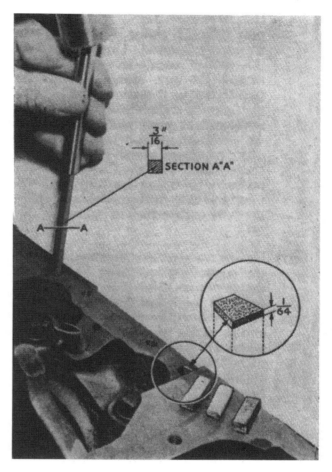

Fig. 35   **Sealing Rear Main Bearing Cap.**

pair of cylinder sleeves and on the mating faces in the cylinder block after ensuring that all components have been thoroughly cleaned of all loose deposits and the machined surfaces in which the cylinder sleeves spigot are clean and free from burrs, the sleeves with their respective piston assemblies can now be fitted to the block.

(xii) Locate the cylinder sleeves and piston assemblies in the cylinder block so that the cap of the connecting rod is adjacent to the camshaft side of the engine. The assembly which bears the number 1 on its connecting rod is fitted to the foremost position. The sleeves should stand .003" to .0055" proud of the cylinder block face (Fig. 5).

(xiii) It is essential that means are employed to prevent the cylinder sleeves from moving in the block. Messrs. V. L. Churchill & Co. Ltd. have manu-

33

factured special retainers for this purpose (Fig. 22) and it is suggested that these are employed. Until this is done the piston assemblies must not be moved, for any movement will be transferred to the sleeve and damage the figure of eight washers. If damage is undetected, water leakage will result. An alternate method is to insert the cylinder sleeves **alone** into the block, clamp them with the Churchill sleeve retainers to ensure no further movement and then fit the piston assemblies similarly as described in paras. (**x**) and (**xi**).

(**xiv**) Having the sleeve retainers in position, the connecting rods may be fitted to the crankshaft, Nos. 1 and 4 cylinders, followed by 2 and 3 cylinders. The caps are fitted to their respective rods and in such a manner that the tubular dowel will sink into its recess and their identification numbers coincide. It should be noted that the bearing cap, because of this dowel, can only be fitted one way round. The cap is secured by two bolts and a locking plate. Tighten the bolts to the correct torque loading and turn over the tabs of the locking plates.

(**xv**) Push the oilite bush into the centre of the crankshaft at its rear end and tap the flywheel locating dowel into position in the flange.

(**xvi**) Fit flywheel located by the dowel so that the arrow marked on its periphery lines up with the centre of the cylinder block with Nos. 1 and 4 pistons at T.D.C. Secure flywheel with the four setscrews and two locking plates, then turn over the tabs of the locking plates when the setscrews have been tightened to their correct torque loading.

(**xvii**) Utilising jointing compound affix the front plate packing and locating the engine plate on the two dowels secure with the five bolts and locking washers. Fit the engine mountings secured by two nyloc nuts.

(**xviii**) To the forward end of the crankshaft fit the sprocket locating shims, the Woodruffe key and the sprocket wheel.

(**xix**) Lubricate the camshaft and feed into the cylinder block and secure the front bearing with two setscrews. Check the end float as described on page 17. Rest the camshaft chainwheel on the camshaft spigot and turn the chainwheel about the camshaft until the identification punch mark on the end of the camshaft can be seen through the punch marked hole in the chainwheel. Secure the chainwheel to the camshaft leaving the two setscrews finger tight. If a replacement chainwheel is being fitted, see "12. To set Valve Timing in the Absence of Markings" (page 22). Check the alignment of the chainwheel with that of the sprocket on the crankshaft, taking into consideration the end float of the camshaft. The alignment can be adjusted by altering the thickness of the shim between the crankshaft sprocket and the abutment on the crankshaft.

(**xx**) Turn the camshaft chainwheel until the scribe line thereon lines up with the scribe line on the crankshaft sprocket. **Without moving** the camshaft remove the chainwheel and when removed fit the timing chain to this wheel and the one on the crankshaft. Reposition the camshaft chainwheel and check by simulating pressure of the chain tensioner that the timing marks have retained their positions and re-adjust if necessary. Tighten bolts to correct torque loading and turn over tabs of locking plates. Lubricate tappets and place in tappet chambers.

(**xxi**) Fit the chain tensioner to its pin and secure with washer and split pin. Screw in timing cover support bolt to the engine plate and fit the oil deflector to the crankshaft so that the raised edge faces the timing cover.

(**xxii**) Press the oil seal with its lip inwards into the timing cover and fit this cover with its packing to the engine plate utilising one nut, eleven bolts with four nuts.
NOTE—See that the short earth bonding strip from engine to chassis frame is attached under the head of the bolt

which aligns with L.H. rubber mounting attachment nut.

(xxiii) The machined faces on the combustion head and the upper flanges of the cylinder sleeves, which contact the combustion head gasket, should be lightly coated with " Wellseal " sealing compound. A substitute compound, which retains its plasticity, may be used if " Wellseal " is not available. This sealing is necessary to ensure a proper life for the gasket.

(xxiv) Assemble the valves and springs to the combustion head (see " To De-carbonise," page 25) and fit the assembly to the block, tightening the ten nuts and washers down as shown in Fig. 28. Fit push rods in the chambers.

(xxv) Assemble the rocker shaft as follows : To the rocker shaft fit No. 4 rocker pedestal in such a manner that the oil-feed holes coincide and secure with setscrew. To the shorter end of the shaft, fit No. 8 rocker, a double coil spring washer and a collar. Secure the collar to the shaft with a mills pin. On the longer end of the shaft feed the remaining rockers, springs and pedestals (see Fig. 36). After fitting No. 1 rocker, fit the double coil spring and collar securing the latter with a mills pin.

(xxvi) Loosen the ball pins and fit rocker shaft assembly to combustion head securing the pedestals to the studs with four nuts and spring washers. Before exerting any pressure on the nuts it is recommended that the adjusting pins are slackened off to prevent them coming into too hard a contact with the push rods. Tighten down the nuts progressively to the correct torque loading (see "General Data" Section A).

(xxvii) Adjust valve clearances. See "11. To Set Valve Clearances" (page 22).

(xxviii) Fit the oil pump assembly and packing secured by three nuts and lockwashers to the inside of the cylinder block.

(xxix) Fit the sump and packing to the cylinder block and secure with nineteen bolts and lock washers. The shorter

bolt is fitted through the front flange of the sump into the sealing block. The rearmost bolt on the left-hand side accommodates the breather pipe clip and the bolt in front of this accommodates the clutch slave cylinder stay. When an aluminium sump is fitted, two packings are used, one either side of the tray.

(xxx) Fit the breather pipe to the cylinder block and secure the clip to the sump plate by the bolt, nut and lock washer with a distance piece between the two plates.

(xxxi) Fit ignition coil to side of cylinder block with two nuts and lock washers.

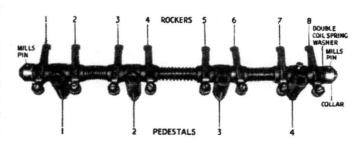

Fig. 36   **The Rocker Gear Assembly.**

(xxxii) Fit distributor and adapter as described in "13. Ignition and Distributor Timing" (page 24).

(xxxiii) To the pulley hub and hub extension assemble the fan pulley in such a manner that the T.D.C. indicating hole in the pulley is diametrically opposite the key way in the pulley hub centre ; secure with six nuts and bolts locked in pairs with locking plates. On later production cars with engine numbers after T.S. 4145E the locking plate and nut was replaced by a plain washer and nyloc nut.

(xxxiv) Fit the Woodruff key to the crankshaft, offer up the pulley assembly and secure with the extension bolt. Shims are placed behind the head of this bolt, which incorporates the starting handle dogs, to provide the correct relation with the starting handle and the engine compressions. This position is obtained with Nos. 1 and 4 pistons at T.D.C. and the dog faces corresponding to "10 minutes to 4 o'clock" (Fig. 37).

(**xxxv**) To the fan assembly fit the split rubber bushes (four front and four to the rear) and slide into the bushes the four metal sleeves. Place on top of the rubber bushes four larger diameter plain washers, the lockwasher for the starting dog extension bolt followed by the balance piece placed in such a manner that the drilled holes coincide with the drill spot on the hub extension To the securing bolts fit the locking plates and smaller diameter plain washers and feed through the holes in the fan blade assembly, and offer up the hub assembly to the crankshaft and secure, finally turning over the tab washers.

(**xxxvi**) Using a new joint washer and sealing compound, offer up the water pump housing to the cylinder block and secure with two bolts and lock washers and tighten to the correct tightening torque. Affix a joint washer to the housing with sealing compound and offer up the water pump impeller. This is secured by two nuts with lock washers and a bolt with lock washer, the purpose of this bolt is twofold, it secures the impeller to the housing and the housing to the cylinder block. Attach the adjusting link with a bolt and tab washer to the right-hand side of the water pump housing but leave the bolt finger tight at this juncture.

(**xxxvii**) Fit the "U" dynamo bracket to cylinder block utilising three setscrews and lock washers. Fit the dynamo pedestal to the front engine plate and secure with nyloc nut; offer up dynamo and secure finger-tight to the pedestal with a setscrew and lock washer and to the bracket at the rear by nut and bolt with lock washer. Secure the front of the dynamo by its second fixing point to the adjusting link (already attached

to the water pump) utilising one set-screw with a plain washer either side of the dynamo.

(**xxxviii**) Fit the fan belt and adjust to give ¾" play either side of a centre line. Tighten up all nuts and bolts securely including the bolt of the adjusting link and turn up tab of tab washer.

(**xxxix**) Fit thermostat housing and packing to combustion head and secure with two bolts and lock washers, leaving finger tight at this juncture. Connect

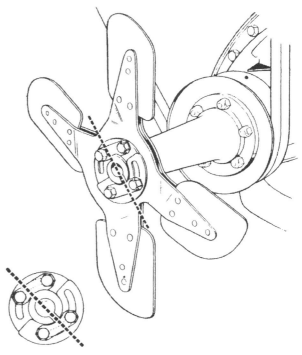

Fig. 37 **Setting the Hand Starter Dog at "ten minutes to four". Note also the hole in fan pulley and pointer on timing cover, which when aligned bring Nos. 1 and 4 pistons to T.D.C.**

the water pump and thermostat housing with the by-pass hose and tighten hose clips.

(**xl**) Assemble the inlet manifold to the exhaust manifold leaving the two nuts finger tight. Position the manifold gaskets on the eight studs fitted in the cylinder head. Fit the manifold assembly to the cylinder head, positioning the four short clamps on the upper row of studs and the longer pair on the two inner studs of the bottom row. Fit the eight nuts and spring washers and tighten to 20—24 lbs. ft. Finally tighten the two nuts attaching the inlet to the exhaust manifold to 16—18 lbs. ft..

(**xli**) Fit the Purolator oil filter with packing to left-hand side of cylinder block. It is located by a tubular stud and secured by three bolts with lock washers. The tubular stud accommodates the oil pressure gauge pipe. This part is fitted to the stud with a copper washer either side of the banjo connection and secured by a cap nut. The pipe is also attached by a clip to the rear stud of the fuel pump.

(**xlii**) Fit fuel pump and packing and secure with two nuts and lock washers. The rearmost stud of this mounting also accommodates the clip steadying the oil pressure pipe.

(**xliii**) Connect fuel pipe from pump, clipping it to the thermostat housing, also the suction tube to the distribution union. The latter, a narrow section tube, is strapped to the fuel pipe.

(**xliv**) Apply oil to the rocker arms and valve tops. Ensure that the rocker cover seal is in position and is in good order and secure cover to top of engine by the two nyloc nuts, each bearing on a fibre and plain washers. Ensure that the rocker cover does not foul the cylinder head nuts at the right-hand side of the engine.

(**xlv**) Offer up the clutch driven plate and housing to flywheel, ensuring first that they are in good condition and the release levers of the housing are correctly adjusted. (See "Clutch" Section.) Settle the housing on the two dowels and secure the flywheel with six set-screws and lock washers, centralising the clutch driving plate with a dummy constant pinion shaft or mandrel.

(**xlvi**) Ascertain that the gearbox, clutch release bearing and clutch operating shaft are in working order before assembling to engine. Offer the gearbox up to the engine, locating it on two dowels and three studs, and secure with six bolts, nuts and lock washers, and three nuts and washers on the studs.

(**xlvii**) The engine and gearbox can be fitted to the chassis with the use of a derrick or moveable crane. Allow the rear extension of the gearbox to be lower than the sump and by slowly lowering the whole unit the mounting points can be found ; utilise a rope sling fitted as shown in Fig. 31.

(**xlviii**) The attachment of the engine and gearbox to the chassis is the reversal of the detachment procedure.

(**xlix**) The engine and the gearbox must be refilled with oil and the radiator with water before the car is used.

22. IGNITION SYSTEM
Notes on Sparking Plugs

(**a**) When sparking plugs are removed from the engine, remove their gaskets with them. Place the plugs and gaskets in a suitable holder, identifying each plug with the cylinder number. The tray shown in Fig. 38 is a simple construction with holes drilled to admit the upper ends of the plugs. Place a new plug of the proper type beside the others to afford a comparison of relative condition of the plugs in use, to the new plug.

(**b**) Look for signs of oil fouling, indicated by wet, shiny, black deposit on the insulator (Fig. 39). Oil pumping is caused by worn cylinders and pistons or gummed-up rings. On the suction stroke of the piston, oil vapour from the crankcase is forced up past the worn rings, where it fouls the plugs and causes sticking valves, with resultant waste of petrol. On the compression stroke, the mixture of oil and petrol vapour is forced past the rings into the crankcase again, contaminating the oil and turning it black with carbon. Carbon deposists in the combustion chamber are formed from burning oil vapour and cause " pinking."

(**c**) Next, look for petrol fouling indicated by a dry fluffy, black deposit (Fig. 40) This is caused by many things— faulty carburation, ignition system, defect in battery, distribution coil or condenser, broken or worn-out cable.

The important thing is for the petrol consumption to be improved. If plugs show suitability for further use, proceed to clean and test.

(d) In preparing for cleaning, remove plug gaskets, and in doing so ascertain their

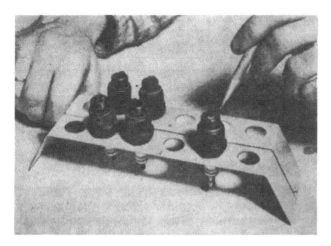

Fig. 38   Sparking Plugs in a tray ready for comparison.

condition. Note the gaskets illustrated in Fig. 41. Upper left shows a gasket not properly compressed. A large proportion of the heat from the insulator is dissipated to the cylinder head by

Fig. 39   Oil fouling indicated by a wet shiny black deposit on the Insulator.

means of the copper gasket between the plug and the cylinder head. Plugs not down tight can be easily over-heated, throwing them out of the proper heat range, causing pre-ignition, short plug life and bringing about so-called " pinking." Don't tighten plugs too much—but be reasonably sure a good seal is made between plug and cylinder head. Lower left shows a gasket on which the plug was pulled down too tight, or had been too long in service. Note the distorted condition. Note evidence of blow-by, also a cause of plug over-heating and resulting dangers. Upper right shows a reasonably compressed gasket giving the plug adequate seal and a good path for heat dissipation. All may be compared with the new gasket, at lower

Fig. 40   Petrol fouling indicated by a dry fluffy black deposit on the Insulator.

right. If gaskets are at all questionable they should be replaced by new gaskets.

(e) Occasionally a blistered insulator or baldy burned electrode may be noted when examining plugs (Fig. 42). If the plug is the type normally recommended for the engine and was correctly installed, i.e., down tight on the gasket—the condition may have been brought about by a very " lean " mixture, or overheated engine. It is well to remember that plugs operating in the condition described above are often the cause of poor engine performance and extravagant petrol con-

Fig. 41   **Sparking Plug Gaskets in various conditions.**

sumption. It may be, however, that a plug of a " colder " type is required.

(**f**) After cleaning, examine plugs for cracked insulators or insulator nose worn away through continued previous cleaning. In this case we should recommend that the plugs have passed their point of useful life and new plugs should be installed. Look for a deposit on the insulator, under side electrode, which may accumulate heat and act as a " hot spot " in service.

(**g**) After cleaning and blowing surplus abrasive out of shell recesses and off plug threads by means of " blow out " nipple—examine threads for carbon accumulation. Use a wire brush to remove carbon and clean the threads. A wire buffing wheel may also be utilised ; however, use reasonable care

Fig. 42   **A Blistered Insulator.**

in both operations in order not to injure electrode or insulator tip. The threaded section of plug shell is often neglected in plug cleaning, even though, like the gaskets, these threads form a means of heat dissipation. When threads are coated with carbon, it retards the even flow of heat to the cooling medium, thereby causing overheating. (When installing plugs, this simple procedure will ensure no binding of threads and avoid unnecessary use of plug spanner.) Screw the plug down by hand as far as possible, then use spanner for tightening only. Always use a box spanner to avoid possible fracture of the insulator.

Fig. 43   **Champion Series " 700 " Cleaner and Tester Unit.**

(**h**) Next, we are ready for resetting the electrodes (Fig. 44). Remember that electrode corrosion and oxides at gap area vitally affect spark efficiency. The cleaner can remove the oxides and deposits from the insulator, but because of gap location, the cleaner stream cannot always reach this area with full effect, also, the tenacious adhesion of corrosion, etc., would require too much subjection to clean blast for removal. Therefore, when plugs are worthy of further use, it is sometimes good practice to dress the gap area, on both centre and side electrodes, with a small file before resetting to correct gap.
Resetting of electrodes should be part of service during useful life of the

39

plugs. However, the strains of intense heat, pressure, mechanical shock, electrical and chemical action, during miles of service, wreak such havoc on the electrodes that molecular construction is affected. Plugs reach a worn out condition and resetting can serve a good purpose only for a time.

Fig. 44   **The Champion Spark Plug Gap Tool.**

When gaps are badly burned, it is indicative the plug is worn to such an extent that further use is unwarranted and wasteful. When resetting, bend the side wire only, never bend centre electrode as this may split the insulator tip.

(i) Inspect for leakage after testing, by applying oil around the terminal (Fig. 45). Leakage is indicated by the presence of air bubbles, the intensity of which will serve to show degree of leakage. Leakage throws the plug out of its proper heat range, as the hot gas escaping has a " blow torch " effect on the plug, causing compression loss, pre-ignition, rapid electrode destruction and overheating of the insulator tip.

(j) New gaskets have been fitted to the plugs and the general improvement in appearance is apparent now that the plugs are ready to be installed in the engine (Fig. 46). It requires no imagination to know that improved engine performance, better petrol consumption and satisfaction will result. The use of the stand (as illustrated) is

evidence of your careful handling of the plugs.

(k) The top half of the insulator is often responsible for causes of poor plug

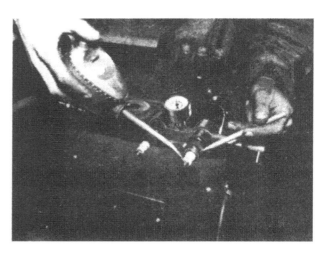

Fig. 45   **Testing for Leaks.**

performance (Fig. 47), namely, paint splashes, accumulation of grime and dust ; cracked insulators caused by slipping spanner, or overtightening of terminals. Examine for cracked insulators at shoulder and terminal post. Remove grime and dust. Recommend inspection, cleaning and testing every 3,000 miles (Fig. 48).
Clean and replace sparking plugs periodically as necessary. The correct gap for the TR2 plugs should provide

Fig. 46   **Sparking Plugs ready to fit to Engine. Note the New Gaskets and the use of the Stand.**

a gap of .032", the Champion L10S $\frac{1}{2}''$ reach plug being specified for normal road work, the L11S for high speed work. The normal efficient life of a sparking plug is 10,000 miles, after which, if full efficiency and economy is desired, the plugs should be replaced by new ones of the type specified.

a piece of carborundum stone, so that when the points are closed they fit flush against each other. If the points have become seriously worn they should be replaced by new items. The points should be properly set to provide a gap of .014" to .016" when fully open.

Fig. 47   Sparking Plugs in various conditions.

Fig. 48   An unretouched photograph of a CHAMPION Sparking Plug after 25,000 miles of service, compared with a new plug. The weak spark given by the former can readily be imagined and amply justifies our recommendation that to save petrol, plugs should be changed before such a stage of wear, as that shown in the photograph is reached.

The distributor cap and rotor should be periodically examined for cracks which will allow electrical leakages.

The contact breaker points should be examined each 5,000 miles, when normal lubrication of this part of the car is recommended, and where these have become burnt or pitted, they should, if possible, be squared up with

The condenser wiring and the low and high tension circuits should be ensured, as should the automatic advance and retard mechanism. Similarly the coil should be ensured.

ENGINE

LIST OF DISTRIBUTORS BEING SERVICED FROM

CHAMPION SPARK PLUG COMPANY, TOLEDO, 1, OHIO, U.S.A.
CHAMPION SPARK PLUG COMPANY OF CANADA LTD., WINDSOR, ONTARIO.
CHAMPION SPARKING PLUG CO. LTD., FELTHAM, MIDDLESEX, ENGLAND.

ADEN PROTECTORATE
Cowasjee Dinshaw & Brothers,
Steamer Point, Aden.

AFGHANISTAN
Afghan Motor Service & Parts Co.,
Shirkate Service, Kabul, Afghanistan.

ALGERIA
A. Sabatier & R. des Cilleuls,
3 Rue Jean Rameau, Algiers, Algeria.

ANDORRA
Etabs. Pyrennes,
Andorra la Vieja, Andorra.

ARGENTINE REPUBLIC
Representative :
    George Dombey,
    Avda. Corrientes 1373,
    Buenos Aires, Argentine Republic.

AUSTRALIA
New South Wales
Bennett & Barkell Ltd.,
G.P.O. Box 3876,
Sydney, N.S.W., Australia.

Bennett & Wood Pty. Ltd.,
G.P.O. Box 4255,
Sydney, N.S.W., Australia.

Queensland
Martin Wilson Bros. Pty. Ltd.,
G.P.O. Box 665 K,
Brisbane, Queensland, Australia.

Engineering Supply Co. of Australia Ltd.,
Box 1411 T, G.P.O.,
Brisbane, Queensland, Australia.

South Australia
Duncan & Co. Ltd.,
Box 1429 J, G.P.O.,
Adelaide, South Australia.

Harris Scarfe Ltd.,
Box 385 A, G.P.O.,
Adelaide, South Australia.

A. G. Healing Ltd.,
G.P.O. Box 645 F,
Adelaide, South Australia.

Tasmania
W. & G. Genders Pty. Ltd.,
Box 98,
Launceston, Tasmania.

Branch at : Hobart.

E. A. Machin & Co. Ltd.,
529-541 Elizabeth Street,
Melbourne C.1, Victoria, Australia.

Branches at : Launceston and Hobart.

Wm. L. Buckland Pty. Ltd.,
139-141 Franklin Street,
Melbourne C.1, Victoria, Australia.

Branches at : Launceston and Hobart.

Victoria
Brooklands Accessories Ltd.,
G.P.O. Box 2030 S,
South Melbourne, S.C.4,
Victoria, Australia.

Keep Brothers & Wood Pty. Ltd.,
200 Latrobe Street,
Melbourne C.1, Victoria, Australia.

Western Australia
Atkins (W.A.) Limited,
Mazda House, 894-6 Hay Street,
Perth, Western Australia.

AUSTRIA
Adolf Riedl,
Turkenstrasse 25,
Vienna IX/66, Austria.

Branches at : Linz and Graz.

AZORES ISLANDS
Varela & Ca. Lda.,
Apartado 29, Ponta Delgada,
S. Miguel, Azores Islands.

BAHREIN (Persian Gulf)
Khalil Bin Ebrahim Kanoo,
P.O. Box 31,
Bahrein, Persian Gulf.

BELGIUM and LUXEMBOURG
Societe de Distribution et d'Agences
    Commerciales,
167 Avenue Brugmann,
Brussels, Belgium.

BERMUDA
Masters Limited,
Hamilton, Bermuda.

BOLIVIA
Cia. Imp. de Automotres,
M. Czapek S.A.,
Casilla 440, La Paz, Bolivia.

BRAZIL
Representative :
    Onorato Rubino,
    Caixa Postal 33-LAPA,
    Rio de Janeiro, Brazil.

BRITISH EAST AFRICA
(Uganda, Kenya, Tanganyika, Zanzibar and Pemba)
The Uganda Co. (Africa) Ltd.,
P.O. Box 1,
Kampala, Uganda.

Branches at : Jinja and Mbale.

Car & General Equipment Co. Ltd.,
P.O. Box 1409,
Nairobi, Kenya.

Branches at : Dar-es-Salaam, Mombasa
    and Zanzibar.

Riddoch Motors Ltd.,
P.O. Box 40,
Arusha, Tanganyika.

Branches at : Dar - es - Salaam, Lindi,
    Tanga, Moshi, Iringa and
    Zanzibar.

BRITISH GUIANA
Bookers Stores Limited,
49-53, Water Street,
Georgetown, British Guiana.

BRITISH HONDURAS
Hofius Hildebrandt.
Albert Street,
Belize, British Honduras.

BRITISH SOMALILAND
K. Pitamber & Co.,
Berbera, British Somaliland.

BRITISH WEST INDIES
George W. Bennett Bryson & Co. Ltd.
St. Johns, ANTIGUA.

Kelly Motor Company,
P.O. Box 365,
Nassau, BAHAMAS.

City Garage Trading Co. Ltd.,
Victoria Street,
Bridgetown, BARBADOS.

McIntyre Bros. Ltd.,
St. George's, GRENADA.

Jamaica Traders (Agency) Limited,
P.O. Box 443,
Kingston, JAMAICA.

J. E. C. Theobalds,
P.O. Box 51,
Castries, ST. LUCIA.

George L. Francis-Lau Limited,
18, Abercromby Street,
Port-of-Spain, TRINIDAD.

BURMA
Representative :
    M. Hasan Behbahany,
    P.O. Box 934,
    115, 38th Street,
    Rangoon, Burma.

The Bombay Motor Company,
115-117, Sule Pagoda Road,
Rangoon, Burma.

The United Motors,
186, Phayre Street,
Rangoon, Burma.

N. B. Mody Brothers,
272, Phayre Street,
Rangoon, Burma.

The Kothari Motor Co. (Burma) Ltd.,
P.O. Box 640,
239, Phayre Street,
Rangoon, Burma.

Globe Automobile Company,
206, Phayre Street,
Rangoon, Burma.

ENGINE

CANARY ISLANDS
Representative :
Francisco Flores,
Espinardo,
Murcia, Spain.

CAROLINE ISLANDS (W. Pacific)
K. Hatoka,
Ponape, Caroline Islands.

CEYLON
Representative :
Rajandrams Limited,
Maharaja Building,
Bankshall Street,
Pettah, Colombo, Ceylon.

A.S.S. Sangaralingham Pillai & Co. Ltd.,
213—215 Norris Road,
Colombo, Ceylon.

Colonial Motors Limited,
297 Union Place,
P.O. Box 349,
Colombo, Ceylon.

CHILE
Representative :
G. Dombey, Avenida Corrientes 1373,
Buenos Aires, Argentine.

Sociedad Anonima Commercial, del Sur,
P.O. Box 30-D,
Punta Arenas, Chile.

Florentine Poblete Perez & Cia.,
Casilla 149-D,
Santiago, Chile.

Vicente Camilio di Biase,
P.O. Box No. 305,
(Calle Roca 981),
Punta Arenas, Chile.

CHINA
Representative :
Dodge & Seymour Limited,
Dodge Building,
53 Park Place,
New York 7, N.Y., U.S.A.

COLOMBIA
Sager & Co.,
Calle 11, No. 219,
Cali, Colombia.

Branch at : Pasto.

Cias Unidas de Combustibles Ltd.,
Apartado Nacional 236,
Madellin, Colombia.

Branches at : Barranquilla, Bogota, Cali,
Manizales.

CONGO
R. J. Franco,
P.O. Box 32,
Elisabethville, Congo.

L. E. Tels & Co's. Handelmaatschappij N.V.,
in Leopoldville, Elisabethville and Madadi.

COSTA RICA
Amalcen Koberg S.A.,
Apartado 1323,
San Jose, Costa Rica.

CUBA
Representative :
C. H. Mackay,
Apartado 1167,
Havana, Cuba.

CYPRUS
Nicos D. Solomonides & Co. Ltd.,
P.O. Box 210,
Limassol, Cyprus.

DENMARK
F. Bulow & Co.,
Polititorvet,
Copenhagen, Denmark.

Branch at : Odense.

DOMINICAN REPUBLIC
Casa Nadal C. por A.,
Apartado 1172,
Ciudad Trujillo, Dominican Republic.

DUTCH GUIANA
A. Van der Voet & Trading Company,
P.O. Box 220,
Paramaribo, Dutch Guiana.

DUTCH WEST INDIES
Martijn-Stokvis N.V.,
P.O. Box 146,
Willemstad, CURACAO.

Rodolfo Pardo,
Madurastraat No. 7,
Willemstad, CURACAO.

Bonaire Trading Company,
BONAIRE.

ECUADOR
Almacenes Comerciales Gonzales Rubio
S.A.,
P.O. Box 54,
Guayaquil, Ecuador.

Alvarez Barba Hnos. y Cia.,
P.O. Box 567,
Quito, Ecuador.

EGYPT
North East Africa Trading Company,
P.O. Box 1800,
43 Rue Kasr-el-Nil,
Cairo, Egypt.

ERITREA
Vrajlal Zaverchand,
P.O. Box 1017,
Asmara, Eritrea.

ETHIOPIA
Edward Achkar & Company,
P.O. Box 250,
Addis Ababa, Ethiopia.

FIJI ISLANDS
Morris Hedstrom Limited,
Suva, Fiji Islands.

FINLAND
Atoy O/Y.,
Mikonkatu 13 A,
Helsinki, Finland.

A. B. Maritim O/Y.,
S. Kajen 14,
Helsinki, Finland.

FRANCE
Bougie Champion S.A.,
5 Square Villaret de Joyeuse,
Paris 17e, France.

FRENCH GUIANA
Yves Massel,
Cayenne, French Guiana.

FRENCH INDO-CHINA
Ets. Jean Comte,
34 Boulevard Norodom,
Saigon, French Indo-China.

FRENCH MOROCCO
Auto-Hall,
Boulevard de Marseille,
Casablanca, French Morocco.

FRENCH SOMALILAND
Maison Ph. Norhadian,
Djibouti, French Somaliland.

FRENCH WEST INDIES
Gaston Lubin,
Basse Terre, GUADELOPE.

Guy de Jaham,
4 Boulevard Allegre,
Fort-de-France, MARTINIQUE.

GERMANY
Automobil-und-Industrie-Artikel,
Hirschstrasse 53,
Karlsruhe, Germany.

Hanko Industrie und Handelsgesellschaft
m.b.h.,
Moselring 27/29,
Koblenz-Neuendorf, Germany.

Branch at : Katharinenstrasse 11—12,
Berlin-Halensee.

GIBRALTAR
A. M. Capurro & Sons Limited,
20 Line Wall Road,
Gibraltar.

GREECE (Incs. Dodecanese & Crete)
The Trading & Commission Agency
(Hellas),
P.O. Box 143,
1 Santarosa Street,
Athens, Greece.

GUAM
James Garland Little,
P.O. Box No. 40,
Agana, Guam.

GUATEMALA
Alfredo S. Clark,
7a, Avenida Sur, No. 105,
Guatemala City, Guatemala.

Fisher, Faeh & Cia. S.C.,
7a, Avenida Sur Prol,
Guatemala City, Guatemala.

HAITI
Jules Farmer,
P.O. Box A-95,
Port-au-Prince, Haiti.

HASHEMITE JORDAN
Ets. F. A. Kettaneh, S.A.,
P.O. Box 485,
Amman, Hashemite Jordan.

HAWAII
The Schuman Carriage Company Limited,
P.O. Box 2420,
Honolulu, Hawaii.

HOLLAND
R. S. Stokvis & Zonen N.V.,
Westzeedijk 507,
Rotterdam, Holland.

Branches at : The Hague, Amsterdam,
Groningen, Utrecht, Leeu-
warden, Deventer, Arnhem,
Breda and Maastricht.

HONDURAS
Walter Brothers,
Tegucigalpa, Honduras.

43

HONG KONG
Representative :
Dodge & Seymour Ltd.,
318, Prince's Building,
P.O. Box 77,
Hong Kong.

ICELAND
Egill Vilhjalmsson H/F.,
P.O. Box 457,
Reykjavik, Iceland.

INDIA
Dodge & Seymour (India) Ltd.,
P.O. Box 144,
Bombay 1, India.

Dodge & Seymour (India) Ltd.,
P.O. Box 457,
P-21, Mission Row Extension,
Calcutta 13, India.

Dodge & Seymour (India) Ltd.,
Lakshmi Insurance Building,
Circular Road,
New Delhi 1, India.

Dodge & Seymour (India) Ltd.,
100, Armenian Street,
Madras, India.

INDONESIA
Javastaal-Stokvis N.V.,
Kramat 4—6,
Djakarta, Indonesia.

Branches at : Surabaia, Semerang,
Medan,Bandung, Makasser,
Pontianak, Padang, Palem-
bang, Bandjermasin.

L. E. Tels & Co's. Handelmaatschappij N.V.,
Gedong, Pandjang 12,
Djakarta, Indonesia.

Branches at : Semerang, Surabaia,
Makasser, Bandjermasin,
Palembang, Medan and
Menado.

IRAQ
Ets. F. A. Kettaneh S.A.,
Al Rashid Street,
Baghdad, Iraq.

ISRAEL
Kaplan Brothers Ltd.,
55 Kingsway,
Haifa, Israel.

Branch at : 3 Hagalilstr., Tel-Aviv.

ITALIAN SOMALILAND (Somalia)
Somalilands Trading Company,
P.O. Box 9,
Via Cardinal G. Massaia N.50,
Mogadishu, Somalia.

ITALY
AD gia Impresa Forniture Industrialiifi,
5 Via Lovanio,
Milan, Italy.

Wilfred Van Singer,
36 Via Barberini,
Rome, Italy.

JAPAN
Empire Motor Co. Ltd.,
2—2 Nihonbashi-Tori Chuo-Ku,
Tokyo, Japan.

Kokusai Kogyo Co. Ltd.,
3—3 Makicho,
Chuo-Ku, Tokyo, Japan.

J. Osawa & Co.,
Sanjo Kaboshi,
Kyoto, Japan.

The Central Automobile Industry Co. Ltd.,
44 Sozecho Kitaku,
Osaka, Japan.

KUWAIT
Sayid Hamid el Nakib,
Kuwait, Persian Gulf.

LEBANON
Ets. F. A. Kettaneh S.A.,
P.O. Box 242,
Beyrouth, Lebanon.

LIBYA
The Automobile Trading Co. Ltd.
29 Via Frosinone,
P.O. Box 353,
Tripoli, Libya.

Beniamino Haddad,
P.O. Box 168,
5/9 Giaddat Omar el Muktar,
Tripoli, Libya.

MADAGASCAR
Edwin Mayer & Co. Ltd.,
Boite Postale 170,
Tananarive, Madagascar.

MALAYA (including Singapore)
L. E. Tels & Co's. Handelmaatschappij N.V.,
P.O. Box 649,
Singapore.

Branches at : Penang and Port Swet-
tenham.

Maclaine-Stokvis (Malaya) Ltd.,
135, 137, 139 Middle Road,
Singapore.

Branches at : Penang and Kuala Lumpur.

MALTA G.C.
Auto Sales Co. Ltd.,
287, Kingsway,
Valletta, Malta, G.C.

Mizzi Brothers Ltd.,
283, Kingsway,
Valletta, Malta, G.C.

MAURITIUS
Manufacturers' Distributing Station Ltd.,
P.O. Box 71,
Quay Square,
Port Louis, Mauritius.

MEXICO
Auto Repuestos, S.A.,
Balderas 36—901,
Mexico D.F., Mexico

Importadora de Artivulos Para Auto-
moviles, S.A.
Calle Barcelone No. 11,
Mexico City, Mexico.

Angel de Caso, Jnr.,
Ave. Bucareli Num 5,
Mexico City, Mexico.

NEW CALEDONIA
Ets. Ballande,
Noumea, New Caledonia.

Maison Barrau,
Noumea, New Caledonia.

NEW ZEALAND
Hope Gibbons Ltd.,
P.O. Box 2197,
Wellington C.1, New Zealand.

Branches at : Auckland, Christchurch,
Dunedin and Invercargill.

E. W. Pidgeon & Co. Ltd.,
228, Tuam Street,
Christchurch, C.1, New Zealand.

Branches at : Auckland, Wellington and
Dunedin.

NICARAGUA
L. M. Richardson en Comandita,
Roosevelt Ave. No. 101,
Managua, Nicaragua.

NORTH BORNEO (British Sarawak)
L. E. Tels & Co's. Handelmaatschappij N.V.,
Kuching, Sarawak,
British North Borneo.

NORWAY
Sorensen og Balchen A/S.,
Box 2261 MJ.,
Oslo, Norway.

NYASALAND
African Lakes Corporation Ltd.,
122, Ingram Street,
Glasgow, C.1, Scotland.

Branch at : Blantyre, Nyasaland.

PAKISTAN
Representatives :
  West Pakistan :
  Raziki Limited,
  P.O. Box 4804,
  Madha Chambers, Bunder Road,
  Karachi-2, West Pakistan.

  East Pakistan :
  Metropolitan Trading Co. (Pak) Ltd.
  76, Lyall Street, Patuatully,
  Dacca, East Pakistan.

PANAMA
The Wholesale Tire & Supply Co.,
Apartado 3270,
Panama City, Panama.

PARAGUAY
Artaza Hermanos,
Comerciale e Industriale, S.A.,
Casilla Postal 235,
Asuncion, Paraguay.

PERSIA
Auto-Teheran, S.A.
Avenue Bargh,
Teheran, Persia.

PERU
E. S. DeLaney, S.A.,
Avenida Grau 290,
Lima, Peru.

PHILIPPINE REPUBLIC
Manila Auto Supply,
1054—56, Rizal Ave.,
Manila, Philippine Islands.

Motor Service Co. Inc.,
Boston St., Port Area,
Manila, Philippine Islands.

ENGINE

Square Auto Supply Co.,
625, Juan Luna,
Manila, Philippine Islands.

PORTUGAL
C. Santos Lda.,
29—41, Avenida da Liberdade,
Lisbon, Portugal.

Soc. de Comal. C. Santos Lda.,
160—168, Rua Santa Catarina,
Porto, Portugal.

**PORTUGUESE EAST AFRICA
(Mozambique)**
Auto Sobressalentes,
P.O. Box 693,
Lourenco Marques, P.E.A.

Emporium Grandes Armazens da Beira,
P.O. Box 200,
Beira, P.E.A.

Adolfo Matos Ltda.,
P.O. Box 11,
Manpula, P.E.A.

**PORTUGUESE WEST AFRICA
(Angola)**
Robert Hudson & Sons Ltd.,
Raletrux House,
Meadow Lane,
Leeds, England.

Branches at : P.O. Box 1210,
Luanda, Angola.

P.O. Box 101,
Lobito, Angola.

PUERTO RICO
Julio T. Rodriguez,
206, O'Donnel Street,
San Juan 6, Puerto Rico.

RHODESIA (Northern and Southern)
Duly & Co. Ltd.,
P.O. Box 131,
Bulawayo, Southern Rhodesia.

Branches at: Salisbury, Umtali, Gatooma,
Gwelo, Fort Victoria, Ndola
and Kitwe.

Motor Car Equipment (Sby.) Ltd.,
P.O. Box 1394,
Salisbury, Southern Rhodesia.

SALVADOR
Frenkel & Co.,
Apartado 63,
San Salvador, El Salvador.

Duran Hermanos,
3A, Avenida Norte No. 17,
San Salvador, El Salvador.

Ernesto McEntee,
Santa Ana, El Salvador.

SAUDI ARABIA
F. A. Kettaneh,
Jeddah, Saudi Arabia.

SEYCHELLES
Temooljee & Co.,
P.O. Box 9,
Mahe, Seychelles.

SOUTH WEST AFRICA
Representative :
J. B. Steele & J. D. Matson (Pty.) Ltd.,
P.O. Box 130,
Knysna,
Cape Province, Union of South Africa.

S. Gorelick's Garage,
P.O. Box 200,
Windhoek, S-W. Africa.

S. Cohen Ltd.,
P.O. Box 215,
Windhoek, S-W. Africa.

SPAIN
Francisco Flores,
Espinardo,
Murcia, Spain.

Branch at : Bilbao.

SPANISH GUINEA
Representative :
Vda. de Jose Penate Medina,
Leon y Castillo 15,
Las Palmas, Canary Islands.

SPANISH MOROCCO
Francisco Flores,
Zoco Grande 53,
Tangiers.

SUDAN
Sudan Mercantile Co. (Motors) Ltd.,
P.O. Box 97,
Khartoum, Sudan.

Branches at: Port Sudan and Wad Medani.

SWEDEN
A.B. Amerikanska Motor Importen,
Stockholm 6, Sewden.

Branches at : Malmo, Gothenburg, Hal-
singborg, Kristianstad and
Sodertalje.

SWITZERLAND
S.A.F.I.A.,
Avenue Pictet de Rochemont, 8,
Geneva, Switzerland.

Branches at : Berne and Zurich.

Max Gromann A.G.,
Solothurnerstrasse 60,
Basle, Switzerland.

SYRIA
Ets. F. A. Kettaneh S.A.,
P.O. Box 242,
Beyrouth, Lebanon.

TAHITI (Society Islands)
Etablissements Donald,
Papeete, Tahiti.

Lionel L. Bambridge,
P.O. Box 88,
Papeete, Tahiti.

TANGIERS
Francisco Flores,
Zoco Grande 53,
Tangiers.

THAILAND (Siam)
Sombat Phanich,
New Road,
Bangkok, Thailand.

TONGA (Friendly Islands)
Morris, Hedstrom Ltd.,
Nukualofa, Tonga.

TUNISIA
S.I.S.A.A.,
42, Rue Thiers,
Tunis.

TURKEY
Etablissements Archmidis, S.A.T.,
Boite Postale 1832,
Galata, rue Okcu Musa 39—51,
Istanbul, Turkey.

G. & A. Baker Limited,
Prevuayans Han Tahtakale,
Postbox 468,
Istanbul, Turkey

UNION OF SOUTH AFRICA
Representative :
J. B. Steele & J. D. Matson (Pty.) Ltd.,
P.O. Box 130,
Knysna, Cape Province.

Branch at : Yorkshire House, Smith
Street, Durban.

URUGUAY
Representative :
George Dombey,
Avda Corrientes 1373,
Buenos Aires, Argentine Republic.

VENEZUELA
Sres. Francisco Sapene e Hijo,
Apartado de Correos 1528,
Caracas, Venezuela.

RAM-MAC,
Apartado de Correos 21,
Calle 99 (Comercio) 9-63,
Maracaibo, Venezuela.

VIRGIN ISLANDS
Virgin Islands Corporation,
St. Croix, Virgin Islands.

Auto Sales & Parts Co.,
St. Thomas, Virgin Islands.

Christiansted Utilities Co.,
Christiansted,
St. Croix, Virgin Islands.

WEST AFRICA
Compagnie Francaise de l'Afrique Oc-
cidentale,
Royal Liver Building,
Liverpool 3, England.

Societe Commerciale de l'Ouest Africain,
5 & 7, Hall Street,
Oxford Street,
Manchester 2, England.

THE ABOVE DISTRIBUTORS COVER
THE FOLLOWING TERRITORIES
IN WEST AFRICA AND HAVE BRAN-
CHES IN THOSE TOWNS SHOWN
IN PARENTHESIS.

GAMBIA—(Bathurst).
**GOLD COAST — (Accra, Kumasi,
Takoradi).**
**NIGERIA — (Lagos, Kano, Onitsha,
Port Harcourt, Sokoto, Warri).**
SIERRA LEONE—(Freetown).
CAMEROONS—(Duala).
DAHOMEY—(Lome and Cottonou).
FRENCH EQUATORIAL AFRICA
FRENCH GUINEA
FRENCH SENEGAL—(Dakar).
FRENCH SUDAN
FRENCH IVORY COAST—(Abijan).
TOGOLAND.

YUGOSLAVIA
Progres General Trade Agency
Knez Mihajlova 1,
Belgrade, Yugoslavia.

Autocentar,
Marticeva Ul. 8,
Zagreb, Yugoslavia.

Auto Srbija,
Bulevar Jogisl,
Armije 61,
Belgrade, Yugoslavia.

23. ENGINE NOISES

(a) Main Bearing Knock

This knock can usually be identified by its dully heavy metallic note which increases with frequency as the engine speed and load rises. A main bearing knock is particularly noticeable when the engine is running very slowly and consequently unevenly, it is more pronounced with advanced ignition.

When this bearing knock is experienced it can be explained by one of the following faults and should be treated accordingly.

(i) Unsuitable grade of oil or badly diluted oil supply.

(ii) Low oil pressure.

(iii) Insufficient oil in sump.

(iv) Excessive bearing clearance caused by worn journal and/or bearings.

(b) Crankshaft End Float

When a knock is being caused by the development of end float, it will be found most noticeable when the engine is running at idling speeds. This knock can temporarily be eliminated by operating the clutch.

(c) Big End Bearing Knock

A big end bearing knock is lighter in note than that experienced with a main bearing. It will be evident at idling speeds and will increase with engine speed.

The best test for this noise is to detach the lead from each sparking plug in turn and reconnecting the lead whilst flicking the throttle open. On reconnection of the lead, a light thud will be audible where the bearing looseness or correcting misalignment exists, further investigation can be carried out to that particular rod or rods.

In addition to the knock being caused by excessive bearing clearance it is sometimes caused by :—

(i) Unsuitable grade of oil or badly diluted supply.

(ii) Insufficient supply of oil.

(iii) Low oil pressure.

(d) Small End Knocks

As the gudgeon pin used in this model is able to float in the piston and the bearing in the connecting rod, a knock may arise owing to slackness in the small end bush or the piston bosses. The knock will make itself audible under idling conditions or at road speeds between 20—30 m.p.h. (32—48 km.p.h.).

To test for a gudgeon pin knock, cut out each cylinder one at a time by disconnecting the plug leads. The offending gudgeon pin will be identified by the fact that a double knock is caused when the disconnection of the plug lead is made.

With complaints of this nature, the following possible causes should be examined.

(i) A too tight gudgeon pin.

(ii) A gudgeon pin slack in the connecting rod bush or piston boss (see page 2 for gudgeon pin clearance).

(iii) Misalignment of connecting rod allowing connecting rod bush to foul the piston bosses.

(e) Piston Knock (Piston Slap)

This will increase with the application of load up to 30 m.p.h. (48 km.p.h.) but only in very bad cases will it continue to be audible over that speed. In some cases piston knock will only be evident when the engine is started from cold and will disappear as the engine warms. In such cases it is suggested that the engine is left untouched.

A suggested method of locating the offending piston is to engage a gear and with the hand brake hard on, just let the clutch in sufficiently to apply a load with the engine at a moderate speed. By detaching a spark plug lead and thus putting a cylinder out of action, it is possible to cut out the knock and so determine the offending piston.

Faults in the engine components listed hereafter often contribute to piston

46

knock (piston slap) and should therefore be examined.

(i) Excessive clearance between piston and cylinder sleeve due to fair usage or to an unsuitable replacement part.

(ii) Pistons or rings striking ridge at the top of the sleeve after fitting a replacement. Such ridges should be removed before replacement parts are fitted.

(iii) Collapsed piston.

(iv) Broken piston ring grooves or excessive clearance in grooves (see page 2).

(v) Connecting rod misalignment.

(f) Noisy Valve Rockers or Tappets

Noise due to valve rockers can be identified fairly easily owing to the fact that these are operated by the camshaft which revolves at half engine speed, the noise will seem to be slower than other engine noises. Valve rocker noise has a characteristic clicking sound which increases in volume as the engine speed rises.

Where rocker noise is caused by excessive tappet clearance, it can be eliminated by the insertion of a feeler gauge between the stem of the valve and the rocker toe whilst the engine is idling.

When this complaint is experienced and is found to be caused by incorrect tappet clearance the rockers should be adjusted as described on page 22.

Push rod noise may be caused by worn or rough rocker ball pins or push rod cups and can be cured by replacing the worn or damaged parts.

(g) Ignition Knock (Pinking)

An ignition knock is recognised by its metallic ringing note, usually occurring when the engine is labouring or accelerating.

The knock can be caused by either detonation or pre-ignition. Detonation is the result of a rapid rise in pressure of the explosive mixture, thus causing the last portion of the charge in the cylinder to be spontaneously ignited, resulting in this striking the cylinder wall with a ringing sound; this noise being familiar to motorists as "pinking."

Pre-ignition may arise as a result of detonation owing to heat generated thereby but may also be caused by sharp edges or points in the combustion space, and where it arises should be treated accordingly.

When "ignition knock" is audible, the following possible causes should be investigated.

(i) Excessive carbon deposits in head and on piston crowns.

(ii) Incorrect or faulty spark plugs causing incandescence.

(iii) Sharp edges or pockets in combustion space.

(iv) Engine overheating.

(v) Too weak carburettor mixture, causing delayed combustion.

(vi) Unsatisfactory grade of fuel.

(vii) Too early ignition timing.

(viii) Faulty automatic advance and retard mechanism due to incorrect or weak centrifugal control springs.

(ix) Hot engine valves due to incorrect seating width, insufficient valve rocker clearances, valve edges thinned by excessive refacing. Valve of unsuitable material.

(h) Back Firing into Carburettor

It is in order that with a cold engine back firing into the carburettors may occur, but this should cease when the engine attains normal working temperature.

If back firing still persists in spite of warming up, the following possible causes should be investigated.

(i) Incorrect ignition timing.

(ii) Incorrect wiring of sparking plugs.

(iii) Centrifugal or suction advance and retard mechanism not functioning correctly.

(**iv**) Incorrect valve timing.

(**v**) Poor quality fuel.

(**vi**) Mixture is too weak or excessively rich.

(**vii**) Pre-ignition due to various causes.

(**viii**) Air leak into induction system giving rise to a weak mixture.

(**ix**) Valves, particularly inlet, not seating correctly.

(**x**) Defective cylinder head gasket.

(**i**) **Excessive Oil Consumption**

Excessive oil consumption is usually associated with a very worn engine, but can arise as a result of external leakages and due to other factors with comparatively new engines.

If excessive oil consumption is established, before commencing to dismantle the engine a check for external leakage should be carried out.

When an engine is burning oil it will be indicated by the emission of bluish grey smoke from the exhaust when the engine is " raced up " after a period of idling.

A check for external leakage can be conveniently carried out by spreading paper on the ground under the forward part of the car, and running the engine at a moderate speed for a few minutes.

In this way it is possible to locate the position of leaks which, without the engine running, would not be evident. External leaks are caused by one or more of the following :—

(**i**) Cracked sump or poor sump packing.

(**ii**) Flange faces of sump not true.

(**iii**) Drain plug loose or defective packing washer.

(**iv**) Defective filter packing, poor joint faces or loose attachment bolts.

(**v**) Oil pressure pipe line leaking.

(**vi**) Defective petrol pump packing, poor joint faces or attachment nuts loosened.

(**vii**) Defective rocker cover packing, poor joint faces or attachment nuts loosened.

(**viii**) Defective front engine plate packing or poor joint faces.

(**ix**) Timing cover oil seal defective.

(**x**) Timing cover cracked, defective packing or loose mounting bolts.

(**xi**) Leakage round camshaft welch plug.

(**xii**) Unsuitable grade of oil or excessively diluted, arduous driving conditions, excessively high pressure or crankcase temperatures.

(**xiii**) Excessive clearance between piston and sleeve or incorrect replacements, damaged rings, rings stuck in grooves, insufficient piston ring end gap, piston rings exercising insufficient radial pressure.

(**xiv**) Excessive diameter and axial clearance due to wear associated with the possibility of oval and worn crankpins.

(**xv**) Excessive diameter clearance in main bearings and/or worn journals. (See page 1 for dimensions and clearances.)

(**j**) **Low Oil Pressure**

The correct oil pressure is 40—60 lbs. per sq. in. for top gear for road speeds between 30 — 40 m.p.h. (48 — 64 km.p.h.). With complaints of low oil pressure the following possible causes should be investigated :—

(**i**) Insufficient oil in sump.

(**ii**) Unsuitable grade of oil or a very badly diluted supply.

(**iii**) Suction oil filter restricted by dirt in sump.

(**iv**) Oil pump loose on mountings.

(**v**) Very badly worn or damaged oil pump. (See " Oil Pump " on page 13.)

48

(vi) Oil release valve in exterior oil filter head out of adjustment, dirt on valve seating, broken or weak release valve spring. Filter loose on bracket, damaged joint packing, poor joint faces.

(vii) Loose connections on pressure gauge pipe or defective pipe line and/or flexible connections.

(viii) Incorrect oil pressure gauge.

(ix) Worn engine bearings and/or crankshaft journals and pins.

(k) High Oil Pressure

(i) Using too heavy a grade of oil.

(ii) Faulty adjustment of oil relief valve, too heavy a relief valve springs.

(iii) Faulty oil pressure gauge.

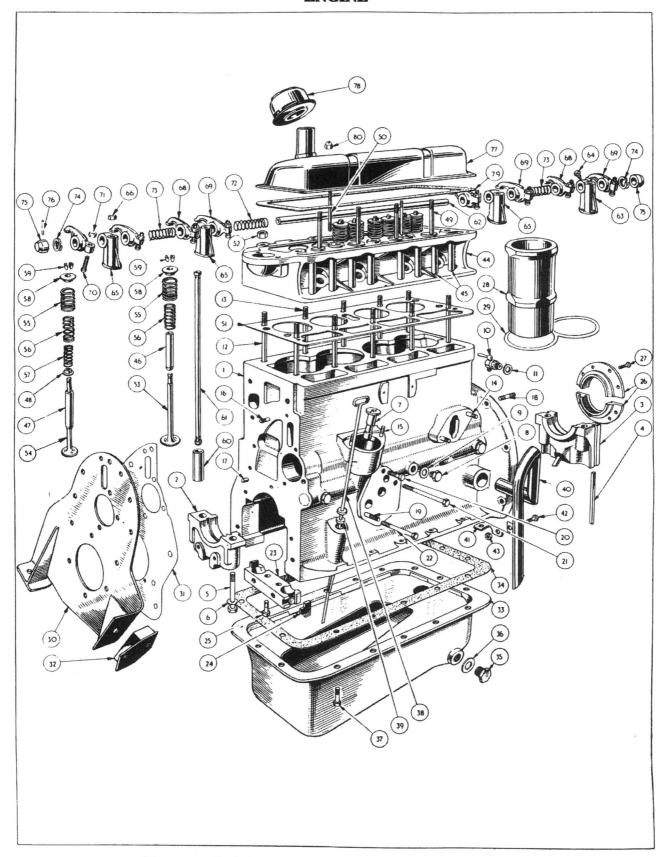

Fig. 49   Exploded view of Engine. Cylinder Block Details.

ENGINE

NOTATION FOR FIG. 49

Exploded view of Cylinder Block Details

Ref. No.	Description	Ref. No.	Description
1	Cylinder Block.	40	Breather Pipe.
2	Front Bearing Cap.	41	Breather Pipe Clip.
3	Rear Bearing Cap.	42	Breather Pipe Clip Bolt.
4	Rear Bearing Cap Felt Packing.	43	Nut for Pipe Clip Bolt.
5	Bearing Cap Bolts.	44	Combustion Head.
6	Spring Washer for Bearing Cap Bolts.	45	Push Rod Sealing Tubes.
7	Oil Pump Drive Bush.	46	Inlet Valve Guide.
8	Oil Gallery Blanking Screw.	47	Exhaust Valve Guide.
9	Washer for Blanking Screws.	48	Exhaust Valve Guide Collar.
10	Drain Tap.	49	Rocker Pedestal Stud.
11	Washer for Drain Tap.	50	Rocker Cover Stud.
12 13 }	Combustion Head Studs.	51	Combustion Head Gasket.
14	Petrol Pump Studs.	52	Combustion Head Securing Nut.
15	Distributor Studs.	53	Inlet Valve.
16	Front Engine Plate Stud.	54	Exhaust Valve.
17	Front Engine Plate Locating Dowel.	55	Outer Valve Spring.
18	Gearbox Stud.	56	Inner Valve Spring.
19	Oil Filter Stud.	57	Auxiliary Exhaust Valve Spring.
20 21 22 }	Oil Filter Attachment Bolts.	58	Valve Spring Collars.
		59	Split Cones.
23	Front Bearing Sealing Block.	60	Valve Tappet.
24	Sealing Block Pads.	61	Push Rod.
25	Screw for Sealing Block.	62	Rocker Shaft.
26	Rear Oil Seal (always a mated pair).	63	Rocker Pedestal (with oil passage drilled).
27	Setscrews for Rear Oil Seal.	64	Rocker Pedestal Screw.
28	Cylinder Sleeve.	65	Rocker Pedestal.
29	Figure of Eight Joint.	66	Rocker Pedestal Attachment Nut.
30	Front Engine Plate.	68	No. 1 Rocker.
31	Engine Plate Joint Washer.	69	No. 2 Rocker.
32	Engine Front Mounting.	70	Ball Pin.
33	Oil Sump.	71	Ball Pin Locking Nut.
34	Joint Washer for Oil Pump.	72	Rocker Centre Spring (coil).
35	Oil Drain Plug.	73	Rocker Intermediate Spring (coil).
36	Washer for Drain Plug.	74	Rocker Outer Spring (flat coil).
37	Oil Sump Bolts.	75	Shaft End Collars.
38	Dipstick.	76	Mills Pins for Shaft End Collars.
39	Felt Washer for Dipstick.	77	Rocker Cover.
		78	Oil Filler Cap.
		79	Rocker Cover Joint.
		80	Nyloc Nuts.

51

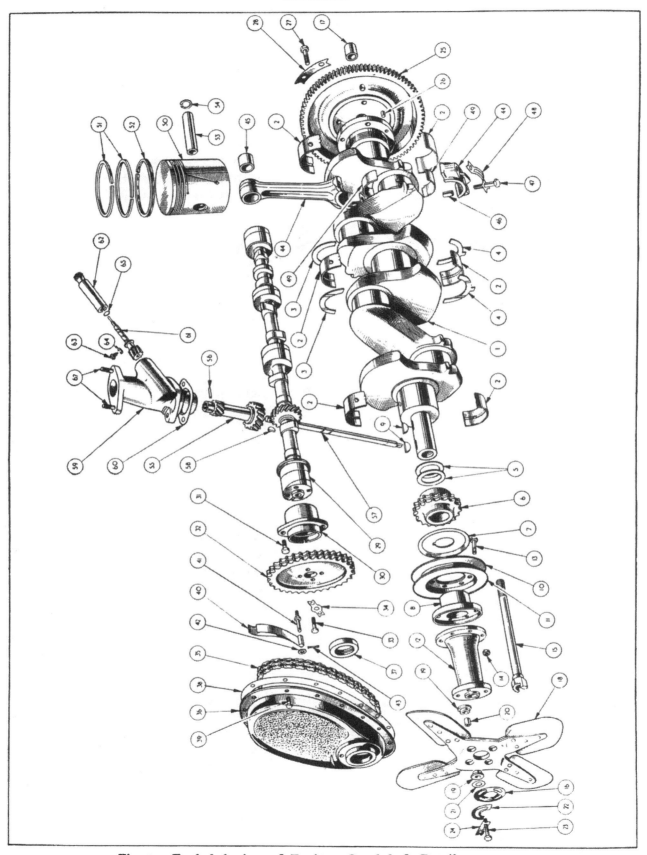

Fig. 50. Exploded view of Engine. Crankshaft Details.

ENGINE

NOTATION FOR FIG. 50

Exploded View of Crankshaft Details.

Ref. No.	Description	Ref. No.	Description
1	Crankshaft.	34	Chain Wheel Bolt Locking Plate.
2	Crankshaft Main Bearings.	35	Timing Chain.
3	Top Thrust Washers.	36	Timing Cover.
4	Lower Thrust Washers.	37	Crankshaft Oil Seal.
5	Sprocket Locating Shims.	38	Timing Cover Joint Washer.
6	Crankshaft Sprocket (Timing Chain).	39	Timing Cover Attachment Bolt.
7	Oil Deflector.	40	Chain Tensioner.
8	Fan Pulley Hub.	41	Chain Tensioner Fulcrum Pin.
9	Woodruffe Keys.	42	Washer for Chain Tensioner Pin.
10	Rear Half of Fan Pulley.	43	Split Pin for Chain Tensioner Pin.
11	Front Half of Fan Pulley.	44	Connecting Rod.
12	Fan Pulley Hub Extension.	45	Small End Bearing.
13	Fan Pulley Bolt.	46	Hollow Dowel.
14	Nyloc Nut for Fan Pulley Bolt.	47	Connecting Rod Bolt.
15	Extension Bolt with Starter Dog Head.	48	Lock Plate for Connecting Rod Bolts.
16	Lock Washer for Extension Bolt.	49	Connecting Rod Bearing.
17	Constant Pinion Pilot Bush.	50	Piston.
18	Cooling Fan Assembly.	51	Compression Ring.
19	Rubber Bushes.	52	Oil Scraper Ring.
20	Metal Sleeves for Rubber Bushes.	53	Gudgeon Pin.
21	Plain Washer.	54	Circlip for Gudgeon Pin.
22	Balance Piece.	55	Distributor and Tachometer Driving Gear.
23	Fan Attachment Bolt.	56	Mills Pin.
24	Locking Plate for Fan Attachment Bolts.	57	Oil Pump Drive Shaft.
25	Flywheel.	58	Woodruffe Key.
26	Flywheel Locating Dowel.	59	Distributor Pedestal.
27	Flywheel Attachment Bolt.	60	Pedestal Joint Washer.
28	Flywheel Bolt Locking Plate.	61	Tachometer Drive Gear.
29	Camshaft.	62	Bearing for Tachometer Drive Gear.
30	Front Camshaft Bearing.	63	Locating Screw for Bearing.
31	Camshaft Bearing Attachment Bolt.	64	Lock Washer for Locating Screw.
32	Camshaft Chain Wheel.	65	Oil Seal.
33	Chain Wheel Securing Bolt.	67	Distributor Stud.

53

FAULT LOCATION

SYMPTOM.	CAUSE.	REMEDY.
Difficulty in Starting Engine.	1. Fault in fuel supply.	(a) Check tank and leaking unions. (b) Clean fuel line, pump and carburettor. (c) Check fuel pump cam lever for bend, weak diaphragm or spring failure. (d) Check carburettor float level. (e) Flooding caused by damaged float or dirty needle and valve.
	2. Sluggish starter motor.	(a) Check battery strength and connection. (b) Dirty bushes. (c) Motor needing overhaul.
	3. Failure of starter pinion to engage with flywheel.	(a) Dirty or bent shaft. (b) In and out of mesh clearance too great.
	4. Faulty ignition.	(a) Condensation on plugs, leads or distributor cap. (b) Plugs dirty or have wrong gap. (c) Dirty or incorrectly set distributor points. (d) Cracked distributor or broken wire. (e) Defective coil or faulty condenser.
Stalling Engine	1. Incorrect carburation.	(a) Dirty jets, mixture and throttle control setting. (b) Air leaks in manifold joints.
	2. Incorrect ignition timing.	Reset timing.
	3. Poor compression.	Decarbonise engine and check for sticking or badly seating valves.
Lack of Power.	1. Choked silencer and/or tail pipe.	Examine the components for carbon deposits.
	2. Binding brakes.	Check brake mechanism.
	3. Slipping clutch.	Check adjustment then overhaul if necessary.

54

FAULT LOCATION (CONTINUED)

SYMPTOM	CAUSE	REMEDY
	4. Incorrect ignition settings.	Check type of plug and spark gap, distributor gap, condenser, seized automatic advance mechanism. Incorrect timing.
	5. Incorrect tappet clearance.	Adjust tappets.
	6. Poor compression.	Check individual compressions with three spark plugs fitted and a compression gauge in the fourth cylinder, throttle set at tick-over using 20 S.A.E. oil and operating the electric starter. Average reading should be 120 lbs. per sq. in. Grind in valves if necessary.
Engine Misfiring.	1. Faulty carburation.	(a) Incorrect float level. (b) Dirty jets. (c) Badly fitting throttle valve or air leaks in joints and manifold connections. (d) Dirty or clogged air filter.
	2. Faulty ignition.	(a) Incorrect ignition timing. (b) Defective plugs or leads. (c) Defective ignition coil or distributor condenser.
	3. Valve condition.	Valves sticking in their guides.

Service Instruction Manual

COOLING SYSTEM

SECTION C

COOLING SYSTEM

INDEX

ILLUSTRATIONS

COOLING SYSTEM

1. DESCRIPTION

The cooling system is pressurised and thermostatically controlled, with an impeller pump to ensure efficient circulation of water at all times. The **capacity** is 13 pints or 14 when a heater is fitted. Careful consideration has been given to points where adequate cooling is necessary, such as sparking plugs and valve guides, etc.

To assist cooling when the car is stationary or travelling at low speeds a 12½" diameter four bladed fan attached to the crankshaft draws air through the radiator.

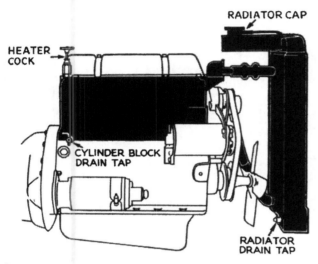

Fig. 1   **Draining of Cooling System.**

2. TO DRAIN THE COOLING SYSTEM

(a) Open the bonnet and remove the radiator filler cap, this is necessary as the system is pressurised. If a heater is fitted ensure that the water shut-off cock is open.

(b) Open both drain taps (Fig. 1), one situated at the lower extremity of the radiator block and a second in the right hand side of the cylinder block below No. 4 inlet and exhaust manifold.

3. FAN BELT ADJUSTMENT

Fan belt adjustment is effected by repositioning the dynamo as follows :—

(a) Loosen the three dynamo attachments.

   (i) The nyloc nut and bolt at the rear, attaching it to the dynamo bracket.

   (ii) The bolt attaching the lower portion of the front flange to the dynamo fulcrum.

   (iii) The bolt securing the upper portion of the flange to the adjusting link.

(b) By moving the dynamo to or away from the engine the fan belt is loosened or tightened respectively. When the belt has ¾" "play" in its longest run suitable adjustment is provided.

(c) Tighten the adjusting link bolt, followed by the two lower attachments.

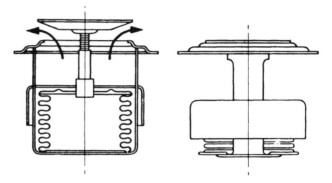

Fig. 2   **The Thermostat in the "Open" and "Closed" condition.**

4. THE THERMOSTAT (Fig. 2)

This is fitted in the cooling system to control the flow of water **before** the engine has reached its normal working temperature.

When the engine is started from cold, water is circulated around the cylinder block by action of the water pump impeller through matched apertures in the impeller pump housing and the cylinder block. The water circulates round the block and cylinder head into the thermostat housing. If the water has not reached a temperature of 158°F. the thermostat will remain closed and the water will pass into the by-pass passage and down to the impeller pump housing to be recirculated through the block by the rotation of the impeller, being driven by a belt at twice crankshaft speed (Fig. 3).

When the water temperature rises above 158°F. (70°C.) the thermostat will commence to open and allow the water to pass into the radiator. This new circulation of water allows the impeller pump to draw water from the lower part of the radiator. The thermostat is fully open at 197°F. (92°C.)

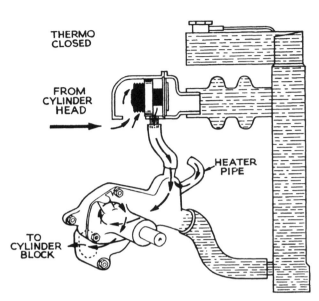

Fig 3 Circulation of Water before the Thermostat has opened.

and at this stage the by-pass is sealed off, this sealing off avoids loss of cooling efficiency when it is most required (Fig. 4).

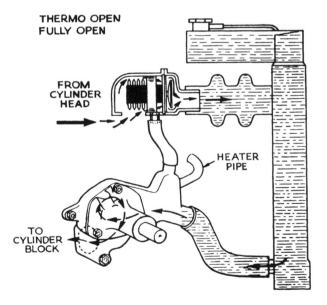

Fig. 4 Circulation when the Thermostat is open.

The radiator temperature for normal motoring should not exceed 185°F. (85°C.).

NOTATION FOR FIG. 5

Ref. No.	Description	
1	Thermostat Housing.	
2	Studs for Top Plate.	
3	Studs for Outlet Cover.	
4	Thermostat.	
5	Outlet Cover.	
6	Outlet Cover Joint Washer.	
7	Nut for securing Outlet Cover.	Up to Commission No. TS.1201
8	Lock Washer for Nut	
9	Top Plate.	
10	Top Cover Joint Washer.	
11	Nut for securing Top Plate	
12	Lock Washer for Nut.	
13	Thermo Housing Joint Washer.	
14	Thermostat Housing.	
15	Studs for Outlet Cover.	
16	Thermostat.	
17	Outlet Cover.	
18	Outlet Cover Joint Washer	
19	Nut for securing Outlet Cover.	From Commission No. TS.1201
20	Lock Washer for securing Nut.	
21	Thermo Housing Attachment Bolt.	
22	Top Hose.	
23	Supergrip Hose Clip.	
24	By-Pass Hose.	
25	Supergrip Hose Clip.	
26	Lower Hose.	
27	Lower Hose Connecting Pipe.	
28	Supergrip Hose Clip.	

5. **TO REMOVE THE THERMOSTAT HOUSING (with thermostat) (Fig. 5)**

 (a) Drain the cooling system. See page 1.

 (b) Disconnect the top and by-pass hoses.

 (c) Loosen the nuts of the thermostat cover, and remove the lower nut to release the petrol pipe clip.

 (d) Remove the thermo gauge capillary tube by withdrawing the gland nut at the left hand side.

 (e) The thermostat housing can be removed by withdrawal of the two bolts attaching it to the combustion head.

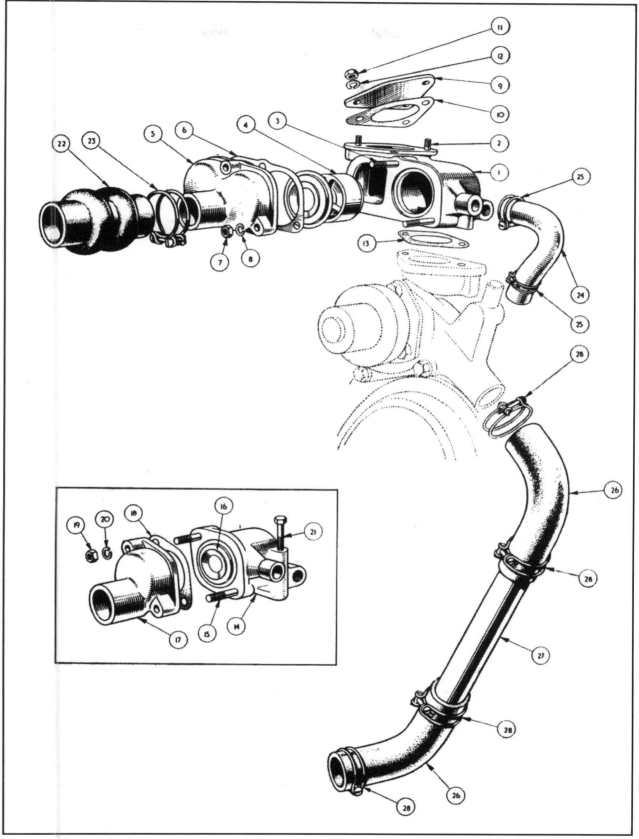

Fig. 5 Exploded details of Thermostat Housings (the housing in the insert is that fitted to current production cars). Cooling System hoses are also shown.

3

(f) The thermostat can be removed from housing by removing the remainder of the front cover nuts (already loosened in para. c) but **after** the removal of the joint washer.

6. TO REPLACE THERMOSTAT HOUSING

The replacement is the reversal of the removal but care should be taken concerning the following points.

(a) That the contact surfaces of the housing and the cover are perfectly clean and do not bear traces of the old joint washer. Failure to observe this point may lead to water leakages.

(b) The thermostat is fitted to the housing first and followed next by the joint washer. In no circumstances should the joint washer be fitted first.

7. TO REMOVE THE THERMOSTAT ONLY (Fig. 5)

(a) Drain the cooling system. See page 1.

(b) Disconnect the top hose.

(c) Withdraw the thermostat housing front cover by removing the three nuts and lock-washers. Remove the petrol pipe clip on the lower right hand stud.
On cars from Commission No. TS.1201 onwards there are only two front cover attachment studs. The lower one accommodating the petrol pipe clip.

(d) Remove the joint washer **before** removing the thermostat.

8. TO REPLACE THERMOSTAT

The replacement is the reversal of the removal but care should be taken concerning the following the points.

(a) That the contact surfaces of the housing and the cover are perfectly clean and do not bear traces of the old joint washer. Failure to observe this point may lead to water leakages.

(b) The thermostat is fitted to the housing first and followed next by the joint washer. In no circumstances should the joint washer be fitted first.

9. TESTING THE THERMOSTAT

Remove the thermostat from its housing as described on page 2. It should be tested in water, at a suitable temperature employing a thermometer to ascertain that the valve does commence to open at the correct temperature 158°F. There is no need to check the temperature at which the valve is fully open as this follows automatically.

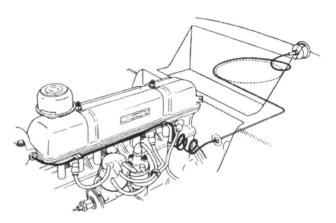

Fig. 6   **The run of the Water Temperature Capillary Tube. The dotted circle indicates the position of the heater.**

10. WATER TEMPERATURE GAUGE

The capillary of this instrument is secured in the thermostat housing by a gland nut and a dial on the instrument panel registers the temperature of the water on the engine side of the thermostat.

Care should be taken that the tubing is not "kinked" for this is liable to fracture the capillary tube thus rendering the instrument unserviceable. Fig. 6 illustrates a suitable "run" for the capillary tube.

11. TO TEST WATER TEMPERATURE GAUGE

When doubt exists concerning the accuracy of the gauge readings, the efficiency of the instrument can be checked by immersing the capillary tube in hot water and checking the gauge reading with that of an accurate thermometer also immersed in the same water adjacent to the bulb.

To effect this test it is merely necessary to remove the gland nut at the left hand side of the thermostat housing.

The instrument is not adjustable or repairable and when a test shows inaccuracies or damage on inspection it will be necessary to replace the complete instrument.

12. THE RADIATOR

The radiator is of the finned pipe type and is secured to the chassis and body of the car at four points. The upper extremity is attached by two nuts and bolts with lock washers to the steady rods, which are in turn secured to the body of the car by jam nuts. The lower attachment is by two pointed shanked bolts with $\frac{1}{8}''$ thick composition packings between the radiator brackets and the chassis frame at either side.

The radiator is pressurised, a relief valve being incorporated in the radiator cap. The spring loaded rubber valve is lifted off its seating when the pressure in the cooling system exceeds 4 lbs. per sq. inch letting the excess pressure escape through the overflow pipe.

To relieve the vacuum when the system cools a small spring-loaded relief valve is incorporated in the centre of the pressure valve unit which will open to admit atmospheric pressure.

The overflow pipe is a rubber tube and is attached to the filler pipe, clipped at the right hand steady attachment, and after running downward it is clipped to the lower right hand wing valance.

13. TO REMOVE RADIATOR

(a) Remove the front cowling as described in the Body Section.

(b) Drain the cooling system as described on page 1.

(c) Remove top and bottom hoses and overflow pipe from radiator.

(d) Remove the nuts and bolts from the two steady rods, one either side at the top of the radiator.

(e) Remove the two bolts and lock washers from the brackets at the sides of the block. The packing between bracket and chassis frame can be removed **after** the radiator has been lifted.

14. TO REPLACE RADIATOR

The replacement of the radiator is the reversal of the removal.

15. FLEXIBLE HOSE CONNECTIONS (Fig. 5)

Four hoses are used in the system and all are moulded rubber with a fibre insert. They are secured to their mating parts by " Supergrip " hose clips.

The smaller diameter curved hose is the by-pass hose for the thermostat—water pump housing connection, the larger diameter straight corrugated hose connects the thermostat housing to the radiator.

The two large diameter curved hoses are assembled to a metal connecting pipe so that their ends are 90° removed from one another. This assembly connects the water pump housing to the radiator outlet.

The overflow pipe is attached to the filler pipe, clipped at the top right hand upper corner of the radiator and again on its run down at a point on the wing valance just above the chassis frame.

16. THE WATER PUMP ASSEMBLY (Fig. 7)

This assembly is attached to the cylinder block by three bolts of unequal length. The longer bolt is situated in the upper right hand position and its purpose is two-fold. In addition to attaching the pump assembly to the cylinder block it also secures the bearing housing to the pump body. The head of this bolt is trapped by the belt pulley and the bolt cannot be removed until this pulley is first removed. The two remaining bolts are of equal length and are situated in the lower extremities of the impeller body.

17. TO REMOVE THE WATER PUMP BEARING HOUSING (Fig. 7)

(a) Loosen the two lower dynamo attachments, remove the upper fixing bolt with the two plain washers and then remove the fan belt.

(b) Loosen the two nuts and the bolt securing the bearing housing to the pump body progressively until the bearing housing can be lifted away with its joint washer.

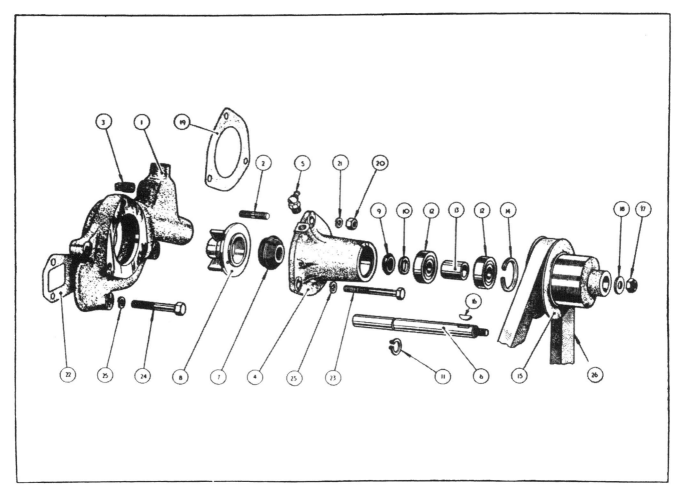

Fig. 7   **Exploded details of Water Pump Housing Assembly.**

NOTATION FOR WATER PUMP HOUSING ASSEMBLY (Fig. 7)			
Ref. No.	Description	Ref. No.	Description
1.	Water Pump Body.	15.	Water Pump Pulley.
2.	Bearing Housing Attachment Stud.	16.	Woodruffe Key.
3.	Plug (removed when heater is fitted).	17.	Nyloc Nut.
4.	Bearing Housing.	18.	Plain Washer.
5.	Grease Nipple.	19.	Water Pump Joint Washer.
6.	Spindle.	20.	Nut.
7.	Water Pump Seal.	21.	Lock Washer.
8.	Impeller.	22.	Water Pump Housing Joint Washer.
9.	Synthetic Rubber Spinner.	23.	Bearing Housing to Cylinder Block Attachment Bolt.
10.	Abutment Washer.	24.	Pump Housing to Cylinder Block Attachment Bolt.
11.	Circlip.		
12.	Bearings.	25.	Lock Washer.
13.	Distance Collar.		
14.	Circlip.		

(c) It will be noted that the bolt is trapped between the bearing housing and the pulley. Mark the position of the bolt on the bearing housing so that during assembly it can be returned to its original position.

18. TO REPLACE THE WATER PUMP BEARING HOUSING

(a) The replacement of this assembly is the reversal of the removal, but the following points should be noted.

(b) The attachment bolt must be fitted before the fan pulley is attached to the shaft. Looking at the pulley end of the assembly with the grease nipple positioned at 11 o'clock, the bolt will occupy the hole at approximately 7 o'clock.

(c) Ensure that the contact surfaces of both components are perfectly clean and a replacement joint washer is used. Failure to observe this point may lead to water leaks.

19. TO DISMANTLE THE BEARING HOUSING ASSEMBLY

(a) Remove nyloc nut and washer from the belt pulley spindle.

(b) Withdraw pulley with the Churchill Universal puller tool No. 6312 and remove the Woodruffe key from its key way.

(c) Utilising the Churchill Tool No. FTS. 127 remove the impeller and rubber seal as shown in Fig. 8.

(d) Remove the bearing locating circlip and gently tap out bearing and spindle assembly.

(e) The bearings and spacer can now be pressed off the spindle, the washer, circlip and synthetic rubber bearing seal can also be removed at this juncture.

20. TO ASSEMBLE THE BEARING HOUSING ASSEMBLY

The assembly is the reversal of the dismantling but the following points must be observed:—

(a) On fitting the bearings to the spindle, ensure that the grease seal incorporated in these bearings face *away* from one another.

(b) The attachment bolt must be fitted before the fan pulley is attached to the shaft. Looking at the shaft end of the assembly with the grease nipple at 11 o'clock, the bolt will occupy the position at 7 o'clock.

(c) The impeller must be a tight fit on the spindle and if it appears to have lost its interference fit with the spindle a replacement must be fitted. It must be pressed on as shown in Fig. 9 and soft

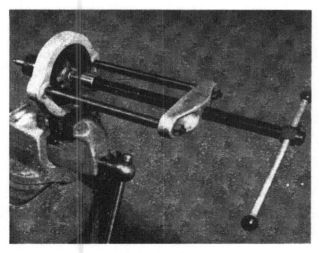

Fig. 8   Utilising the Churchill Tool No. FTS 127 to remove the Water Pump Impeller.

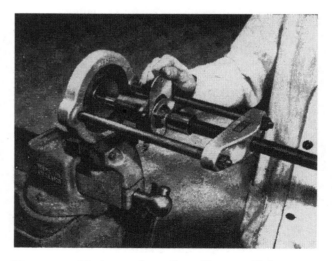

Fig. 9   Fitting the Impeller, utilising the Churchill Tool No. FTS 127.

7

solder run round the end face to ensure a water-tight joint (Fig. 10).

Fig. 10   Showing the correct clearance between Water Pump Impeller and Bearing Housing.

21. RECUTTING THE WATER PUMP SEALING FACE

When servicing the water pump it is sometimes necessary to re-cut the water seal abutment face. The Churchill Tool No. 6300 and bush S.126 is designed for this operation (Fig. 11) and carried out as follows :—

(a) The bearing housing is dismantled as described on page 7.

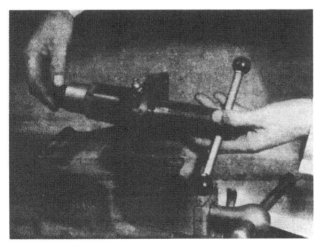

Fig. 11   Refacing Water Seal Face with Churchill Tool No. 6300 and Bush S.126.

(b) Feed the pilot shaft of the Churchill Tool No. 6300 in from the seal seating of the bearing housing. On to the protruding end of the pilot feed the bush S.126, followed by the tool bearing and knurled nut (Fig. 11).

(c) Turn the knurled nut until the cutter contacts the seal face and turn the tool round by the tommy bar, apply firm and steady pressure.

(d) Tightening the knurled nut slightly continue to turn the tool until the seal face is free from score lines and has attained a polished surface.

(e) Whilst carrying out this operation it will be necessary to remove the tool and clean the cutter with a blast from a compressed air line. Do not remove more than .030″ from the seal surface, if the score marks are not removed at this figure a replacement bearing housing should be fitted.

22. TO REMOVE WATER PUMP BODY (When bearing assembly has been removed)

(a) Disconnect the by-pass hose, also the heater pipe if the car is so fitted.

(b) Remove dynamo adjusting link which is secured to the pump body by a set-screw locked by a tabwasher.

(c) Remove the remaining two bolts securing the pump body to the cylinder block.

(d) Remove the body complete with its joint washer.

23. TO REPLACE WATER PUMP BODY

The replacement is the reversal of the removal, but care should be taken concerning the following point.

That the contact surfaces of the housing and the cover are perfectly clean and do not bear traces of the old joint washer. Failure to observe this point may lead to water leakages.

24. THE FAN ASSEMBLY

The fan is built up on a hub and hub extension, then balanced as a unit. When this operation has been completed the balancing plate is drilled right through and the drill allowed to touch the hub extension.

If, for any reason, the fan is dismantled all that is necessary on re-assembly is to line up the component parts so that the drill holes are all in line with the dimple in the hub extension and the re-assembled unit is in balance. Only when replacement parts are fitted will it be necessary to re-balance the unit.

The hub extension is attached to the hub, the latter being keyed to the crankshaft by six nyloc nuts and bolts and the whole assembly is secured to the crankshaft by the extension bolt, the head of which acts as the starting handle dog and on re-assembly it will be necessary to place sufficient shims under the head of the extension bolt to bring it into such a position that when the starting handle is in use compression is felt just after the handle has left B.D.C. as shown in Fig. 37 in Engine Section.

25. TO REMOVE THE FAN ASSEMBLY FROM ENGINE UNIT

(a) Remove the front cowling as described in the Body Section " N ".

(b) Remove the radiator as described on page 5.

(c) Scribe a mark on the balancing plate and fan assembly to ascertain the front of these components for re-assembly.

(d) Turn back the tabs of the locking plates and withdraw the four bolts together with lock plates, plain washers, the balance plate (if one is fitted) and the extension bolt locking plate. The fan assembly, together with split rubber bushes, metal sleeves and larger diameter plain washer can now be removed.

(e) Remove the extension bolt and shims from the hub extension.

(f) By tapping the front flange of the hub extension remove the hub extension, hub and fan belt pulley from the crankshaft. Collect Woodruffe key.

(g) By releasing the tabs of the locking plates the nuts and bolts can be removed. On engines after Engine No. TS.4145E nyloc nuts and plain washers were fitted in place of lock plates and plain nuts. The hub extension can be removed and the hub withdrawn from the pulley pressings.

26. TO FIT FAN ASSEMBLY TO ENGINE UNIT

(a) Fit the Woodruffe key to the crankshaft and slide on the hub and hub extension assembled as described in operations a, b and c of " To assemble fan for balancing," hereafter.

(b) Fit the two shims under the head of the extension bolt and insert through the centre of hub extension and tighten until the abutment of the starting dog jaws, incorporated in the head of the extension bolt, assume a " 10 to 4 o'clock " position to ensure correct relationship with compression when the starting handle is in use.

(c) On to one pair of fan securing bolts feed one lock plate followed by one plain washer per bolt.

(d) Offer up the fan assembly in such a manner that the hole in the web is over the dimple in the hub extension face. Fit the extension bolt locking plate with the larger diameter plain washer between it and the rubber bushes. Secure the extension bolt locking plate with the bolts built up as described in operation (c above) utilising the two tappings opposite those with the $\frac{5}{32}''$ drill hole.

(e) The remaining pair of bolts are made up in a similar manner to those already mentioned, but with the balancer fitted. These bolts are assembled to the remaining tappings in the hub extension. Before tightening, the balancer is moved until the hole aligns with those in the fan assembly ; after tightening the tabs of the locking plates are turned over.

(f) Replace the radiator and hoses.

(g) Replace the front cowling as described in the Body Section.

27. TO ASSEMBLE FAN FOR BALANCING

Check that the four fan blades riveted to the fan webs are free from movement. If for any reason replacement parts have been fitted the fan unit should be re-balanced. The dimple in hub extension face should be filled in with solder to avoid confusion during re-assembly.

(a) Place the two pulley pressings together, the flatter one with the drilled hole uppermost and the second pressing on top; feed the hub through the pressings with its keyway lowermost. It is necessary that this procedure is followed for it ensures a visual check of setting the engine at T.D.C. on Nos. 1 and 4 cylinders.

(b) Position the six bolts and secure the hub extension with the nyloc nuts. On early production cars, nuts and locking plates were used.

(c) Insert the rubber bushes in the fan assembly and locate the metal sleeves through the centres of these bushes.

(d) Feed the four fan attachment bolts through the larger diameter plain washers and metal sleeves of the fan assembly and secure the latter to the hub extension.

(e) Using a jig, ascertain the lighter side of the assembly and fit the balancer to that side. This can be moved to obtain perfect balance.

(f) When the balanced condition is attained a $\frac{5}{32}''$ drill hole should be put through the thinner edge of the balancer and fan assembly webs until it makes a small dimple in the face of the hub extension. Withdraw the four bolts and remove fan assembly from hub extension.

28. ANTI-FREEZE PRECAUTIONS

During frosty weather it is necessary to protect the engine from damage and this can be effected by draining the cooling system by opening the tap at the lowermost portion of the radiator, and the second tap at the right hand side of the cylinder block.

In severe frosty weather an anti-freeze additive to the cooling system is strongly recommended, for it is possible for the lower portion of the radiator to become frozen, even when the car is being driven, restricting the circulation of the water as well as causing possible damage to the radiator itself. Before adding the anti-freeze compound thoroughly flush out the radiator and cylinder block, and ascertain that all hoses and connections are in perfect condition. Check also that the cylinder head nuts are tight, for if due to leaks, any anti-freeze solution finds its way into the cylinder bores serious damage may result.

The anti-freeze solution itself does not usually evaporate, thus apart from leakage, it should only be necessary to top up with water as the level in the radiator head drops.

This Company uses and recommends Smiths "Bluecol", and for protection from various degrees of frost the following proportions are recommended.

Degrees of Frost (Fahrenheit)	15	25	35
Proportion (per cent)	10	15	20
Amount of Bluecol (pints)	2	3	4

Water capacity 13 pints, 14 pints with heater. Other reputable anti-freeze compounds are available and the compound chosen should be used in accordance with the manufacturer's instructions.

It is a very wise precaution when using anti-freeze in the cooling system to employ some method of indicating the fact for the enlightenment of repairers who may be called upon to carry out adjustments or the replacement of parts.

COOLING SYSTEM

SERVICE DIAGNOSIS.

OVERHEATING.

This difficulty may arise owing to one or more of the causes listed below :—

CAUSE	REMEDY
Ignition timing too late or auto advance and retard mechanism or suction not operating correctly.	Check ignition timing, automatic advance and retard mechanism and the suction pipe for the carburettor.
Fan belt slipping.	Adjust to give belt $\frac{3}{4}''$ play by moving dynamo outwards along adjusting link.
Insufficient water in cooling system.	Check all joints for leaks including combustion head gasket
Radiator and/or cylinder block restricted by the accumulation of sludge, dirt or other solid matter.	Flush out system with a detergent and refill, using clean, softened or soft water.
Thermostat not operating correctly.	Remove and test as described on page 4.
Weak mixture caused by incorrect carburettor setting or air leaks in induction manifold.	Check carburettor manifold and carburettor joints, ensure tightness of manifold.
Initial tightness after an engine overhaul or insufficient clearance of replacement parts during an overhaul.	If due to the former, run-in engine most carefully and overheating should disappear. If overheating is caused by the latter it will not disappear, it can even get worse. The engine should be examined for badly fitting parts.
Overheating from bad lubrication, incorrect oil level or incorrect grade of oil. The use of certain brands of anti-freeze compound which have a lowering effect on the boiling point during warm weather.	Check oil level, grade and circulation, flushing system and refilling if necessary. Smiths " Bluecol " has a tendency to raise the boiling point.

Service Instruction Manual

CLUTCH

SECTION D

CLUTCH

INDEX

ILLUSTRATIONS

CLUTCH

1. GENERAL DATA

Model A 6 G 9".

Hydraulically operated from twin bore master cylinder which incorporates the brake master cylinder.

Ball bearing release bearing.

Clearance between ball bearing release bearing and release levers—.0625".

Nine, 120—130 lb. cream thrust springs.

Single dry plate with six springs. All six springs cushion the driving torque, whilst three (grey in colour) cushion the over run.

Free travel on clutch pedal = .820".

Clearance between piston rod and master cylinder piston = .030".

End float in Slave Cylinder fork assembly = .079".

Height of release lever tip from face of flywheel = 1.895".

Long portion of hub towards Gearbox.

2. TOOL DATA

Borg and Beck Gauge Plate No. CG.192.

Land Thickness = .330" (see page 13).

Churchill Tool Spacers .... No. 3
Churchill Tool Adapters ...... No. 7
Churchill Tool base plate position D

3. CLUTCH OPERATION

The clutch is hydraulically operated and has a twin bore master cylinder (see Brake Section "R", for full explanation) attached to the bulkhead under the bonnet and a slave cylinder secured to the gearbox bell housing by a support plate, these are connected together by a length of Bundy-tubing and a flexible hose.

When pressure is applied to the foot pedal of the master cylinder it is transmitted through the pipe line to the slave cylinder. The piston of this cylinder operates a rod attached to the lever of the clutch operating shaft, a fork mounted on the latter engages in an annular groove of the release bearing mounting sleeve and moves the release bearing into engagement with the release levers.

4. TWIN BORE MASTER CYLINDER

The unit consists of an integrally cast body with a common fluid reservoir for the two identical bores, one connected to the brakes and the second to the clutch. Each bore

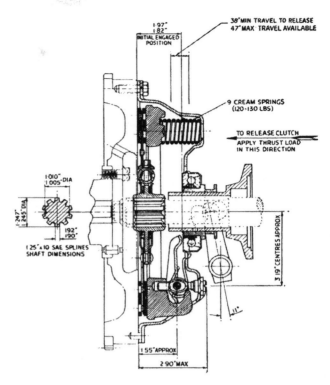

Fig. 1   **Sectional view of Clutch.**

accommodates a piston having a main cup loaded on to its head by a return spring. In order that the cup shall not tend to be drawn into the holes in the piston head, a piston washer is interposed between the main cup and the piston head.

Unlike the brake cylinder bore, with that for the clutch, there is no check valve fitted at the delivery end of the return spring and this spring uses the body as an abutment.

The absence of this check valve precludes the risk of residual line pressure which would tend to keep the release bearing in contact with the release levers, causing excessive wear on the bearing and possible clutch slip.

5. CLUTCH SLAVE CYLINDER

The slave cylinder is mounted on a support plate which is attached by the two lower bell housing bolts to the left-hand side of the engine unit. A steady bracket, attached at its forward end to the engine unit by one of the sump bolts, forms the slave

1

cylinder and plate upper attachment by means of a jam nut and a nyloc nut. The lower attachment being effected by nut and bolt with washer. A return spring is fitted to a plate on the clevis pin of the fork assembly to the lower portion of the support plate.

The inner assembly of the slave cylinder is made up of a coil spring, cup filler, rubber cup and a piston. The piston moves in the highly polished bore when hydraulic pressure is applied through the pipe line.

6. THE CLUTCH OPERATING SHAFT

This shaft is carried in the bell housing in two " Oilite " bushes, it is positioned by a fixing screw, the shank of which locates the reduced diameter portion of the shaft. A short coil spring is placed between the shaft lever and the bell housing which steadies the shaft and prevents rattle.

Mounted on the shaft is the release bearing operating fork, being secured thereto by a tapered pin, the shank of which passes into the shaft, whilst its head is locked to the fork by a short length of wire.

The shaft is lubricated by grease nipples and over-lubrication must be avoided (see page 4).

7. THE RELEASE BEARING

This is a ball bearing housed in a cover. A sleeve pressed into the inner race of this bearing, is grooved externally to accommodate the pins of the clutch operating fork mounted on its shaft in the bell housing of the gearbox. The sleeve, pressed into the bearing, moves on an extension of the front gearbox cover which ensures its correct angular engagement with the three release levers.

The ball bearing is grease packed during its manufacture and does not require re-greasing.

8. COVER ASSEMBLY

This assembly consists of a steel pressing to which the component parts are assembled, being attached with the Driven Plate Assembly to the flywheel.

NOTATION FOR CLUTCH ASSEMBLY (FIG. 2)

Ref. No.	Description
1	Clutch Cover.
2	Pressure Plate.
3	Thrust Springs.
4	Release Lever Eye Bolt.
5	Release Lever Pin.
6	Release Lever.
7	Release Lever Strut.
8	Anti-Rattle Spring.
9	Adjusting Nut.
10	Driven Plate Assembly.
11	Driven Plate Facings.
12	Ball Bearing, Release Bearing and Pressed-in Sleeve.
13	Clutch Operating Fork.
14	Taper Pin.
15	Clutch Operating Shaft.
16	Spring on Operating Shaft.
17	Grease Nipple (one each end of shaft).
18	Shaft Locating Bolt.
19	Locking Washer for Locating Bolt.
20	Slave Cylinder Body.
21	Bleed Screw.
22	Cup Filler Spring.
23	Cup Filler.
24	Rubber Cup.
25	Piston.
26	Rubber Boot.
27	Small Circlip for Rubber Boot.
28	Large Circlip for Rubber Boot.
29	Fork Assembly Rod.
30	Fork End.
31	Clevis Pin.
32	Clevis Pin Spring.
33	Fork End Locking Nut.
34	Clutch Shaft Return Spring.
35	Anchor Plate for Return Spring.
36	Slave Cylinder Support Bracket.
37	Lower Attachment Bolt.
38	Nut.
39	Lock Washer.
40	Slave Cylinder Stay.
41	Nyloc Nut.

The cover assembly contains a cast iron pressure plate loaded by nine cream thrust springs (120—130 lbs.). Mounted on the pressure plate are three release levers which pivot on floating pins retained by eye bolts. Adjusting nuts are screwed on the eye bolts, which pass through the cover pressing these nuts being secured by staking.

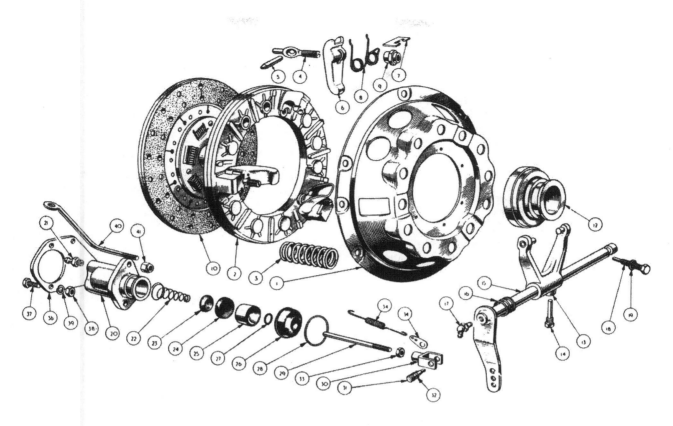

Fig. 2   **Exploded details of Clutch Assembly with Slave Cylinder.**

Struts are interposed between the lugs on the pressure plate and the outer ends of the release levers. Anti-rattle springs are fitted between the release levers and the cover pressing.

9. DRIVEN PLATE ASSEMBLY

This is the Borglite spring type, having a splined hub and a disc adapter fitted with nine cushioned segments which carry two facings attached by rivets.

The hub flange and disc adapter are slotted to carry six springs (3 red, 3 grey) positioned by a retaining plate which is secured to the disc adapter by stop pins. This flange is drilled to carry three steel balls positioned by the two friction plates located by tabs in holes in the hub flange.

A spacer is fitted between the disc adapter and one friction plate and another spacer is fitted between the retaining plate and the second friction plate.

10. MAINTENANCE

It is essential that the master cylinder is at least half full of Lockheed Brake Fluid at all times, and should be checked every 5,000 miles (8,000 kms.).

Only Lockheed Brake Fluid should be used in this system. This fluid has been selected as it has no injurious effects on the rubber seals and flexible hoses used.

Before removing the filler cap, wipe the top of the master cylinder and the cap clean with a non-fluffy material. Cleanliness is particularly important and every precaution should be taken to ensure no dirt or foreign matter is allowed to enter the system. Failure to observe this point may lead to blockages; damage to the highly polished bores and pistons, resulting in expensive replacements.

Ensure also that the breather hole in the filler cap is not restricted and that the sealing washer and pipe lines are in good order.

11. BLEEDING THE HYDRAULIC SYSTEM

Bleeding is only necessary when a portion of the system has been disconnected or if the level of the fluid has been allowed to fall so low that air has been allowed to enter the system. If bleeding is carried out for the latter reason the brake system will need to be bled also, as they share the same reservoir.

(a) Fill the reservoir with Lockheed Brake Fluid and keep at least half full throughout the operation. Failure to observe this point may lead to air being drawn into the system and the operation of bleeding will have to be repeated.

(b) Attach a length of rubber piping to the bleed screw and allow the free end to be submerged in a little Lockheed Brake Fluid contained in a *clean* glass jar, open the bleed port by giving the screw one complete turn.

(c) Depress the clutch pedal with a slow full stroke and before the pedal reaches the end of its travel the bleed screw is tightened sufficiently to seat it.

(d) Repeat the operation (c) until air bubbles cease to appear from the end of the tube.

(e) Ensure that there is sufficient fluid in the reservoir, at least half full, and replace cap first, ensuring that its seal is in good order and its vent is unobstructed.

12. GREASING OF THE CLUTCH OPERATING SHAFT

Hand grease gun lubrication should be used when greasing this shaft. Two strokes of the gun to each nipple after 5,000 miles (8,000 kms.) of running will provide adequate lubrication

Over lubrication, from generous use of pressure lubricating may lead to grease finding its way on to the clutch facing.

13. ADJUSTING THE CLUTCH

The adjustment connection between pedal and master cylinder is set on initial assembly and should not need re-adjustment.

During complete overhauls or the repair of accidental damage the master cylinder may have to be disturbed. Its replacement is dealt with in the Brake Section "R" and the adjustment is described in this section below.

The clutch pedal will provide no sensitive indication of loss of release bearing clearance ($\frac{1}{16}''$) consequent upon wear of the facings. Adjustment at the slave cylinder fork assembly must therefore be checked periodically, at whatever intervals the operating conditions may dictate. The adjusting sequence is described below.

The adjustment is said to be correct when there is .079″ end float in the slave cylinder fork assembly.

14. ADJUSTING THE MASTER CYLINDER

It is important to provide .030″ free travel of the push rod before it reaches the piston. This clearance is necessary to ensure that the piston will return to its stop in its cylinder and thus prevent the possibility of the lip of the main cup covering the by-pass port. If such a condition were to exist the excess fluid drawn into the cylinder during the return stroke of the piston will find no outlet and pressure will build up in the system causing the clutch to "slip".

(a) Loosen the jam nut of the clutch pedal stop at the forward end of the master cylinder support bracket.

(b) Turn the adjuster screw inwards and testing the push rod eliminate all end float. Tighten jam nut finger tight, holding adjuster screw.

(c) Unscrew the adjuster together with the jam nut until a .030″ feeler can be placed in between the jam nut and the master cylinder bracket.

(d) Holding adjuster screw, lock jam nut to the bracket.

15. ADJUSTING THE SLAVE CYLINDER

(a) Unlock the jam nut on the slave cylinder fork assembly.

(b) Turn the rod until ALL end float is just eliminated.

(c) Hold the push rod and turn the jam nut until a .079″ feeler gauge will pass in between the nut and the fork end.

(d) Screw the rod together with the jam nut to the fork and lock. Check by moving the fork assembly and readjust if necessary.

Fig. 3   The Slave Cylinder and support bracket.

16. TO REMOVE THE FLEXIBLE HOSE

(a) Drain the hydraulic system.

(b) Holding the hexagon of the flexible hose, withdraw the Bundy tubing by first removing the union nut.

(c) Still holding the hexagon of the flexible hose, remove the locking nut and shake proof washer.

(d) Withdraw the flexible hose from its bracket and disconnect it from the slave cylinder.
Ensure that its whole length is turned whilst unscrewing as any twist will impair the life of the hose.

17. TO FIT THE FLEXIBLE HOSE

Ensure that all connections are perfectly clean. Dirt being allowed to enter the system may cause blockages, or damage to the highly polished bores and pistons resulting in expensive replacements.

(a) Utilising a new copper gasket, attach and secure the flexible hose to the lower port of the slave cylinder.

(b) Feed the hose into the bracket welded on the left hand chassis member. Gripping the hexagon of the hose with a spanner set the hose in such a manner that it will have a free run, away from all obstructions and rubbing contacts.

(c) Still holding the hexagon of the hose secure it to the chassis bracket with the shakeproof washer and lock-nut.

(d) Insert the Bundy tubing into its housing and check that it is correctly seated before securing with the union nut.

(e) Bleed the clutch system as described on page 4.

18. REMOVAL OF THE SLAVE CYLINDER (with fork-rod assembly) Fig. 3.

(a) Remove the flexible hose as described on this page.

(b) Unhook the spring from the slave cylinder support plate. Remove the split pin and the clevis pin exercising care not to mislay the spring between the fork and the clutch shaft lever. Remove the spring attachment plate.

(c) Remove the nyloc nut from the slave cylinder stay and the nut, bolt and lock washer from the lower cylinder fixing point and withdraw slave cylinder from its support plate.

(d) Withdraw the fork assembly from the slave cylinder together with the rubber boot by first removing the wire clip from the exterior of the boot and slave cylinder.

19. TO REPLACE SLAVE CYLINDER

(a) Seat the slave cylinder in the support bracket with the bleed screw uppermost.

(b) Secure at the uppermost point by a nyloc nut on the threaded end of the stay and at the lowermost point with nut, bolt and lock washer.

(c) Fit the small coil spring and spring anchor plate either side of the clutch operating lever, followed by the fork assembly. Secure with the clevis pin and lock with split pin.

(d) Attach the return spring to the spring anchor plate of the fork end assembly and anchor the other end to the slave cylinder support bracket.

(e) Fit the flexible hose as described on this page.

(f) Adjust the clutch at the fork end assembly as described on page 4.

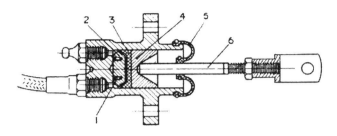

Fig. 4   Sectional View of Slave Cylinder. 1 Spring.
2 Cup Filler. 3 Rubber Cup. 4 Piston.
5 Rubber Boot. 6 Fork assembly.

20. DISMANTLING THE SLAVE CYLINDER (Fig. 4)

(a) Remove the slave cylinder assembly from its mounting as described on page 5. Remove bleeder screw.

(b) Remove the wire circlip from the rubber boot and ease the rubber boot from the alloy body.

(c) The rubber boot can be removed from the fork end assembly by first removing the wire circlip. The assembly can now be drawn through the rubber.

(d) By applying low air pressure through one of the tapped holes the piston can be removed from the cylinder bore followed by the rubber cup, the cup filler and spring.

(e) The components should be washed in Lockheed Brake Fluid and any component that shows excess wear should be replaced. Particular attention must be paid to the cylinder bore and piston.

21. ASSEMBLY OF THE SLAVE CYLINDER

(a) Give the component parts a liberal coating of Lockheed Brake Fluid and also the bore of the cylinder.

(b) Assemble the spring to the cup filler and insert both, spring first, into the bore of the cylinder.

(c) Fit the rubber cup, lip first, into the bore, exercising great care that the edges do not curl up inside the bore. After assembly it will be noticed that the flat surface of the rubber cup is uppermost and will accommodate the piston.

(d) Slide the piston into the cylinder, flat side first, the piston may be assisted in the travel by the rod of the fork end assembly.

(e) Insert the push rod of the fork assembly into the rubber in such a manner that the push rod end is nearer to the lips of the boot. Secure the rubber boot to the rod with a small circlip.

(f) Fit the fork end assembly and rubber boot to the slave cylinder body and secure with the large wire circlip.

(g) Fit the bleed screw to one of the ports in the slave cylinder body.

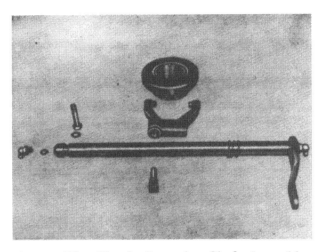

Fig. 5   The Clutch Operating Shaft Assembly.

22. TO REMOVE RELEASE BEARING AND CLUTCH OPERATING SHAFT (Fig. 5)

(a) Remove the gearbox from the car as described in the Gearbox Section "E".

(b) Break and remove the wire locking the taper pin to the clutch bearing operating fork, remove taper pin.

(c) Withdraw the release bearing and sleeve from the front end cover of the gearbox.

(d) Remove grease nipple and fibre washer from right hand end of clutch operating shaft.

(e) Withdraw the shaft locating bolt and lock washer from right hand side of bell housing.

6

(f) Holding the clutch operating fork withdraw the shaft from the left.

(g) Remove spring and grease nipple with fibre washer from lever end of shaft.

NOTE—To effect the removal of the shaft from cars prior to Commission No. TS. 411, there is no necessity to remove the grease nipple (operation **d**) and the shaft locating bolt (operation **e**) is situated on the left hand side of the bell housing.

23. TO REPLACE CLUTCH OPERATING SHAFT AND RELEASE BEARING

The replacement of the clutch operating shaft and release bearing is the reversal of the removal. It will be found, however, that light pressure will be necessary to compress the spring on the operating shaft to insert and tighten the shaft locating bolt.

When fitting the ball bearing release bearing, locate the pegs of the operating fork in the groove of the bearing. Secure the operating fork to the shaft with the taper pin and lock the head with wire.

24. REMOVAL OF THE CLUTCH FROM FLYWHEEL WITH GEARBOX REMOVED

(a) Slacken the six holding bolts, in the outer rim of the cover pressing, a turn at a time by diagonal selection until the thrust spring pressure is relieved.

(b) Remove the six bolts and lift away the cover assembly and driven plate assembly from the two locating dowels.

(c) Inspect the two dowels in the flywheel for looseness and burrs and replace if necessary.

25. REPLACEMENT OF CLUTCH TO FLYWHEEL (Fig. 6)

(a) Place the driven plate assembly on the flywheel with the larger portion of the splined hub towards the gearbox. Centralise this plate with the Churchill Tool No. 20S. 72 or the splined portion of a constant pinion shaft.

(b) Fit the cover assembly over the driven plate and locate it on the two dowels in the face of the flywheel.

Fig. 6   Showing Constant Pinion Mandrel in position. Churchill Tool No. 20S.72.

(c) Secure the cover assembly to the flywheel with six bolts and lock washers, tightening them a turn at a time by diagonal selection to the correct tightening torque, 20 lbs. ft.

(d) Remove the driven plate centraliser only when the cover assembly is attached to the flywheel.

It is essential that the driven plate assembly is central at all times during the assembly of the cover to flywheel. Failure to observe this point may lead to difficulty in attaching the gearbox, for the constant pinion shaft may not have a free passage to the pilot bearing bush in the rear end of the crankshaft.

26. DISMANTLING THE COVER ASSEMBLY USING THE CHURCHILL FIXTURE No. 99A (Fig. 7)

(a) Before dismantling the clutch, suitably mark the following parts so that they can be re-assembled in the same relative positions to each other and so preserve the balance of the clutch cover assembly.

(i) Cover pressing.

(ii) Lugs on the pressure plate.

(iii) Release levers.

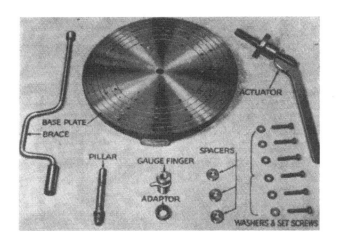

Fig. 7   Clutch Assembly Fixture (Churchill Tool No. 99A) as used with 9″ clutch.

(**b**) Determine from the code card in the Churchill Fixture No. 99A, the reference numbers of the adapter, the spacers, and the spacers position letter on the Churchill base plate. For this clutch they are 7, 3 and D respectively.

(**c**) Clean the top of the base plate and place the three number 3 spacers (Fig. 8) on the positions marked " D ".

Fig. 8   Spacers in position on Base Plate.

(**d**) Place the cover assembly on the base-plate so that the release levers are situated directly above the spacers and the bolt holes in the rim of the cover pressing are in line with the tapped holes in the base plate.

(**e**) Screw the actuator into the centre hole, and press the handle down to clamp the cover housing to the base plate.

(**f**) Insert through the cover pressing six bolts and secure cover assembly to base plate (Fig. 9). Remove the actuator.

Fig. 9   Securing Cover assembly to Base Plate.

(**g**) Remove the three adjusting nuts, considerable torque will be necessary as the staking of these nuts has to be overcome.

(**h**) Remove the bolts clamping cover assembly to base plate by diagonal selection to release load on springs (Fig. 10).

Fig. 10   Adjusting nuts have been removed, cover securing bolts being removed.

(**i**) Take off cover pressing and remove the nine thrust springs and anti-rattle springs.

(j) Lift up inner end of release lever and disengage the strut. Repeat procedure for 2nd and 3rd levers.

(k) Gripping the tip of the release lever and the eye bolt lift out the assembly from the pressure plate. Repeat procedure for 2nd and 3rd levers.

(l) Remove the eye bolts from release levers and take out pins. Remove the struts from pressure plate.

27. ASSEMBLY OF COVER PLATE ASSEMBLY USING THE CHURCHILL FIXTURE No. 99A

Before assembling a smear of Lockeed Expander Lubricant or Duckham's Keenol K.O. 12 should be applied to the release lever pins, contact faces of the struts, eyebolt seats in the cover pressing, drive lug sides on the pressure plate and the plain end of the eye bolts.

Assembly is to be made with strict regard to the markings on certain parts and so ensure that the unit remains in balance.

(a) Place strut in position in lug of pressure plate.

(b) Assemble pin to eye bolt and feed threaded portion through release lever.

(c) By holding the strut in the pressure plate to one side, feed the plain end of the eye bolt (assembled to release lever) into the pressure plate (Fig. 11).

Fig. 11   Fitting Release Levers to Pressure Plate.

(d) Place the strut into groove in the outer end of the release lever.

(e) Repeat operations (a) to (d) for the remaining two release levers.

(f) Place the pressure plate and the assembled release levers, with the latter over the spacers, on the base plate of the Churchill Fixture.

(g) Place the cover pressing over the pressure plate laying on the base allowing the lugs to protrude through the cover. Should the holes in the cover pressing fail to line up with those in the base plate the cover and pressure plate must then be turned to allow alignment. Remove the cover pressing without disturbing the position of the pressure plate. Fit the anti-rattle springs.

(h) Place springs on their seats on the pressure plate, followed by cover pressing (Fig. 12).

Fig. 12   Cover pressing with anti-rattle springs fitted ready for final assembly.

(i) Insert bolts through cover pressing into base plate. Tighten bolts by diagonal selection, checking that the pressure plate lugs protrude through the cover and the anti-rattle springs contact the release levers.

(j) Screw on adjuster nuts until their heads are flush with the tops of the eye bolts.

(k) Fit the actuator into the centre hole of the base plate and pump handle up and down half a dozen times to settle the assembled mechanism, remove actuator.

(l) Secure pillar firmly into centre of base plate. Place on No. 7 adapter, recessed side downward, followed by gauge finger.

(m) Screw the adjusting nuts to raise or lower the release levers sufficiently to just contact the finger gauge (Fig. 13).

Fig. 13   Adjusting the release levers.

(n) Exchange the finger gauge and pillar for the actuator and operate the clutch a dozen or so times. Check again with finger gauge and make any adjustments necessary.

(o) Lock the adjusting nuts by peening over the collars into the cuts of the eye bolts.

(p) Remove cover assembly from base plate and it is ready to be fitted to the flywheel (with the driven plate assembly).

28. DISMANTLING THE COVER ASSEMBLY (Fig. 14) WITHOUT CHURCHILL FIXTURE

In the event of the Churchill Fixture not being available the following method is suggested.

This method utilises a fly or hydraulic press and suitable size wooden blocks ; two blocks on which to stand the pressure plate and allow the cover pressing downward movement. Before dismantling the cover assembly suitably mark the following parts so that they can be re-assembled in the same relative positions to each other and so preserve the balance of the cover assembly :—

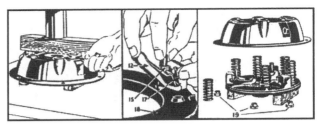

Fig. 14   Dismantling the Cover Assembly utilising a ram press. 12 Release Lever. 15 Eye Bolt. 17 Strut. 18 Pressure Plate. 19 Adjusting Nuts.

(i) Cover pressing.

(ii) Lugs on the pressure plate.

(iii) Release levers.

(a) Lay the assembly on the bed of the press with the pressure plate resting on the two wooden blocks so arranged that the cover pressing is free to move downwards when pressure is applied.

(b) Lay another wooden block on top of the cover pressing in such a manner that it will contact the ram of the press and will also move downward between the release levers.

(c) Lower the ram of the press sufficiently to bring the cover pressing in contact with the bed of the press. Secure the ram and remove the three adjusting nuts, considerable torque will be necessary as the staking of these nuts has to be overcome.

(d) Release the pressure of the press slowly to prevent the thrust springs from flying out.

(e) Remove the cover pressing and collect the component parts.

29. TO ASSEMBLE COVER ASSEMBLY WITHOUT CHURCHILL FIXTURE

Before assembly note the markings on the various components and return them to their original positions. Grease the components slightly at their contact faces with Lockheed Expander Lubricant or Duckham's Keenol K.O. 12.

(a) Fit the pins to the eye bolts and locate these parts within the release levers. Hold the threaded end of the eye bolt and the inner end of the lever as close together as possible and, with the other hand, engage the strut within the slots in a lug on the pressure plate and the other end of the strut push outwards to the periphery of the pressure plate. Offer up the lever assembly, first engaging the eye bolt shank within the hole in the pressure plate, then locate the strut in the groove of the release lever. Fit the remaining levers in a similar manner.

(b) Place the pressure plate on the wooden blocks on the base of the press and position the thrust springs on the bosses on the pressure plate.

(c) Place the cover pressing, with the anti-rattle springs fitted, over the pressure plate ensuring that the lugs protrude through the cover slots.

(d) Arrange a wooden block across the cover and apply pressure to compress the whole assembly. Screw the adjusting nuts on to the eye bolts sufficiently so that pressure can be released.

30. INSPECTION OF COVER ASSEMBLY

Before re-assembling the clutch unit the parts should be cleaned and inspected. Any components which show considerable wear on its working surface should be replaced. The thrust springs and anti-rattle springs should be checked against new ones of the correct strength, and any found to be obviously weak should be replaced. The anti-rattle springs should be assembled to the cover pressing. The working face of the cast iron pressure plate should also be inspected and if the ground face is deeply scored or grooved it should be either re-ground or replaced by a new plate.

If any parts are changed or a new pressure plate fitted, it is essential it should be statically balanced.

31. ADJUSTING THE RELEASE LEVERS

In service, the original adjustments made by the clutch manufacturer, will require no attention and re-adjustment is only necessary if the cover assembly has been dismantled.

There are three methods by which the release levers may be adjusted.

    (i) Churchill No. 99A Clutch Fixture.

    (ii) Borg and Beck No. CG 192 gauge plate. (If available).

    (iii) In the absence of the above the Driven Plate Assembly may be used.

(a) Churchill No. 99A Clutch Fixture. Both this Company and the Clutch manufacturers recommend this method. Details can be found on page 9.

(b) Borg and Beck No. CG 192 Gauge Plate method (Fig. 15).

    (i) Utilising the actual flywheel lay the Gauge Plate in the position normally taken by the driven plate assembly. Mount the cover plate

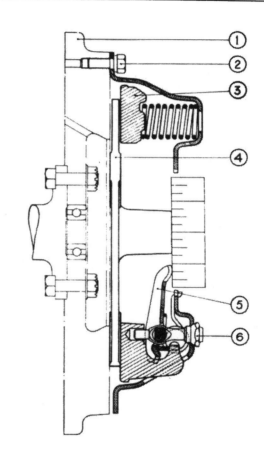

Fig. 15   **Adjusting the Release Levers utilising the Borg and Beck Gauge plate No. CG 192.**

Notation for Fig. 15.

Ref. No.	Description
1	Flywheel.
2	Cover assembly attachment bolts.
3	Pressure plate.
4	Borg and Beck gauge plate No. CG 192.
5	Release lever.
6	Adjusting nut.

11

assembly on the flywheel so that the ground lands of the gauge plate are situated under the release levers.

(ii) Turn the adjusting nuts to bring the release lever tips to contact a short straight edge resting upon the boss of the gauge plate.

(iii) Having made preliminary setting, operate the mechanism several times in order to settle the mechanism. The press used for assembling the cover assembly will perform this operation.

(iv) Carry out a check with the straight edge and re-adjust if necessary. Lock the adjusting nuts.

(c) **Utilising the Driven Plate Assembly.**

This method of setting the levers is not highly accurate and should only be used when the Churchill Fixture or the Borg and Beck Gauge Plate No. CG 192 are not available.

The draw back to this method is that although the driven plate is produced to close limits, it is difficult to ensure absolute parallelism. Although the error in the plate is small it becomes magnified at the lever tip due to lever ratio.

(i) Utilising the actual flywheel, lay the driven plate in position and clamp the cover plate assembly over it. The driven plate can be centralised by the Churchill Tool No. 20S. 72 (or similar tool).

(ii) By turning the adjusting nut adjust the height of the lever tips to 1.895″ from the flywheel face utilising a suitable depth gauge.

(iii) Operate the Clutch by using a small press several times in order to settle the mechanism.

(iv) Check the height of the release lever tips and re-adjust if necessary.

(v) Slacken the cover assembly and turn the drive plate 90°. Reclamp the cover assembly to the flywheel and check the height of the release lever tips as a safeguard against any lack of truth in the driven plate.

32. **CONDITION OF CLUTCH FACINGS**

The possibility of further use of the driving plate assembly is sometimes raised, because the clutch facings have a polished appearance after considerable service. It is perhaps natural to assume that a rough surface will give a higher friction value against slipping, but this is not correct.

Since the introduction of non-metallic faces of the moulded asbestos type, in service, a polished surface is a common experience, but it must not be confused with a glazed surface which is sometimes encountered due to conditions discussed hereafter.

The ideal smooth polished condition will provide a normal contact, but a glazed surface may be due to a film or a condition introduced, which entirely alters the frictional value of the facings. These two conditions might be simply illustrated by the comparison between a polished wood and a varnished surface. In the former the contact is still made with the original material, whereas in the latter instance, a film of dried varnish is interposed between the contact surfaces.

The following notes give useful information on this subject.

(a) After the clutch has been in use for some little time, under perfect conditions, with the clutch facings working on a true and polished or ground surface of correct material, without the presence of oil, and with only that amount of slip which the clutch provides for under normal condition, then the surface of the facings assumes a high polish, through which the grain of the material can be clearly seen. This polished facing is of a mid-brown colour and is then in perfect condition, the co-efficiency of friction and the capacity for transmitting power is up to a very high standard.

NOTE : The appearance of Wound or Woven type facings is slightly different but similar in character.

(b) Should oil in small quantities gain access to the clutch in such a manner as to come in contact with the clutch facings it will burn off, due to the heat generated by slip which occurs during normal starting conditions. The burning off of the small amount of lubricant, has the effect of gradually darkening the clutch facings, but providing the polish on the facing remains such that the grain of the material can be clearly distinguished, it has very little effect on clutch performance.

(c) Should increased quantities of oil or grease attain access to the facings, one or two conditions or a combination of the two, may arise, depending on the nature of the oil etc.

(i) The oil may burn off and leave on the surface facings a carbon deposit which assumes a high glaze and causes slip. This is very definite, though very thin deposit, and in general it hides the grain of the material.

(ii) The oil may partially burn and leave a resinous deposit on the facings, which frequently produce a fierce clutch and may also cause a "spinning" clutch due to a tendency of the facings to adhere to the flywheel or pressure plate face.

(iii) There may be a combination of 1 and 2 conditions, which is likely to produce a judder during clutch re-engagement.

(d) Still greater quantities of oil produce a black soaked appearance of the facings, and the effect may be slip, fierceness or judder in engagement etc., according to the conditions. If the conditions under (c) or (d) are experienced, the clutch driven plate assembly should be replaced by one fitted with new facings, the cause of the presence of oil removed and the clutch cover housing assembly and flywheel thoroughly cleaned.

33. RECONDITIONING OF DRIVEN PLATE ASSEMBLY

Whilst a much more satisfactory result is obtained by the complete replacement of this assembly, circumstances may force the renewal of the clutch facings. The after-mentioned notes will prove useful.

(a) Ensure that the metal components of the assembly are in good condition and pay particular attention to the following:—

(i) Uneven spline wear.
(ii) Cracked segments.
(iii) Springs are not broken.
(iv) Test the drive and over run.

(b) Drill out the rivets securing the facings to the plates.

(c) Rivet the new facings onto the plate assembly. It is suggested that an old flywheel is used as an anvil and the rivets supported by short pieces of $\frac{3}{16}''$ dia. mild steel rod.

(d) Mount the driven plate assembly on a mandrel between the centres of a lathe and check for "run out" with a dial test indicator set as near to the edge of the assembly as possible.

Where the run-out exceeds .015″ locate the high spot and true the assembly by prizing over in the requisite direction. Care must be taken not to damage the facings.

NOTE : When offering up the driven plate assembly to the flywheel, the LONGER side of the splined hub must be nearer to the gearbox.

IMPORTANT

The Borg and Beck Gauge Plate No. CG 192.

Mention of this Gauge Plate is made on Pages 1 and 11, but this plate can no longer be purchased. It is possible however that some dealers have an existing gauge of this type and for this reason instruction as to its use is included.

13

CLUTCH

SERVICE DIAGNOSIS.

SYMPTOM	CAUSE	REMEDY
1. Drag or Spin.	(a) Oil or grease on the driven plate facings.	Fit new facings.
	(b) Misalignment between the engine and gearbox shaft.	Check over and correct the alignment.
	(c) Improper pedal adjustment not allowing full movement to release bearing.	Correct pedal adjustment.
	(d) Warped or damaged pressure plate or clutch cover.	Renew defective part.
	(e) Driven plate hub binding on splined shaft.	Clean up splines and lubricate with small quantity of high melting point grease such as Duckham's Keenol.
	(f) Pilot or operating shaft bearings binding.	Renew or lubricate bearings.
	(g) Distorted driven plate due to the weight of the gearbox being allowed to hang in clutch plate during erection.	Fit new driven plate assy. using a jack to take the overhanging weight of the gearbox.
	(h) Broken facings of driven plate.	Fit new facings.
	(j) Dirt or foreign matter in the clutch.	Dismantle clutch from flywheel and clean the unit, see that all working parts are free. **Caution.** Never use petrol or paraffin for cleaning out clutch.
	(k) Air in hydraulic line or insufficient fluid.	Bleed or replenish.
2. Fierceness or Snatch.	(a) Oil or grease on driven plate facings.	Fit new facings and ensure isolation of clutch from possible ingress of oil or grease.
	(b) Misalignment.	Check over and correct the alignment.
	(c) Binding of clutch pedal mechanisms.	Free and lubricate journals.
	(d) Worn out driven plate facings.	New facings required.
3. Slip.	(a) Oil or grease on the driven plate facings.	Fit new facings and eliminate cause of foreign presence.
	(b) Improper pedal adjustment indicated by lack of the requisite .820″ free or unloaded foot pedal movement—.030″ at master cylinder, .079″ at slave cylinder.	Correct pedal adjustment and/or clearances.
4. Judder.	(a) Oil, grease or foreign matter on the driven plate facings.	Fit new facings and eliminate cause of foreign presence.
	(b) Misalignment.	Check over and correct alignment.

CLUTCH

SYMPTOM	CAUSE	REMEDY
	(c) Pressure plate out of parallel with flywheel face in excess of of permissible tolerance.	Re-adjust levers in plane and, if necessary, fit new eyebolts.
	(d) Contact area of friction facings not evenly distributed. Note that friction facing surface will not show 100% contact until the clutch has been in use for some time, but the contact area actually showing should be evenly distributed round the friction facings.	This may be due to distortion, if so fit new driven plate assembly.
	(e) Bent splined shaft or buckled driven plate.	Fit new shaft or driven plate assembly.
	(f) Unstable or ineffective rubber engine mountings.	Replace and ensure elimination of endwise movement of power unit.
5. Rattle.	(a) Damaged driven plate, i.e., broken springs, etc. (b) Worn parts in release mechanism. (c) Excessive back lash in transmission. (d) Wear in transmission bearings. (e) Bent or worn splined shaft. (f) Ball release bearing loose on operating sleeve.	Fit new parts as necessary.
6. Tick or Knock.	(a) Hub splines badly worn due to misalignment. (b) Worn pilot bearing.	Check and correct alignment, then fit new driven plate. Pilot bearing should be renewed.
7. Fracture of Driven Plate.	(a) Misalignment distorts the plate and causes it to break or tear round the hub or at segment necks in the case of Borglite type. (b) If the gearbox during assembly be allowed to hang with the shaft in the hub, the driven plate may be distorted, leading to drag, metal fatigue and breakage.	Check and correct alignment and introduce new driven plate. Fit new driven plate assembly and ensure satisfactory re-assembly.
8. Abnormal Facing Wear.	Usually produced by overloading and by the excessive slip starting associated with overloading.	In the hands of the operator.

Service Instruction Manual

GEARBOX

SECTION E

GEARBOX

INDEX

GEARBOX

ILLUSTRATIONS

Page

GEARBOX

Dimensions and Tolerances

PARTS AND DESCRIPTION	DIMENSIONS NEW	CLEARANCE	REMARKS
Constant Pinion Shaft			
Spigot External Diameter	.494" .492"	.0058" to .0085"	
Crankshaft Bush Internal Diameter	.5005" .4998"		
Mainshaft			
Spigot External Diameter	.6875" .6870"	.0005" to .0017"	
Internal Diameter of Constant Pinion Bush	.6887" .6880"		
Mainshaft Bushes			
Diameter of Mainshaft	1.2488" 1.2481"	.0007" to .0026"	
Internal Diameter of 2nd Gear Bush	1.2507" 1.2495"		
Internal Diameter of 3rd Gear Bush	1.2495" 1.2488"	.0000" to .0014"	
Mainshaft Bush Float			
Length of 2nd Gear Bush (measured without flange)	1.162" 1.160"	.004" to .008"	End float of .004" to .006" obtained by selective assembly.
Length of 2nd Gear	1.156" 1.154"		
Length of 3rd Gear Bush	1.225" 1.223"	.004" to .008"	End float of .004" to .006" obtained by selective assembly.
Length of 3rd Gear	1.219" 1.217"		
Overall Float of Bushes			
Overall Length of Mainshaft Bushes	2.511" 2.505"		2nd gear bush has .124"—.122" flange.
Thickness of 2nd Gear Thrust Washer	.122" .120"		
Thickness of 3rd Gear Thrust Washer	.124" .122"		
Overall Float of Bushes		.000" .015"	End float of .007" to .012" obtained by selective assembly.

1

GEARBOX

Dimensions and Tolerances

PARTS AND DESCRIPTION	DIMENSIONS NEW	CLEARANCE	REMARKS
Countershaft			
External Diameter of Countershaft	.7913″ .7908″		
Internal Diameter of Countershaft Gear	.8983″ .8978″		
Needle Roller Diameter	.119″		
Countershaft Gear End Float			
Internal Width of Casing	6.771″ 6.769″		
Affected Length of 1st Countershaft Gear	2.2487″ 2.2473″		
Width of Constant Gear	1.3132″ 1.3118″		
Width of Third Gear	1.1882″ 1.1868″		
Width of Second Gear	.7607″ .7593″		
1st Thrust Washer Thickness	.068″ .066″		
Rear Thrust Washer Thickness	.107″ .105″		
Distance Piece	1.0817″ 1.0803″		
Overall Width of :— Countershaft Gears and Two Thrust Washers	6.7675″ 6.7565″		
Overall Float of Countershaft Gears	.0015″ .0145″		Select parts to provide .006″—.010″ end float.
Reverse Idler Shaft			
Diameter of Shaft	.5618″ .5613″	.0007″ to .0012″	
Internal Diameter of Bushes	.5625″		

2

GEARBOX

Dimensions and Tolerances

PARTS AND DESCRIPTION	DIMENSIONS NEW	CLEARANCE	REMARKS
Gearbox Top Cover			
Selector Shaft External	.4985"		
Diameter	.4972"	.0010"	
		to	
Bore in Cover for	.5005"		
Selector Shaft.	.4995"	.0033"	

Gear Synchronisation and Loading Details

2nd Speed Synchro Axial Load for Release	25 to 27 lbs.	
3rd and Top Synchro Axial Load for Release	19 to 21 lbs.	
Gap between Baulk Ring Dog Teeth and Cone Dog Teeth on Mainshaft Synchro Gears	.035" to .040" Engaged. .060" to .075" Free.	

Selector Rod Loading

Selector Rod Axial Load for Release

	1st and 2nd	32 to 34 lbs.
	3rd and TOP	17 to 20 lbs.
	Reverse	21 to 23 lbs.

Load required at Gear Change Knob to Select :—

1st and 2nd Gear	7 to 9 lbs.
3rd and TOP Gear	4 to 6 lbs.
Reverse Gear	6 to 7 lbs.

NOTE : To convert lbs. to Kgs. divide by 2.204.

„ „ ins. to Millimetres multiply by 2.54.

3

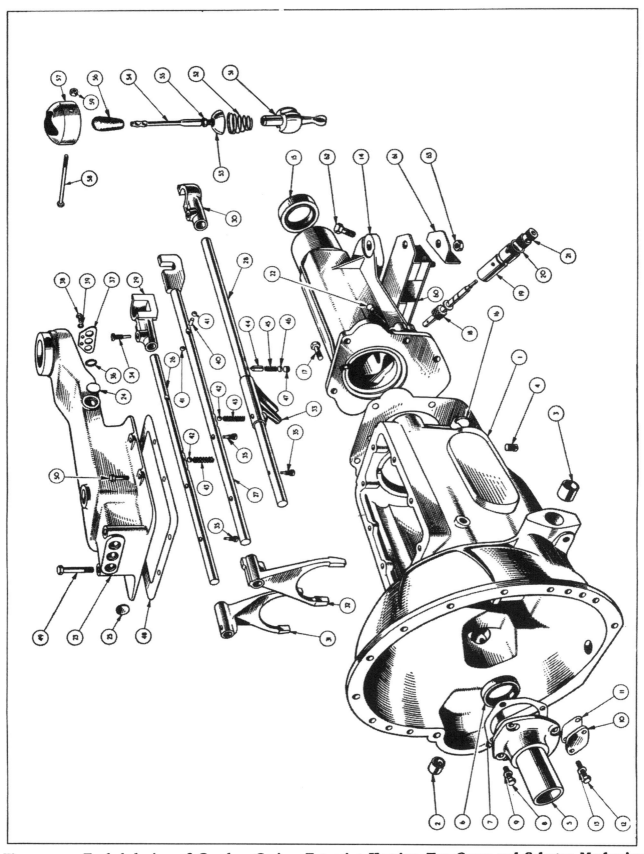

Fig. 1   Exploded view of Gearbox Casing, Extension Housing, Top Cover and Selector Mechanism.

NOTATION FOR FIG. 1			
Ref. No.	Description	Ref. No.	Description
1	Clutch and Gearbox Casing.	33	Reverse Selector.
2	Bush for Clutch Shaft.	34	Taper Screw.
3	Bush for Clutch Shaft.	35	Stop Screw.
4	Drain Plug.	36	Sealing Ring.
5	Front End Cover.	37	Cover Plate.
6	Oil Seal.	38	Setscrew for Cover Plate.
7	Joint Washer.	39	Lock Washer.
8	Setscrew for Cover.	40	Interlock Roller 3rd/Top.
9	Plain Washer for 8.	41	Interlock Balls.
10	Countershaft Cover.	42	Selector Shaft Ball.
11	Joint Washer.	43	Spring for Ball.
12	Setscrew.	44	Reverse Shaft Plunger.
13	Plain Washer.	45	Spring for Plunger.
14	Gearbox Extension.	46	Distance Piece.
15	Oil Seal.	47	Plug.
16	Joint Washer.	48	Joint Washer.
17	Extension Attachment Bolt.	49	Attachment Bolt (long). } Top cover.
18	Speedometer Drive.	50	Attachment Bolt (short). }
19	Speedometer Bearing.	51	Ball End.
20	Washer.	52	Spring.
21	Screwed Adaptor.	53	Spring Retainer.
22	Locating Screw.	54	Lever Assembly.
23	Top Cover.	55	Lever Locknut.
24	Core Plug.	56	Knob.
25	Selector Shaft Welch Washer.	57	Cap.
26	Selector Shaft (1st and 2nd Gear).	58	Bolt.
27	Selector Shaft (Top and 3rd Gear).	59	Nyloc Nut.
28	Reverse Selector Shaft.	60	Rear Mounting.
29	1st/2nd Gear Selector.	61	Steady Bracket.
30	Reverse Gear Selector.	62	Bolt.
31	1st/2nd Selector Fork.	63	Nut.
32	3rd/Top Selector Fork.		

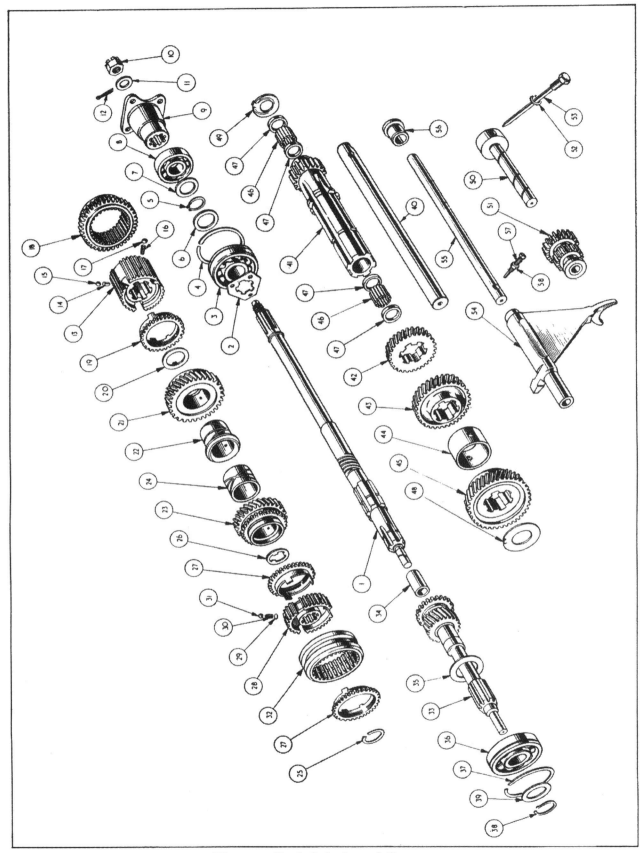

Fig. 2.   Exploded view of Gears.

GEARBOX

Ref. No.	Description	Ref. No.	Description
	NOTATION FOR FIG. 2		
1	Mainshaft.	30	Synchro Spring.
2	Triangular Washer.	31	Synchro Ball.
3	Centre Bearing (Interchangeable with 36).	32	3rd and TOP Gear Sychronising Sleeve.
4	Outer Circlip for Centre Bearing (Interchangeable with 37).	33	Constant Pinion Shaft.
		34	Constant Pinion Bush.
5	Circlip for Centre Bearing.	35	Oil Thrower.
6	Washer for Centre Bearing.	36	Ball Bearing.
7	Washer for Rear Bearing.	37	Outer Circlip for Constant Pinion Bearing.
8	Rear Bearing.	38	Circlip.
9	Driving Flange.	39	Washer between Bearing and Circlip.
10	Slotted Nut.	40	Countershaft.
11	Plain Washer.	41	1st Speed Countershaft Gear.
12	Split Pin.	42	2nd Speed Countershaft Gear.
13	1st Gear Synchro Hub.	43	3rd Speed Countershaft Gear.
14	Interlock Plunger.	44	Distance Piece Countershaft Gear.
15	Interlock Ball.	45	Constant Gear.
16	Synchro Spring.	46	Needle Rollers.
17	Synchro Ball.	47	Retaining Ring for 46.
18	1st Gear Synchronising Sleeve.	48	Front Thrust Washer.
19	2nd Speed Synchronising Cup.	49	Rear Thrust Washer.
20	Washer.	50	Reverse Spindle.
21	2nd Gear.	51	Reverse Wheel.
22	2nd Speed Bush.	52	Lock Washer.
23	3rd Speed Gear.	53	Countershaft Retaining Screw.
24	3rd Speed Bush.	54	Reverse Operating Fork.
25	Circlip.	55	Operating Rod.
26	Washer.	56	Bush on rear end of Rod.
27	3rd and TOP Gear Synchronising Cup.	57	Rod Retaining Screw.
28	3rd and TOP Gear Synchronising Hub.	58	Locknut.
29	Synchro Spring Shim.		

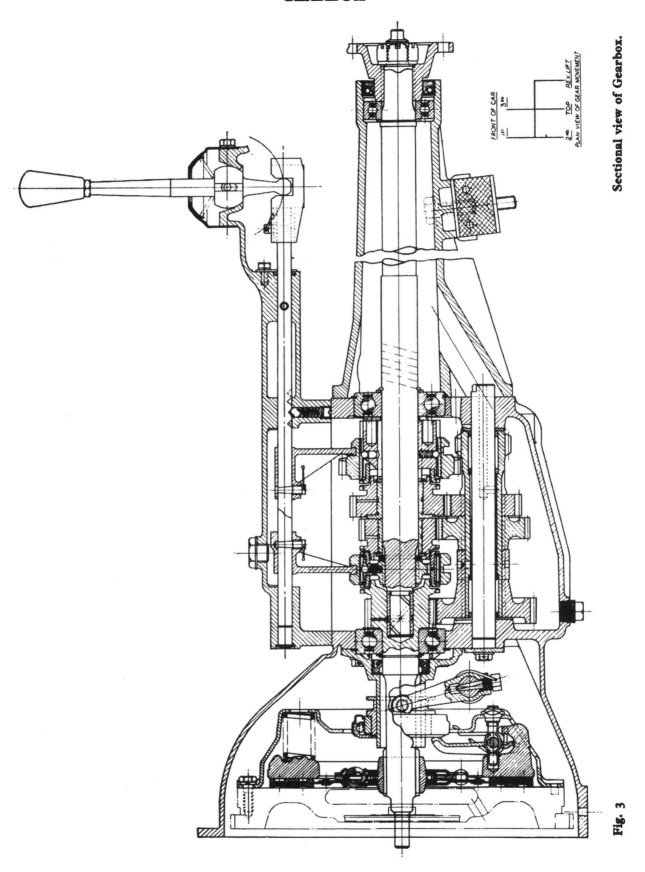

Sectional view of Gearbox.

Fig. 3

8

GEARBOX

Description

Four forward speeds with gear synchronisation on 2nd, 3rd and Top and one Reverse ratio actuated by a compound gear which is disengaged when in Neutral or in any of the forward gears.

1. OPERATION

A remote control lever is carried in a turret formed in the rear end of the top cover, which is at a point approximately halfway down the rear extension housing. The selector forks are mounted on three selector shafts which are carried in the gearbox top cover and both cover and shafts extend rearwards to the control lever turret where gear selection is made by conventional "H" gate movement.

2. RATIOS

	Gearbox	Overall
Overdrive Top	0.82	3.03
Top	1.00	3.70
Third	1.325	4.90
Second	2.00	7.40
First	3.38	12.50
Reverse	4.28	15.80

3. BEARINGS

(a) Constant Pinion Shaft
Bearing (S.M.Co. Part No. 58391):
Fischer Ball Bearing No. MS12 S.G.
Hoffman Ball Bearing No. MS12 K.

(b) Mainshaft Centre
Bearing (S.M.Co. Part No. 58391):
As for Constant Pinion Shaft.

(c) Mainshaft Rear
Bearing (S.M.Co. Part No. SP75 G.):
Fischer Ball Bearing No. 6206.
Hoffman Ball Bearing No. 130.

(d) Countershaft Cluster
Front : 24 needle rollers retained by means of two retaining rings (press fit).
Rear : 24 needle rollers retained by means of two retaining rings (press fit).

4. MOUNTING

Unit assembly with engine which is two point mounted to the chassis at front, the gearbox being mounted on a silent block under the gearbox extension housing to the chassis cross member.

5. OIL CAPACITY

1½ pints (0.8 litres) from dry.
With Overdrive 3½ pints (2.0 litres).
For recommended grades of oil refer to Lubricant Recommendations in the "General Data" Section.

6. NUT AND BOLT DATA AND TIGHTENING TORQUES

For these particulars refer to "General Data" Section.

NOTE—For details regarding Special Tools, please refer to Section "Q" of this Manual.

7. TO REMOVE GEARBOX LEAVING ENGINE IN POSITION

(a) Disconnect battery lead.
(b) Remove both seats by withdrawing sixteen nuts, eight from beneath each seat cushion.
(c) Remove gear lever and grommet, after slackening the locknut and unscrew gear lever from its ball end.
(d) Withdraw floor centre section and carpet after the withdrawal of sixteen setscrews located round the edges of the pressing. Similarly remove the "U" plate (R.H. side) secured with two P.K. screws.
(e) Disconnect the propeller shaft at the front end by withdrawing the four bolts and nyloc nuts.
(f) Disconnect speedometer cable from gearbox by unscrewing the knurled collar from its adaptor.
(g) Remove clutch slave cylinder with its mounting bracket after withdrawing two nuts and bolts from the bell housing and one sump bolt securing the steady rod. The slave cylinder push rod can be removed from the clutch operating shaft after the withdrawal of the split pinned clevis pin from the operating fork to which is attached the clutch return spring.

9

125

(**h**) Disconnect the two wires from their terminals on the solenoid if an Overdrive is fitted.

(**i**) Remove gearbox mounting after the withdrawal of two nuts by jacking up the unit, using a block of wood between jack and sump to avoid damage.

(**j**) Remove starter motor bolts and slide starter motor forwards clear of the bell housing.

(**k**) Remove nuts and bolts from bell housing and withdraw gearbox (Fig. 4).

Fig. 4   Gearbox Unit ready for withdrawal.

TO REPLACE GEARBOX

Carry out the above procedure in reverse, but it is advisable before doing so to check the alignment of the clutch unit with a suitable mandrel (Fig. 5). If this is found to be incorrect slacken the clutch cover assembly bolts until the mandrel slides in freely, then re-tighten the bolts.

8. TO DISMANTLE

(**a**) Remove eight setscrews from the top cover assembly and withdraw complete with selector mechanism. To dismantle top cover assembly see page 18.

(**b**) Remove top cover paper joint.

(**c**) Break locking wire on clutch operating fork positioning setscrew and withdraw.

(**d**) Remove clutch operating shaft positioning bolt and grease nipple with fibre washer from R.H. of clutch shaft.

Fig. 5   Aligning Clutch Floating Plate with Mandrel. Churchill Tool No. 20S. 72.

Then withdraw operating shaft, coil spring, operating fork, clutch throwout bearing and sleeve.

(**e**) Detach the speedometer drive after removal of the special securing setscrew.

(**f**) Withdraw propeller shaft coupling, having first removed split pin, nut and plain washer.

(**g**) Remove gearbox extension and paper joint after the withdrawal of six securing setscrews and spring washers Fig. 6. The oil seal and ball race will remain in position in the housing but can easily be tapped out with a suitable drift.

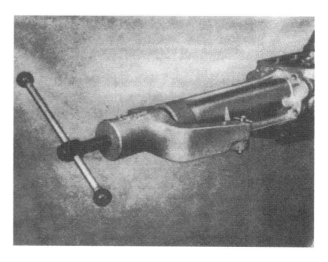

Fig. 6   Showing the removal of Gearbox Extension with Churchill Tool No. 20S. 63.

126

10

(h) Withdraw the countershaft locating setscrew as shown in Fig. 7.

Fig. 7   Countershaft and Reverse Locating Setscrew partially withdrawn.

(i) After removal of the countershaft front end cover plate which is secured by two wired setscrews, plain washers and lead linger drive out the countershaft using a suitable tube as shown in Fig. 8, to retain the 48 needle rollers in position maintaining contact throughout between the tube and countershaft.

Fig. 8   Showing Needle Roller Retainer Tube Tool No. 20SM68 being used to drive out Countershaft.

(j) Remove the gearbox front end cover and paper joint after cutting the wire

in the setscrew heads and withdrawing them complete with their plain washers and lead linger.

(k) Extract the constant pinion shaft assembly as shown in Fig. 9, and remove

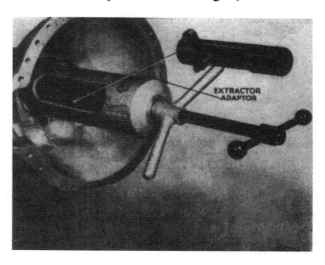

Fig. 9   Extracting Constant Pinion Shaft Assembly with Churchill Tool No. 20SM66A.

the mainshaft spigot bush located in the pinion itself. The further dismantling of this assembly necessitates the removal of the small circlip and thrust washer which fit against the inner ring of the ball race. After extraction of ball race in the fixture shown in Fig. 10, the oil thrower may be withdrawn, but owing to probable damage to this thrower during the dismantling operation a new one may be required when re-assembling the unit.

Fig. 10   Extraction of Constant Pinion Ball Race with Churchill Press No. S4221 and Adapter from Set S.4615.

11

(1) Tap the mainshaft towards the rear with a soft metal drift, as shown in Fig. 11, sufficiently to clear the bearing

Fig. 11   Driving Mainshaft to rear with Tool No. 20SM1 to free Centre Main Bearing.

from the casing.   Next tilt the shaft sufficiently to enable the third and top synchro unit to be withdrawn as shown in Fig. 12.   Note the position

Fig. 12   Removing Top and Third Synchro Unit.

of the short boss on the synchro hub is towards the mainshaft circlip.

(m) Remove mainshaft circlip with the special extractor shown in Fig. 13. The extraction of this circlip is made

somewhat difficult by the adjacent thrust washer which has three lugs,

Fig. 13   Showing the removal of the Mainshaft Circlip with Churchill Tool No. 20SM69.

equally spaced, and engaging alternate splines on the mainshaft.   Quite apart from the necessity to engage the three available splines with the full length prongs, in some cases it may be necessary to tap the circlip round on these prongs, to free it from its recess before it can be withdrawn.   A new circlip should always be used when re-assembling.

(n) Withdraw thrust washer, third mainshaft constant gear and bush, second mainshaft constant gear and bush, thrust washer with three lugs to fit splines and the second speed synchro unit which also incorporates the first mainshaft gear.   The mainshaft can now be withdrawn.

(o) Remove the small seeger circlip and thrust washer which locates the ball race on the mainshaft and extract the race as shown in Fig. 14.   The triangular washer can then be removed from behind the race.

(p) After removal of the lock nut and locating screw the reverse selector shaft and bronze selector fork can be withdrawn.   A steel selector shaft insert located at the rear of the casing and a welch plug at the front can easily be removed.

Fig. 14   Removing Mainshaft Centre Bearing with Churchill Press No. S4221 and adapter from Set No. S4615.

(q) Lift out the reverse pinion (compound gear) after tapping out its spindle through the rear of the casing, the retaining setscrew having been removed in a previous operation (h).

(r) The countershaft assembly can now be lifted out of the casing with the needle roller retaining tube still locating the 24 rollers at each end of the countershaft in their respective recesses. Lay aside the two phosphor bronze thrust washers for re-assembly.

(s) The countershaft gears and distance sleeve can now be removed from the splined portion of the countershaft, noting their position for re-assembly.

(t) If it is desired to examine the needle rollers they can be removed by withdrawing the retaining tube. Note the correct number of 48 for re-assembly (24 at each end) and the needle roller retaining rings can be tapped out with a suitable drift.

9. TO ASSEMBLE

(a) Thoroughly clean out the casing and examine for cracks, ball race housings for wear or other damage.

(b) Fit needle roller retaining rings if necessary, as shown in Fig. 15. Fit 24 needle rollers at each end of the countershaft ensuring first that the locating rings are in position. The chamfer on each retainer ring should be placed towards the bottom of the bore in the case of the inner ones, outwards for the outer ones. The rollers should

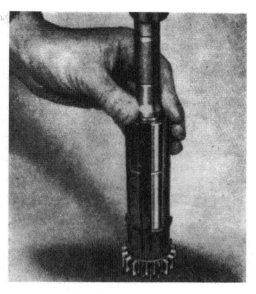

Fig. 15   Fitting Needle Roller Retainer Rings with Churchill Tool No. 20SM68.

be retained in grease and counted after installation to ensure that they have not become displaced before fitting the retainer tube.

(c) Assemble countershaft, noting correct position for the gears, observed in operation (s) when dismantling (see also Fig. 3).

(d) Install the countershaft assembly, positioning the thrust washers on the casing with grease. The correct end float for the countershaft gears should be between .006″—.010″. If there is insufficient end float the distance piece should be reduced as necessary by rubbing it down on a sheet of emery cloth placed on a surface plate. Where too much end float exists new thrust washers and/or distance piece should be fitted.

(e) Fit reverse pinion (compound gear) with smaller gear towards front of box, having first ensured that there is no tooth damage or wear in bushes; leave the fitting of the locating set-screw until the countershaft has been assembled in its normal fitted position.

(f) Install the reverse selector shaft and bronze selector fork position with setscrew and tighten lock nut. The selector shaft steel insert and welch plug can now be fitted.

13

(g) (i) Install the triangular washer on its splines on the mainshaft.

(ii) Press ball race on to mainshaft with Churchill fixture as shown in Fig. 16. Then fit the thrust washer

Fig. 16   Fitting Mainshaft Centre Bearing with Churchill Press No. S4221 and Adapter from Set No. S4615.

and small seeger circlip. A large circlip should be fitted into the annular groove in the outer ring of the bearing.

(h) Before the mainshaft is assembled into the gearbox the following points should be checked:

(i) The 2nd speed constant gear float on its bush (.004″—.006″).

Fig. 17   Checking Second Mainshaft Constant Gear for End Float.

(ii) The 3rd speed constant gear float on its bush (.004″—.006″). **(i)** is checked as in Fig. 17, and **(ii)** as in Fig. 18.

Fig. 18   Checking Third Gear Mainshaft Constant Gear for End Float.

(iii) Overall bush float on mainshaft (.007″—.012″).
To check gear bush end float, fit 2nd speed mainshaft gear thrust washer, ensuring that its three lugs engage in the mainshaft splines, 2nd and 3rd mainshaft gear bushes and 3rd mainshaft gear thrust washer fitted with oil scroll towards the bush. Install the original circlip and measure float with a feeler gauge as shown in Fig. 19.

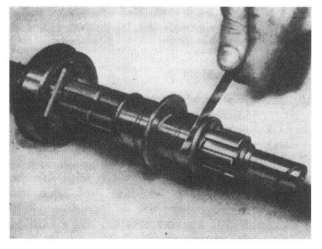

Fig. 19   Checking Mainshaft Gear Bush Overall Float.

(iv) Axial release loading of 2nd speed synchro unit 25—27 lbs.

(v) Axial release loading of 3rd and top speed synchro unit 19—21 lbs.

(iv) and **(v)** can be checked as shown in Fig. 20. If it is found to be in-

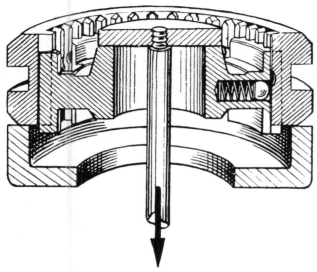

TO SPRING BALANCE

Fig. 20 Fixture which can be readily manufactured to test Synchro Axial loading.

correct, steel shims can be added or removed from below the axial release loading springs to increase or decrease respectively the axial release load as required.

(i) After completion of checks the mainshaft circlip, thrust washers and constant gear bushes can be removed. The mainshaft can then be installed into the gearbox casing, and assembled as follows :

(i) Second speed synchro unit incorporating the first mainshaft gear.

(ii) Thrust washer with three lugs to fit splines.

(iii) Second mainshaft constant gear and bush.

(iv) Third mainshaft constant gear, bush and thrust washer fitted with oil scroll towards gear.

(v) New mainshaft circlip as shown in Fig. 21.

Fig. 21 Fitting Mainshaft Circlip with Churchill Tool No. 20SM. 46.

(vi) Third and top speed synchro unit with the short boss of the synchro hub towards the mainshaft circlip or rear of gearbox as shown in Fig. 3. The mainshaft and ball race can then be driven into the gearbox casing, positioning the gap of the circlip on the outer ring of the bearing in line with the atmosphere hole in the casing as shown in Fig. 7.

(j) Assemble oil thrower on to constant pinion shaft and press ball race on the shaft as shown in Fig. 22, ensuring that this goes right home and that in

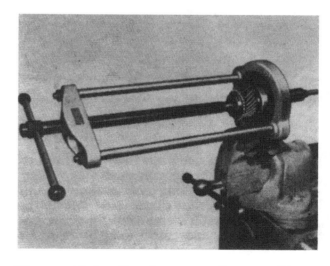

Fig. 22 Fitting Bearing on to Constant Pinion Shaft with Churchill Press No. S4221 and Adapter from Set No. S4615.

15

this position with the correct thrust washer fitted, the small seeger circlip fits properly into its recess. When passing this circlip along the ground portion of the constant pinion shaft, take care not to score the shaft as such damage may cause subsequent leakage of oil. Fit larger circlip into the annular groove in the outer ring of the ball race.

(k) Fit Oilite spigot bush into constant pinion, placing the internally bevelled portion of it towards the mainshaft.

(l) Drive the constant pinion shaft and bearing into the gearbox casing, positioning the gap in the circlip on the outer ring of the bearing in line with the oil hole in the casing.

Utilising a feeler gauge, measure the distance between the dog teeth of all the mainshaft synchro gears, and the dog teeth of their respective baulk rings. (Fig. 23).

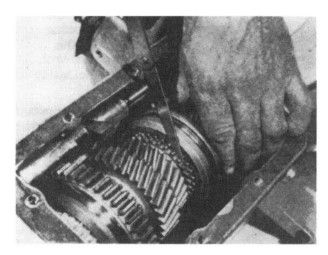

Fig. 23   Measuring the gap between Baulk Ring teeth and Cone.

Move the outer synchro sleeve towards the gear being measured thus forcing the baulking ring on to its cone. In this position the dimension should be between .035″ and .040″ for new components and .005″ to .010″ less for components which have been run-in.

(m) Utilise a pilot to align thrust washers and countershaft gear assembly as

Fig. 24   Inserting Churchill Tool No 20S. 77 preparatory to driving out needle roller retaining tube.

shown in Fig. 24, driving out needle roller retaining tube, subsequently ejecting the pilot tool with the actual countershaft. It is important when carrying out this operation that the pilot tool should maintain contact with the retaining tube or counter-shaft, as appropriate, throughout the operation, alternatively there is danger that the needle rollers may leave their recess.

(n) Install locating setscrew through countershaft, and reverse spindle, first checking the alignment of the holes in the reverse gear spindle and counter-shaft.

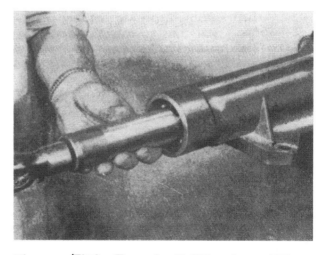

Fig. 25   Fitting Extension Ball Bearing and Thrust Washer with Churchill Tool No. 20S. 87.

(o) Fit countershaft front end cover plate and paper joint securing with two set-screws and washers using lead linger and wiring as necessary.

(p) Assemble gearbox extension and paper joint, securing with six setscrews and washers, using lead linger and wiring as necessary.

(q) Install thrust washer and ball race into gearbox extension with suitable tool as shown in Fig. 25.

(r) Locate gearbox extension oil seal as shown in Fig. 26.

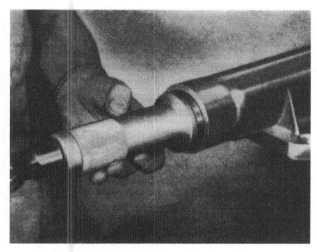

Fig. 26   Fitting Extension Housing Oil Seal with Churchill Tool No. 20S. 87.

(s) Fit plain washer, slotted nut as shown in Fig. 27, tightening to 85—100 lbs. ft., and install split pin.

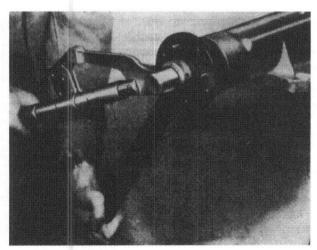

Fig. 27   Tightening Driving Flange Securing Nut with torque spanner and Churchill Tool No. 20SM. 90.

(t) Install speedometer driving gear and accommodating bush, securing with special setscrew.

Fig. 28   Fitting Front Cover Oil Seal with Tool No. 20SM. 73.

(u) Fit front cover, having installed oil seal as shown in Fig. 28, utilising fitting tool to protect oil seal (see Fig. 29).

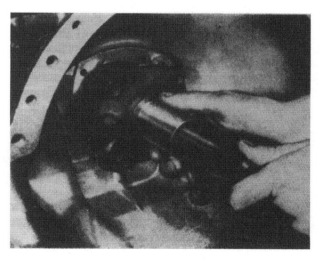

Fig. 29   Assembling Front Cover, utilising Churchill Tool No. 20SM. 47 to protect Seal Face.

Fit four setscrews and plain washers with lead linger after positioning the slot in the face of the front cover horizontally at 9 o'clock and wire setscrew heads.

17

(v) Assemble clutch throw-out bearing and sleeve and install with clutch operating shaft coil spring and clutch operating fork, positioning both with special securing setscrews, wire locking the latter. Install grease nipple with fibre washer into R.H. end of clutch operating shaft.

(w) Fit top cover assembly with selector mechanism, paper joint, securing with eight setscrews.

10. TO DISMANTLE TOP COVER ASSEMBLY

(a) Remove oil level dipstick.

(b) Ensure that the selector mechanism is in the neutral position.

(c) Remove change speed lever positioning bolt, nyloc nut and setscrew. This enables the change speed lever complete with knob, cap, spring retainer, spring and ball end to be removed as an assembly. Further dismantling requires the removal of the knob and/or the removal of the screwed change speed lever ball end.

(d) Remove 1st and 2nd speed selector shaft wire locked stop screw and $\frac{3}{8}''$ dia. positioning ball, spring and retaining screw, then 1st and 2nd speed bronze selector fork wire locked positioning setscrew, and slide selector shaft rearwards clear of the casting to enable the selector fork to be removed.

(e) Remove reverse selector fork and shaft, carrying out procedure as in (d) except that the shaft is positioned by a plunger spring, distance piece and retaining screw instead of the ball, spring and retaining screw.

(f) Remove 3rd and top speed selector shaft and fork, carrying out the procedure as in (d).

N.B. It is important that no attempt is made to move more than one selector shaft at a time otherwise damage will be caused to the selector shaft bores by the interlock mechanism consisting of two $\frac{3}{8}''$ dia. ball bearings located in the top cover casting either side of the 3rd and top speed selector shaft, and the .185″ dia. interlock roller made of key steel which makes contact with these balls being installed, in a hole drilled transversely through the 3rd and top speed selector shaft. (See Fig. 1.)

The interlock roller and steel balls can easily be shaken or pushed out of position if it is desired to examine them.

(g) Further dismantling of the selector shafts only requires the removal of the selector shaft end pieces on the 1st and 2nd and reverse rods, they are located by a wired setscrew; on the 3rd and top they are silver soldered together.

(h) Remove the two setscrews and spring washers from the oil sealing ring cover plate, enabling the plate and three rubber sealing rings at the end of the selector shaft bores to be removed.

(i) The three 16G pressing selector shafts welch plugs located at the front of top cover and the two 14G pressing welch plugs either side of top cover can easily be removed with a suitable drift.

(j) The threaded plug located on the top cover can also be removed.

TO ASSEMBLE, carry out the reverse procedure to that of dismantling, but for ease of assembly install the $\frac{3}{8}''$ dia. interlock mechanism balls after the 3rd and top speed selector shaft has been fitted but before the reverse and 1st and 2nd selector shafts.

Important.
Whilst fitting the selector shafts make sure that the selector shaft or shafts already fitted are in the neutral position.

INSTALLATION OF OVERDRIVE

1. DISMANTLING

Remove the detachable floor pressing from around the gearbox. Remove the four bolts connecting the propeller shaft to the gearbox flange. Disconnect the speedometer drive from the gearbox. Remove the bottom nuts of rear mounting and jack up engine sufficiently to allow removal of rear mounting. Remove the starter motor. Remove the clevis pin from the lever on the clutch operating shaft.

Remove the bolts from around the bell housing and detach the gearbox from the engine.

The gearbox should now be dismantled and the various gears and ball races examined for possible damage. Any parts which are damaged or suspect in any way should be replaced.

The mainshaft originally fitted will be replaced by the special one supplied. To ensure the future life of the Overdrive Unit it is advisable to fit a new mainshaft bearing.

2. ASSEMBLY OF GEARBOX

Proceed to re-assemble in the following manner after ensuring that the gearbox has been drilled as shown in Fig. 31.

(a) Fit 1st and reverse idler pinion and shaft with the smaller gear pointing forward and the hole in the shaft in line with the securing bolt.

(b) Fit the reverse selector fork and shaft with the tapered hole forward. Secure in position by fitting the tapered bolt and locknut.

(c) With heavy grease, assemble the needle rollers into the 1st countershaft gear (24 each end) and slide in a needle retaining tube to retain the rollers during assembly.

(d) With heavy grease, position the front thrust washer with the lip of the washer engaged with a recess in the gear case.

(e) Slide the small, or 2nd speed gear, on to the 1st countershaft gear, following this by the 3rd speed gear with the boss pointing forward. Next slide on the distance piece and finally the constant speed gear with the boss towards the distance piece.

(f) Position the completed countershaft gear assembly in the bottom of the gear case and slide into position the rear thrust washer.

(g) For checking purposes the countershaft should be fitted. The countershaft gears (when new) have an end float of .006″—.010″.

(h) After checking, the countershaft should be removed by pushing the needle retaining tube into the countershaft gears and forcing the layshaft out, after which the gears will drop to the bottom of the gearbox casing.

(i) Fit the triangular washer, ball race, distance washer and circlip to the new mainshaft. Gripping the mainshaft in the protected jaws of a vice, assemble the gears on this shaft up to the main locating circlip, ensuring that the recess for this is free for its eventual entry by checking with half the circlip previously used (a new one will be required when re-assembling). When a new 2nd or 3rd mainshaft gear is to be fitted, ensure that .004″—.006″ end float of the gears is permitted by the length of their bushes, when in their fitted position.

Having ensured that the synchro units are perfectly free on their splines, check the overall float of the constant mesh assemblies by removing the 2nd and 3rd speed constant gears, but leaving their respective bushes in position with the hardened steel thrust washers and the half circlip.

Fig. 30 Checking overall float of mainshaft bushes with feeler gauge.

The end float can then be checked with a set of feeler gauges as shown in Fig. 30. The correct float should be between .007″ and .012″.

(j) Remove the mainshaft details remaining on the shaft and begin the final assembly.

Feed the shaft into the casing and assemble the 2nd gear synchro unit, the hardened steel thrust washer which must be located on the splines, the 2nd constant gear with its bush, the 3rd constant gear with bush, the front hardened steel thrust washer and finally fit the main locating circlip with a special sleeved tool.

(k) Withdraw the gearbox mainshaft, with the gears so far assembled, sufficiently towards the rear to enable the assembly to be tipped upwards, thus permitting the 3rd and top synchro unit to be fitted.

(l) Tap the mainshaft assembly into position and fit the constant pinion assembly.

Fig. 32   Showing correct location of four springs.

3. FITTING THE OVERDRIVE UNIT

(a) Locate the paper washer on the gearbox casing with grease, fit the overdrive adapter plate and wire the six securing bolts, as shown in Fig. 31. The correct positioning of the wiring is important to ensure proper working clearance for the assembled overdrive unit. Ensure that the oil transfer hole is free (see Fig. 31).

(b) Ensure that the eight springs in the overdrive unit are correctly located, as shown in Fig. 32, that is, the long springs on the outside and the short springs nearer the centre.

(c) After placing a paper joint on the adapter plate, fit the gearbox assembly

Fig. 31   Showing the oil transfer hole and method of wiring bolts. The Gearbox casing has to be drilled on early models whereas all present production are already drilled.

Fig. 33   Method of fitting Gearbox to Overdrive Unit.

to the overdrive unit, holding the latter vertically in the vice as shown in Fig. 33. After engaging top gear, turn the constant pinion until the splines in the overdrive unit mesh with those of the mainshaft. The eight springs are now located on their spigots and a nut and washer fitted to each long stud.

These two nuts are now evenly tightened until the pump roller is nearing the pump driving cam. The driving cam should have been assembled on the gearbox mainshaft splines so that the least amount of eccentricity is nearest to the pump roller. It is necessary to depress the pump plunger with a screwdriver to allow the pump roller to pass over the cam. The nuts are now finally tightened.

CAUTION. Do not use undue force in tightening the nuts on the long studs. There are two sets of splines in the overdrive unit and unless these are in line, it is impossible to tighten the overdrive unit home on to the adapter plate face.

The overdrive valve setting should now be checked.

4. VALVE CHECKING

On the R.H. of the overdrive unit and pinned to the valve operating shaft, is a valve setting lever with a $\frac{3}{16}''$ diameter hole. In the casting adjacent to this lever is another $\frac{3}{16}''$ diameter hole. Actuate the solenoid with a 12 volt battery and while the plunger is drawn into the solenoid and if the valve adjustment is correct, it should be possible to insert a $\frac{3}{16}''$ diameter pin through the valve setting lever and into the casting (see Fig. 34).

If this is not possible then the valve must be re-adjusted in the following manner.

5. VALVE ADJUSTMENT (Fig. 35)

Remove the cover plate by unscrewing three cheese headed bolts. Slacken off the clamping bolt on the solenoid lever.

Rotate the valve setting lever until its $\frac{3}{16}''$ diameter hole coincides with the $\frac{3}{16}''$ diameter hole in the casting. Insert a $\frac{3}{16}''$ diameter pin through the hole in the

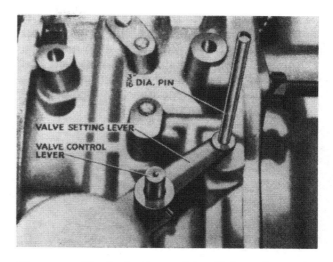

Fig. 34   Method of setting Valve Operating Levers.

setting lever and into the casting thus locking the valve operating shaft.

Actuate the solenoid with a 12 volt battery and while the plunger is drawn into the solenoid, tighten the clamping bolt on the solenoid lever and at the same time ensure that opposite end of the solenoid lever is against the head of the actuating bolt. Repeat the first check and if satisfactory, refit the cover plate.

Fig. 35   Setting Solenoid Lever.

6. FITTING THE ISOLATOR SWITCH

On the lid of the gearbox, and situated near the dipstick, is a plug with a 16 mm. dia. thread. This plug should be removed and replaced by an isolator switch type SS10/1, which is supplied (see Fig. 36). See page 24 for multi-gear overdrive.

Fig. 36 Showing the position of the Isolator Switch on the Gearbox Cover.

7. THE OPERATING SWITCH

L.H. Drive Cars. Two holes are pierced in the facia panel on the L.H. side of the speedometer and covered with fabric. The fabric should be pierced through the extreme L.H. hole and the operating switch fitted. The remaining hole is used for a heater switch when fitted.

R.H. Drive Cars. Two holes are pierced in the facia panel on the R.H. side of the speedometer and covered with fabric. The fabric should be pierced through the extreme R.H. hole and the operating switch fitted (see Fig. 37).
The remaining hole is used for a heater switch when fitted.

The Relay. Reference to Fig. 38 shows the fitted position of the relay.

Wiring. The feed wire to the terminal marked " W1 " on the relay is taken from

Fig. 37 Showing the position of the Overdrive Control.

the " live " side of the starter switch on the facia panel (see Fig. 39).

NOTE—The terminal on the starter switch is " live " only when the ignition is switched " ON."
A wire is connected from " W2 " on the relay to a terminal of the operating switch on the facia panel. The remaining terminal

Fig. 38 Instructions for fitting the Relay Switch.

of the operating switch is connected through a snap connecter to a terminal on the isolating switch situated on the gearbox lid. The remaining terminal of the isolating switch is earthed to one of the bolts securing

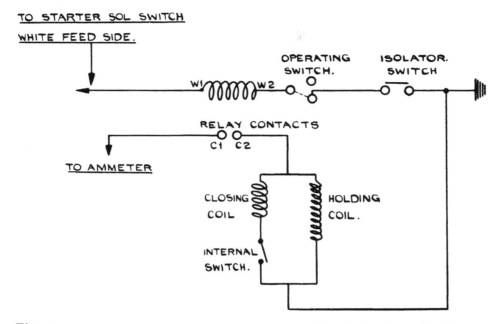

TO STARTER SOL SWITCH
WHITE FEED SIDE.

OPERATING SWITCH.

ISOLATOR. SWITCH

W1 W2

RELAY CONTACTS
C1 C2

TO AMMETER

CLOSING COIL

HOLDING COIL.

INTERNAL SWITCH.

Fig. 39   **Overdrive Control Circuit.**

the gearbox lid. A second feed wire is connected from the negative side of the ammeter to " C1 " on the relay.

To complete the wiring, a wire is connected from " C2 " on the relay through a snap connecter to the solenoid.

Built into the solenoid are two coils, a closing coil and a holding coil. These two coils are connected in parallel with an internal switch connected in series with the closing coil.

When the solenoid is energised, both coils are in circuit until the plunger reaches a pin which operates the internal switch. This switch switches out the closing coil and allows the holding coil to remain in circuit.

The closing current of 15 amperes is of a very short duration. The holding current should be less than one ampere. Fig. 39 shows the theoretical wiring diagram.

23

SUPPLEMENTARY INSTRUCTIONS FOR INCORPORATING OVERDRIVE ON "SECOND" AND "THIRD" GEARS

1. The incorporation of Overdrive on "Second" and "Third" gears has necessitated the following engineering alterations :—

(a) Increasing the diameter of the clutch operation pistons in the overdrive unit from $1\frac{1}{8}''$ to $1\frac{3}{8}''$.

(b) Re-designing the gearbox top cover assembly to permit the selection of overdrive in other gears.

2. OVERDRIVE UNIT

To enable the unit to transmit the maximum available torque in the lower gears, it is necessary to use larger clutch operating pistons than those fitted previously.

From Chassis No. TS.5980 onwards, all Triumph Sports Cars, which have been equipped with overdrive, have been fitted with the re-designed unit, Part No. 301991 : Serial No. 22/1374/— incorporating the larger pistons.

NOTE. A small number of cars with chassis numbers prior to TS.5980 have been fitted with the re-designed overdrive unit.

To establish whether or not a re-designed unit has been fitted, remove the gearbox floor covering and a brass plate can be seen bearing a serial number. The old unit number is 22/1275/—, and the re-designed unit number is 22/1374/—.

Unit Exchange
The Spares Department of The Standard Motor Company Ltd., in conjunction with Messrs. Laycocks, operate an exchange system whereby the old unit can be exchanged for the later type at a cost fixed by the Spares Division of The Standard Motor Company Ltd.

3. GEARBOX TOP COVER ASSEMBLY
Fig. 41

To permit the selection of overdrive in "Second" and "Third" as well as "Top" a new top cover assembly has been designed and the Part No. is 502411.

The new cover assembly has been fitted to Chassis No. TS.6280 and all subsequent Sports Cars.

NOTE. A limited number of cars prior to Chassis No. TS.6280 were fitted with the new cover assembly and can be recognised by the two isolator switch bosses, Fig. 41.

Modification of Top Cover Assembly. To modify the old top cover assembly, thus permitting the selection of overdrive in 2nd, 3rd and top gears necessitates the fitting of certain new parts. The new parts required are detailed under "Top Cover Conversion Pack" on page 27.

Top Cover Assembly—Fig. 40—Dismantling. Proceed as follows :—

(a) Remove the dipstick and ensure that the selector mechanism is in the "Neutral" position.

(b) Disconnect the wires from the isolator switch, where fitted, and remove the top cover assembly from the gearbox.

(c) Remove the change speed lever by :—
(i) Unscrewing and removing the $\frac{1}{4}''$ UNF setscrew (1) which secures the retaining cap to the top cover casting.

(ii) Unscrewing the nyloc nut (2) from the pivot bolt.

SUPPLEMENTARY INSTRUCTIONS FOR INCORPORATING OVERDRIVE ON "SECOND" AND "THIRD" GEARS

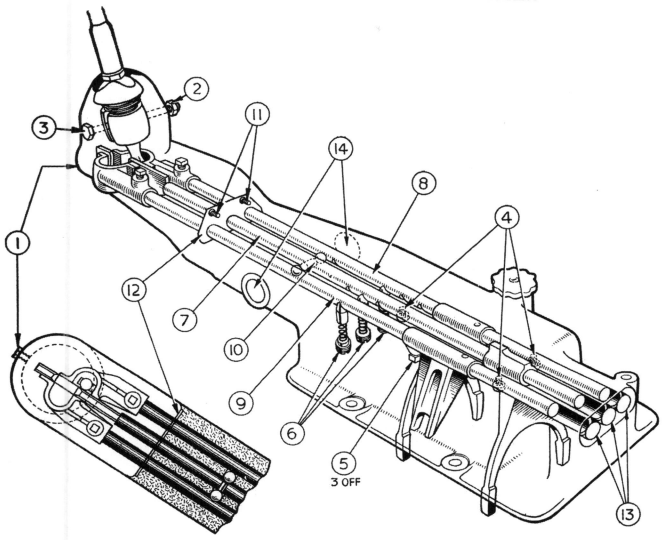

Fig. 40

Ghost view of Top Cover Assembly.

(iii) Withdrawing the pivot bolt (3) to enable the change speed lever assembly to be withdrawn.

Caution. When withdrawing the change speed lever assembly, ensure that the anti-rattle spring and retainer, which is located on the spherical part of the lever, is retained for re-assembling.

(d) Remove the three wire locked stop screws (4).

(e) Unscrew and remove the three wire locked screwed taper pins (5) securing the forks to the selector shafts.

(f) Remove 1st and 2nd speed selector shaft retaining screw (6), spring and $\frac{3}{8}''$ locking ball and slide this selector shaft rearwards clear of the casting to enable the removal of the selector fork.

(g) Remove " Reverse " selector fork and shaft (9) carrying out the procedure as in (f) above, excepting that the shaft is positioned by a plunger, spring, distance piece and retaining screw instead of the ball, spring, and retaining screw.

(h) Remove 3rd and " Top " speed selector shaft (7) and fork, carrying out the procedure used in (f) above.

25

SUPPLEMENTARY INSTRUCTIONS FOR INCORPORATING OVERDRIVE ON "SECOND" AND "THIRD" GEARS

NOTE. It is important that no attempt is made to move more than one selector shaft at a time otherwise damage will be caused to the bores of the top cover and difficulty will be experienced in removing the shafts.

(i) Finally shake out the interlock balls from the casing.

(j) Remove the existing isolator switch.

(k) Remove the two $\frac{1}{4}''$ UNF setscrews (11) from the oil sealing ring cover plate (12), enabling the plate and three rubber sealing rings to be removed.

(l) It being very difficult to remove the welch plugs (13 and 14) without damaging them, it is desirable to replace the old plugs with new ones when re-assembling the new top cover.

Top Cover Assembly—Fig. 40—To Assemble. Assemble the new selector forks into the new top cover by reversing the dismantling procedure, observing the following :—

(a) Ensure before fitting the centre selector shaft that the interlock pin is positioned in the end of the shaft. (See 10).

(b) After fitting and moving the centre shaft to the " Neutral " position, feed the two interlock balls into position from either side. (See 10).

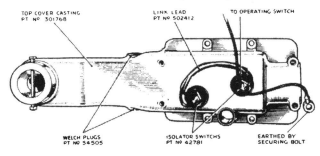

Fig. 41   Top Cover showing Isolation Switches.

Isolator Switches. The isolator switches, Fig. 41 (Part No. 42781), are not included in the top cover assembly (Part No. 502411) and will therefore be required.

Switch Adjustment. Fig. 42. It is important when moving the gear lever to an engaged position, that the switch contacts close at a precise point during the lever's movement.

The correct time for contact closure is when :—

(a) Synchronisation is complete.

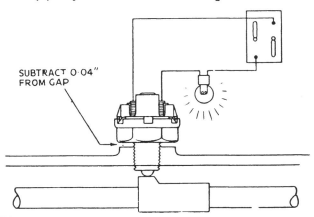

Fig. 42   Adjusting the Isolation Switches.

(b) The synchro sleeve begins to cover the dog teeth of the driving gear.

NOTE. Failure to obtain these conditions will result in noisy and difficult gear changing.

To obtain correct switch adjustment proceed as follows :—

(a) Move the gear lever until " Second " gear is fully engaged.

(b) Wire a bulb in series with the switch contacts and connect to a battery. (Fig. 42).

(c) Screw the switch into the rear switch boss (Fig. 42), until the contacts close. (Indicated by the bulb lighting.)

GEARBOX

SUPPLEMENTARY INSTRUCTIONS FOR INCORPORATING OVERDRIVE ON "SECOND" AND "THIRD" GEARS

(d) Measure with feeler gauges the gap between the switch and boss, that is, the amount the switch would have to be screwed down to be fully home.

(e) From this dimension subtract .040" and make up the remainder with paper packing washers, Part No. 502146.

> **Example.** If the gap measured .090" the subtraction of the .040" would leave .050". By selection (the washers vary in thickness) obtain a pack which measures .050".

(f) Disconnect the switch and remove it from the top cover.

(g) After installing the washer-pack over the screwed portion of the switch, screw the switch securely into the top cover.
Repeat the procedure with the " Third " and " Top " isolator switch.

Wiring. The switches are wired in parallel (Fig. 43) and the necessary link lead from switch to switch is obtainable under Part No. 502412.

One of the link wires is connected to earth (Fig. 41). The remaining link wire is connected through a snap connector to one side of the operating switch.

Top Cover Conversion Pack—Part No. 503219. The following is a list of the parts included in the pack to convert the old type cover assembly, part No. 502078 to 502411.

1	Top Cover Casting	301768
1	1st and 2nd Selector Fork	110753
1	Top and 3rd Selector Fork	110754
2	Welch Plugs	54505
1	Isolator Switch	42781
6	Packing Washers	502146
1	Link Lead	502412
3	Welch Plugs	104449

Overdrive Kit—Part No. 501803 for R.H.
Part No. 502104 for L.H.

These kits may be used either :

(a) Where a car is to be fitted with overdrive on all gears and is already fitted with a top cover (Part No. 502411).

(b) To convert cars fitted with the old type overdrive unit (Serial No. 22/1275/—, in which case either :—

> **(i)** A complete new top cover assembly, Part No. 502411, may also be required, or

> **(ii)** A top cover conversion Pack, No. 503219.

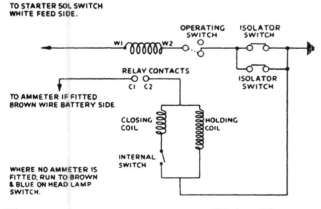

TO STARTER SOL SWITCH WHITE FEED SIDE.

TO AMMETER IF FITTED BROWN WIRE BATTERY SIDE

WHERE NO AMMETER IS FITTED, RUN TO BROWN & BLUE ON HEAD LAMP SWITCH.

OPERATING SWITCH

ISOLATOR SWITCH

RELAY CONTACTS
C1 C2

ISOLATOR SWITCH

CLOSING COIL

HOLDING COIL

INTERNAL SWITCH

Fig. 43   **Wiring Diagram.**

Service Instruction Manual

REAR AXLE

SECTION F

REAR AXLE

INDEX

ILLUSTRATIONS

REAR AXLE

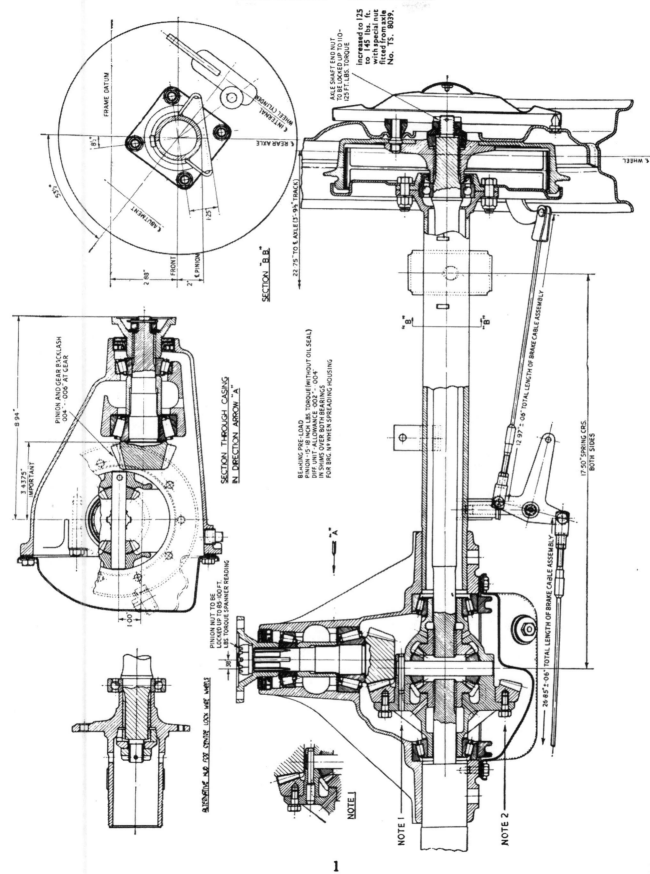

Axle arrangement. (For Notes 1 and 2 See page 4).

Fig. 1

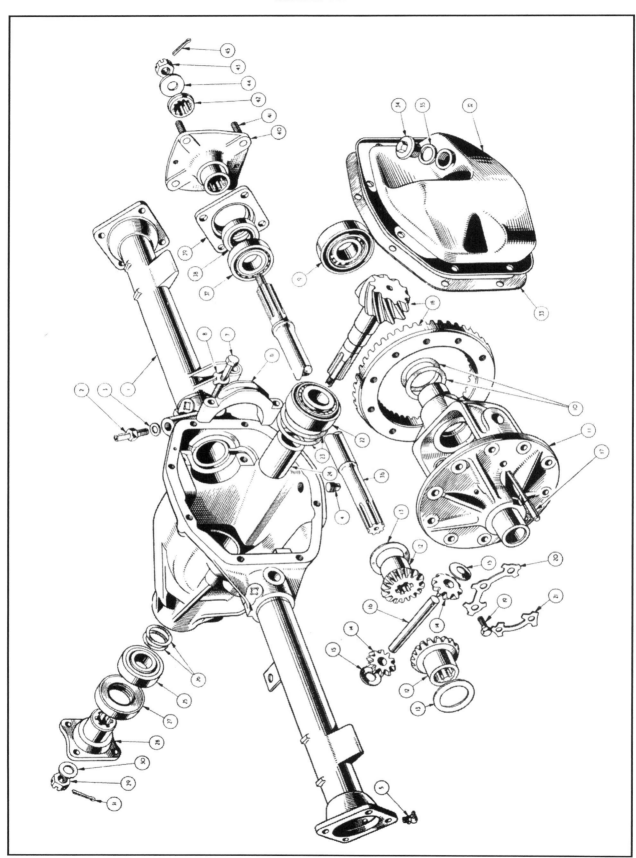

Exploded view of axle details.

Fig. 2

NOTATION FOR REAR AXLE EXPLODED VIEW (FIG. 2)	
Ref. No. — Description	Ref. No. — Description
1. Axle casing assembly	23. Pinion head bearing ring shim.
2. Breather.	24. Pinion bearing spacer.
3. Fibre washer.	25. Pinion tail bearing.
4. Drain plug.	26. Pinion shaft shims.
5. Grease nipple.	27. Pinion shaft oil seal.
6. Bearing cap.	28. Pinion driving flange.
7. Bearing cap setscrew.	29. Castellated nut.
8. Tab washer.	30. Washer.
9. Differential bearing.	31. Cotter pin.
10. Shims.	32. Rear cover.
11. Differential casing.	33. Joint washer.
12. Sun gear.	34. Oil filter plug.
13. Thrust washer.	35. Washer.
14. Planet gear.	36. Rear axle shaft.
15. Thrust washer.	37. Hub bearing.
16. Cross pin.	38. Hub oil seal.
17. Locating pin. (See note 1 page, 4).	39. Bearing housing.
18. Crown wheel and pinion.	40. Hub assembly.
19. Crown wheel bolt. (See note 2, page 4).	41. Wheel stud.
20. Tab washer.	42. Splined collar.
21. Tab washer.	43. Castellated nut.
22. Pinion head bearing.	44. Washer.
	45. Cotter pin.

See also Fig. 35.

DATA

Crown wheel run out	Not more than .003″
Backlash between crown wheel and pinion	.004″ — .006″
Distance from ground thrust face on pinion to centre of crown wheel	3.4375″
Pinion bearing pre-load, measured without oil seal	15 — 18 in. lbs.
Pre-load for differential bearings	Allowance for .002″ to .004″ shims, spread over both bearings
Diameter of differential bearings	2.8446″ — 2.8440″
Later production cars	2.8460″ — 2.8450″
Pinion nut tightening torque	85 — 100 lbs. ft.
Hub securing nut tightening torque	110 — 125 lbs. ft.
125—145 lbs. ft. with special nut fitted to axle No. TS.8039 onwards.	

1. GENERAL DESCRIPTION (Fig. 1)

The rear axle is of the hypoid semi-floating type with shim adjustment for the differential bearings and for the endwise location of the pinion in relation to the crown wheel. The axle sleeves are pressed into the centre casing and each sleeve is located by four pegs.

The centre casing is a casting which accommodates the differential cage and the attached crown wheel, together with the hypoid pinion. A detachable pressed steel cover, at the rear of the centre casing, allows access to the differential unit and crown wheel, the removal of this cover clears the way for the dismantling of the axle. The hypoid pinion is mounted on two taper roller bearings which are separated from one another by a tubular spacer. The pinion's endwise relation with the crown wheel is adjusted by means of shims inserted between the "head" bearing outer ring and the casing. Preloading of bearings is adjusted by means of shims between the spacer and tail bearing.

The differential casing contains two sun and two planet wheels and also carries the crown wheel, which is bolted in position by ten bolts passing through the casing and into tapped holes in the back of the wheel itself.

NOTE 2: Fig. 1. The crown wheel is attached to the differential casing by bolts locked by tab washers. The crown wheel showed a tendency to work loose after exacting rally acceleration and reversing gear tests and to obviate this possibility the $\frac{5}{16}''$ UNF attachment bolts were replaced by $\frac{3}{8}''$ UNF in axles numbered TS.4731 onwards.

The two planet wheels are mounted on a cross spindle, this spindle being provided with a hole at one end and located by a pin passing through the hole and the differential casing.

NOTE 1: Fig. 1. The locating pin used has a "stepped" shape but this is to be changed in the near future to the "parallel" type pin as shown in the main illustration. Incorporated in axle No. TS.6260 onwards.

The axle shafts are splined at both ends. The inner end fitting into the sun wheels and the outer extremity accommodating the wheel bearing and hub. The hub is secured to the splined end of the axle shaft by means of a splined taper collar, a shaped washer and a castellated nut.

The wheel bearing is accommodated in the axle sleeve and a housing which is bolted to the flanged end of each axle tube. The inner portion of the wheel bearing is gripped between the hub and a flange on the axle shaft.

The differential casing is mounted on two taper roller bearings, the position of these being adjusted by means of shims interposed between them and the casing itself. The disposition of these shims decides the crown wheel and pinion depth of engagement and the thickness of these the amount of pre-loading.

2. TO REMOVE HUBS

(a) Remove the nave plate.

(b) Withdraw the split pin from end of axle shaft. Partly release the torque on the castellated hub securing nut.

(c) Jack up the car, remove the castellated nut, the road wheel and by the withdrawal of the two countersunk setscrews remove the brake drum.

(d) Remove the washer and the splined taper collar from the axle shaft.

Fig. 3   **Hub removal utilising the Churchill tool No. S 132/2.**

(e) Fit the Churchill hub removing tool No. M86 or S132/2 and withdraw the hub from the shaft (Figs. 3 and 6).

An alternate method is to withdraw the half shaft with the hub in position (see page 5), this method necessitates the removal of the brake backing plate and the severing of the hydraulic and hand brake connections.

3. TO REPLACE HUBS

The replacement of the hubs is the reversal of the removal but the following notes should be considered.

The axle shafts of the later production cars provided an interference fit with the hub splines. To facilitate the replacement of the hubs—the Churchill hub replacing Tool No. S125 was introduced (see Fig. 5). Should the axle shafts be out of the axle casing it will still be necessary to use the hub replacing tool or a fly press.

4. TO REMOVE HUBS (Centre lock type) (Fig. 35)

(a) Jack up the car and remove the hub cap by tapping the lugs with a copper faced mallet. Remove the road wheel.

(b) Remove the split pin through the aperture in the barrel of the hub.

(c) Remove the hub securing nut from the axle shaft. It may be necessary to replace the wheel and lower the car when torque is applied to the nut. After removing the nut withdraw the washer and splined collar.

(d) By inserting a screwdriver blade into the cut of the split tapered collar, the collar will expand and allow it to be withdrawn from the hub.

(e) Remove the two countersunk brake drum securing screws and withdraw the brake drum.

(f) Fit the Churchill hub removing Tool No. S132 and remove the hub. It should be remembered that the hubs have right or left-hand threads and care must be exercised when selecting the removal rings. (See Fig. 4.)

Fig. 4   The removal of the knock on type hub utilising the Churchill Hub Removing Rings S132 with the S4221 frame and slave ring. Shown with brake assembly removed for photographic purposes.

An alternate method of hub removal is to remove the axle shaft complete as described on this page. This necessitates the severing of the hand brake and hydraulic connections and removing the brake backing plate.

5. TO REPLACE HUBS (Centre lock type) (Fig. 35)

The replacement of the hubs is the reversal of their removal. However the following points should be noted.

(a) The axle shafts of later production cars provide an interference fit with the hubs. To facilitate the replacement of the hubs the Churchill hub replacing Tool No. S125 was introduced and is illustrated in Fig. 5.

Fig. 5   The replacing of the knock on type hub utilising the Churchill Hub Replacing Tool No. S125. Shown with brake assembly removed for photographic purposes.

(b) When the axle shafts are out of the casing it is still necessary to use the hub replacing tool or a fly press.

6. TO REMOVE AXLE SHAFT

(a) Jack up car and remove road wheel.

(b) If the car is equipped with wire wheels remove the split tapered collar by inserting a screwdriver blade into the cut of the ring. It can now be drawn off the barrel of the hub.

(c) Withdraw the two countersunk brake drum securing screws and remove the brake drum.

(d) Drain the hydraulic system, disconnect the pipe line and the hand brake cable at the wheel cylinder.

5

(e) Remove the four bolts and nyloc nuts which secure the brake backing plate and the bearing housing to the axle flange.

(f) Withdraw the axle shaft assembly from the axle casing together with the brake backing plate assembly.

(g) Grip the axle in the protected jaws of a vice and utilising the aperture in the barrels of the centre lock hub, remove the split pin. Remove the castellated nut, washer and splined taper collar. Remove the hub with the Churchill hub remover, Tool No. S132. (See Fig. 4).

(h) To remove the disc wheel hub, first remove the split pin at the axle end followed by the castellated nut, washer and splined taper collar. Remove the hub with the Churchill hub remover, Tool No. M86 (Fig. 6) or S132/2 (Fig. 3).

Fig. 6   Hub removal utilising the Churchill tool No. M86.

The extraction of each hub will release the oil seal and bearing housing but leave the hub bearing on the axle shaft.

(i) Remove the hub bearing from the shafts, utilising the Churchill Tool No. S4615 Codes 8 and 10. (See Fig 7.)

(j) The oil seal can now be drifted out of the bearing housing if it is seen to be unserviceable.

7. **TO REPLACE AXLE SHAFT**

The replacement of the axle shaft is the reversal of their removal. However the following points should be noted.

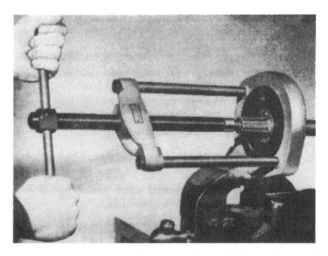

Fig. 7   Removing hub bearing utilising Churchill Tool No. S4615 Codes 8 and 10 with S4221 frame and slave ring.

(a) The replacement of the hub oil seal is shown in Fig. 8, utilising Churchill Tool No. M29.

Fig. 8   Replacing hub oil seal utilising Churchill Tool No. M29.

(b) Replacement of the hub bearing is illustrated in Fig. 9, using Churchill Tool No. M92.

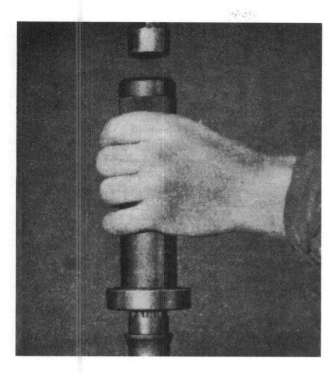

Fig. 9   Replacing wheel bearings utilising Churchill Tool No. M92.

(**c**) On later production cars the axle shaft provided an interference fit with the hub and it is necessary to replace the hub, utilising the Churchill hub replacing Tool No. S125.

(**d**) On completion of the replacement operations it will be necessary to bleed the brakes.

8.   TO REMOVE AXLE

NOTE : As the axle has to be tilted it may be desirable to drain off the oil.

(**a**) Jack up car and remove road wheels.

(**b**) Detach propeller shaft from pinion flange by the removal of four bolts and nyloc nuts.

(**c**) Disconnect hand brake cable from the compensator lever.

(**d**) Drain the hydraulic system and disconnect the line at the front end of the flexible hose. (See Brake Section.)

(**e**) Remove the brake drums after withdrawing the two countersunk setscrews. If wire wheels are fitted the split taper ring will have to be removed first,

this can be effected by inserting the blade of a screwdriver into the split to expand the ring which can then be drawn off the hub.

(**f**) First disconnect the Bundy Tubing and the hand brake cables at the wheel cylinders and then remove the bolts and nyloc nuts securing the brake back plate to the axle casing.

The hubs, together with the half shafts, oil seals, bearings and brake backing plate, can now be removed from the axle. These can be dismantled as described on page 5.

(**g**) Remove the axle check straps by first removing the four nuts and lock washers.

(**h**) Remove the nyloc nuts from the " U " bolts securing the axle to the road spring and swing the shock absorber arm (attached to the spring plate) clear. The " U " bolts may now be removed from the axle.

(**i**) Lift the axle clear of the spring and move it to the left, allow the right-hand side to be lowered when the axle end is clear of the right-hand spring. By moving the axle to the right it can be withdrawn from the chassis. (Fig. 10.)

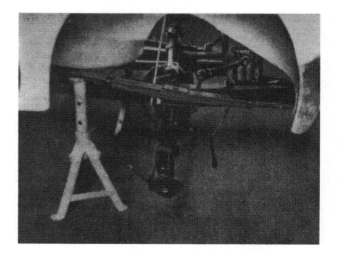

Fig. 10   Axle being removed from car.

7

9. TO REPLACE THE AXLE

If a replacement axle is being fitted it will be necessary to remove the complete brake assemblies at the axle ends.

It is not necessary to remove the hubs, for these can be removed with the half shafts and brake backing plates.

The axle must be tilted during the fitting operations and filling the axle with oil should be delayed until the axle has been fitted to the car.

The fitting is the reversal of the removal. For the bleeding of the hydraulic system see " Brakes—Section R."

10. TO DISMANTLE

(a) Drain oil.

(b) Remove wheel securing cones (wire wheel hubs only). This enables the brake drum securing screws to be removed and the drums withdrawn.

(c) Remove split pins (as shown in Fig. 11)

Fig. 11 Removing split pin from hub securing nut.

and hub securing nuts. Preventing the hubs from rotating by means of a road wheel, the conical washers can then be removed and the hubs, complete with their splined tightening cones, withdrawn with a suitable tool or press. Churchill Tool No. M86 or S132/2.

NOTE: Some difficulty may be experienced in the slackening of the nuts due to rotation of the hubs, but since

the axle is going to be completely dismantled the hubs can be removed at a later stage, which means that the half shafts, hubs, brake backing plates, etc., must be removed as an assembly.

(d) Remove brake shoes and return springs.

(e) Withdraw the brake backing plates after removal of the eight bolts, spring washers and nyloc nuts, four from either back plate. Further dismantling of the brake backing plates only require the removal of the hydraulic wheel cylinders and anchor blocks, the latter being secured by spring washers and two nuts, the former can be withdrawn provided the hydraulic connections, rubber dust sealing boots, etc., have been removed.

(f) The half shafts can now be withdrawn from the axle casing, the bearing housings tapped off the bearings and the bearings withdrawn with a suitable puller. (As shown in Fig. 7.) The grease seal can then be tapped out of the bearing housings.

NOTE: If the hubs have not been previously withdrawn due to difficulties in slackening the hub nuts mentioned in (c) they can now be slackened by gripping the axle shaft in the vice, and the hubs then pressed off the axle shafts with a suitable tool or press.

Fig. 12 Identification numbers on bearing caps and axle casing. Note also the tops of washers laying in groove of bearing cap.

(g) Remove axle centre casing cover and joint after withdrawal of eight setscrews.

(h) Remove the differential bearing caps, noting the markings stamped on the top of these and the correspondingly abutting portions of the casing. The existing relation between the caps and casing must be retained when re-assembling. Fig. 12 shows example of markings.

(i) Apply axle casing spreader as shown in Fig. 13, and lift differential assembly

Fig. 14   Removing split pin from driving flange securing nut.

Fig. 13   Casing spreader in position. Churchill Tool No. S101.

out of the axle centre casing. " Spreading " should be limited to that required to just free the assembly in the casing.

(j) Suitably identify the respective outer portion of the differential bearings with their inner races. The inter-relation of the component parts of these races must be retained when re-assembling the rear axle.

(k) Remove the crown wheel from its mounting flange after the withdrawal of the ten fixing bolts, leaving further dismantling of the differential unit until a later stage.

(l) After removal of split pinned flange nut as shown in Fig. 14, and having removed the flange, drive the pinion out through the casing with a hide faced hammer. Lay aside the shims which are fitted between the spacer

and tail race for possible use when re-assembling. Remove pinion head bearing inner cone as shown in Fig. 15.

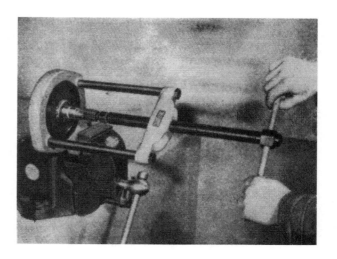

Fig. 15   Removal of pinion head bearing utilising Churchill Tool No. TS1 and S4221 frame with slave ring.

(m) Drive out the pinion outer rings as shown in Fig. 16. The removal of the outer ring of the tail bearing will also eject the oil seal and tail bearing inner cone. The ejection of the head bearing outer ring will uncover the shims fitted between this and the casing.

9

Fig. 16   **Driving out pinion bearing outer rings utilising Churchill Tool No. 20SM FT71.**

These shims should be laid aside with the component parts of this bearing as a guide when re-assembling.

(**n**) Replace the differential assembly in the axle casing and release the tension from the axle casing spreader.

(**o**) Check the " run out " of the crown wheel mounting flange; this should not exceed .003″ (Fig. 17). The crown wheel itself can be checked on a surface table with the aid of a set of feeler gauges. Having satisfactorily completed these checks, the differential

Fig. 17   **Checking the run out of the crown wheel mounting flange utilising a D.T. 1.**

assembly can be removed from the axle casing and dismantled as follows:

(**i**) Drive out the cross pin locating pin and withdraw the cross pin.

(**ii**) Rotate the sun wheels which will in turn rotate the planet wheels until the planet wheels with their respective thrust washers are opposite the cut away portions of the crown wheel carrier from which they can easily be withdrawn.

(**iii**) Remove the sun wheels and their thrust washers, so completing the dismantling of the rear axle.

11   **TO RE-ASSEMBLE**

All parts must be examined carefully and a decision should be made as to which items require renewal. Where it is found necessary to replace the crown wheel or pinion for any reason the gears must be replaced as a pair, as they are "lapped" together in manufacture.

The first consideration, after replacing damaged or worn parts, must be the correct interrelations between the crown wheel and pinion. The assembled relation of these two gears must very closely approximate that used when the gears where "lapped" together after heat treatment during manufacture.

The datum position of the pinion with relation to the crown wheel is specified as 3.4375″ from the ground thrust face on the back of the pinion to the centre line of the differential bearings. It is also important that not only should this datum position be achieved, but that sufficient bearing pre-load should be arranged to ensure the maintenance of the specified relations in service.

Having cleaned the abutment faces and bearing housings thoroughly, and removed any excrescences from these surfaces, the following procedure for re-assembly is recommended.

(**i**) Fit the outer rings of the pinions two bearings, pulling them into place with a special tool. (Fig. 18).

Fig. 18   Fitting pinion bearings outer rings utilising Churchill Tool No. M70.

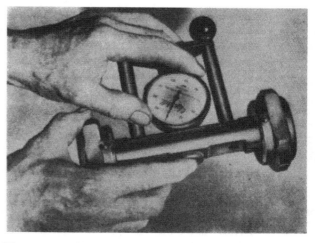

Fig. 20   Utilising the " button ", the Pinion Setting Gauge (Churchill Tool No. M84) is set to zero.

(ii) Fit the dummy pinion (M.84), the pinion bearing inner cones and install into the axle centre casing; tightening the flange nut progressively until the correct pinion pre-load of 15—18in. lbs. is obtained.

(iii) Install the pinion setting gauge in the axle centre casing (Fig. 19), (after zeroing the dial with a ground button held firmly on the gauge plunger, Fig.20) and tighten bearing caps. This gauge is used to assess the shim thickness which is required under the pinion head bearing outer ring, to bring the pinion into its correct datum position mentioned earlier. Due to the fact that the bearing inner cones are a slide fit on the dummy pinion and a press fit on the actual pinion to be used, bearing expansion will undoubtedly take place in the latter case. A pack of shims .002"—.003" below the gauge reading will be required to allow for this expansion and thus ensure the pinion is in its correct datum position.

Fig. 19   Pinion setting gauge, Churchill Tool No. M84 assembled to axle centre casing.

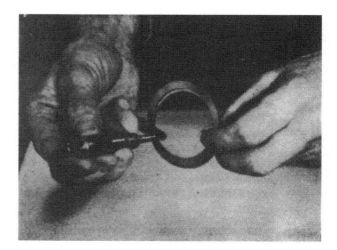

Fig. 21   Measuring the shim pack.

(iv) Although the packing shims are supplied to nominal thicknesses, the dimensions should be measured with a micrometer gauge. It is important that no damaged shims are used and that they are thoroughly cleaned before measurement. (Fig. 21.)

(v) Remove the pinion setting gauge, dummy pinion and pinion bearing outer rings.

(vi) Insert the measured pack of shims on the pinion head bearing outer ring abutment face (Fig. 22) and replace the pinion bearing outer rings, pulling them into place with the special tool shown in Fig. 18.

Fig. 23   Pressing the pinion head bearing on to the pinion shaft utilising the Churchill Tool No. TS1 and S4221 frame with slave ring.

Fig. 22   Shims placed in position on outer ring abutment face.

(vii) Press the pinion head bearing inner cone on to the pinion shaft (Fig. 23).

(viii) The bearing spacer is fed on to the pinion shaft with the chamfer outwards as shown in Fig. 24. The shims previously removed when dismantling the axle are placed in position on the pinion and the assembly fitted into the axle centre casing. The thickness of shims fitted will probably have to be adjusted to provide the correct pre-load figure.

(ix) The inner cone of the pinion tail bearing is tapped into position on the pinion and up against the shims on the distance collar.

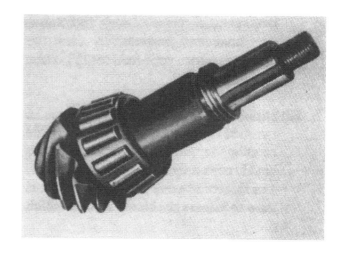

Fig. 24   Showing the pinion bearing spacer chamfer pointing outwards, followed by the previously used shim pack.

(x) The driving flange is fitted on the end of the pinion shaft and firmly secured with the castellated nut and plain washer to a tightening torque of 85—100 lbs. ft. THE OIL SEAL IS NOT FITTED UNTIL THE BEARING PRELOAD HAS BEEN CHECKED AS DESCRIBED IN THE NEXT OPERATION.

Fig. 25 Testing the pre-load of the pinion bearing utilising the Churchill Tool No. 20SM. 98 Note : The oil seal is not fitted at this juncture.

(xi) The fixture shown in Fig. 25 is now applied and the pre-load of the bearings checked. The correct pre-load should fall between 15—18 in. lbs. If the pre-load is inadequate shims must be withdrawn, whereas if an excessive figure is obtained additional shims must be fitted.

(xii) When the correct pinion pre-load is obtained remove driving flange and fit the oil seal (Fig. 26), after which the

Fig. 26 Fitting pinion housing oil seal utilising Churchill Tool No. M100.

flange should be replaced, the castellated nut tightened to the correct torque and split pinned.

(xiii) The differential assembly bearings are now fitted without, as yet, installing any packing shims. A suitable driver such as that which is shown in Fig. 27 should be used for driving the bearings on to the crown wheel carrier.

Fig. 27 Fitting differential bearings utilising the Churchill Tool No. M89 to the differential casing.

(xiv) The axle bearing seats are carefully cleaned and any excrescences removed. The differential casing is positioned and the bearing caps, fitted with regard to the identification markings, are tightened down and then slackened off a ¼ turn. This will prevent the bearings tilting but allow sideways movement. A dial indicator gauge is mounted on the axle centre casing with the plunger resting on the crown wheel mounting flange (Fig. 28). The assembly is forced away from the dial gauge and then the indicator set to zero. The assembly should then be

Fig. 28 Ascertaining the total end float of the differential casing without the crown wheel fitted. The caps should be tight and then slackened ¼ turn.

13

levered in the opposite direction until the taper roller bearings go hard home. The reading on the dial gauge (.062″ for example) will indicate the total side float of the crown wheel carrier and should be noted for later reference.

(xv) The crown wheel carrier is now removed from the axle centre casing so that the sun gears, planet gears and thrust washers can be assembled, the cross pin being used to locate the two planet gears with their respeceive thrust washers temporarily in position (Fig. 29). Subsequently, the planet gears are

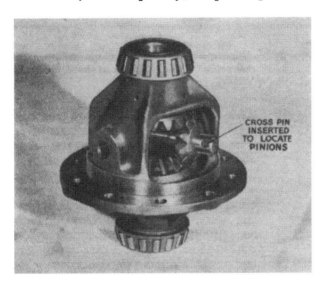

Fig. 29 Location of planet gears for entry into the differential casing.

rotated round the sun wheel through 90 degrees, the cross pin being withdrawn to allow the gears to assume their normal fitted position, and the cross pin finally fitted and secured by its locking pin, this pin being located by "centre popping."

(xvi) The crown wheel is fitted to the crown wheel carrier, the fixing bolts thoroughly tighten to 22—24 lbs. ft. and secured with their respective locking plates.

NOTE: The crown wheel attachment bolts were increased in diameter from $\frac{5}{16}$″ to $\frac{3}{8}$″ at rear axle No. TS.4731. The crown wheel is checked for flush fitting against the flanged face of the

carrier with a feeler gauge, thus ensuring that the crown wheel goes right home and also that there can be no question of casting distortion. The maximum permissible run out of the crown wheel and crown wheel mounting flange is .003″. The flange can be checked before the fitting of the crown wheel by rotating it on its bearings, using a dial indicator, the crown wheel itself on a surface table with the aid of feeler gauges.

(xvii) The differential assembly is installed in the casing in a similar manner to operation (xiv), but in this instance the D.T.I. plunger bears against the back of a crown wheel fixing bolt (Fig. 30).

Fig. 30 Ascertaining the depth of engagement between crown wheel and pinion.

(xviii) The assembly is now forced away from the dial gauge until the teeth on the crown wheel go fully home with those on the pinion. The dial gauge is now set to zero and the assembly levered towards the dial gauge. Let this dimension be .045″.

(xix) The side float of the assembly measured in the last operation, less the crown wheel and pinion backlash specified, will indicate the shim thickness required on the crown wheel side. The backlash is specified as between .004″ and .006″ and an average figure of .005″ should be used for this calculation giving .040″ to be fitted on the crown wheel side.

(xx) To obtain the thickness of the shims required between the other differential bearing and casing, the figure arrived at in previous operation, *i.e.*, .040", should be subtracted from the total side float measured in operation (xiv), plus an allowance of .005" to provide the necessary degree of bearing pre-load. This gives a total shim thickness of .067" and thus shims on two bearings will be .040" already estimated and .067" — .040" = .027" on the other side.

(xxi) Having decided the thickness of shims required behind each differential bearing, these bearings are extracted with the special tool shown in Fig. 31. The

Fig. 31   Removal of the differential bearing utilising the Churchill Tool No. S103 and S4221 frame.

respective shim packs are measured with a micrometer gauge after ensuring that the shims are clean and undamaged and allocated to their respective sides of the crown wheel carrier.

(xxii) As each bearing is extracted, the two portions of each must be laid aside for refitting in the same relation and position as that used during initial assembly. Failure to fit these bearings in their original positions will upset the measurements made in previous operations.

(xxiii) Having fitted the two packs of shims in their respective positions the bearing inner cones are driven on to the carrier with a suitable sleeve tool as shown in Fig. 27 and the outer rings applied.

(xxiv) The differential assembly is now fitted into the axle centre casing and, owing to the pre-loading of the bearings, a certain amount of casing spreading is desirable to complete this operation. THE CASING SPREADER SHOWN IN FIG. 13 SHOULD BE USED AND THE SPREADING OF THE CASE LIMITED TO THAT JUST REQUIRED TO ENABLE THE DIFFERENTIAL ASSEMBLY TO ENTER THE CASING.

(xxv) The bearing caps are then fitted in their respective positions so that the number stamped on the caps coincide with those stamped on the axle casing, tightening them to their correct torque of 34—36 lbs. ft.

(xxvi) The pinions and crown wheel backlash is checked with a dial gauge as shown in Fig. 32, and should be .004"—.006": an average should be taken of several teeth.

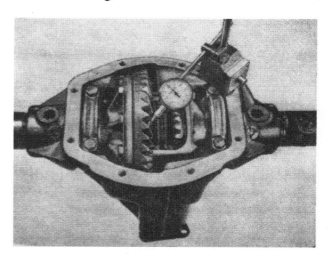

Fig. 32   Checking the back lash of the differential unit utilising a DTI.

Should the backlash be incorrect, the transfer of shims from one side of the differential carrier to the other will be necessary. If the backlash is too great, then a shim or shims will have to be taken from the side opposite the crown wheel and the same shims added to the crown wheel side, always maintaining the same overall total. Should the backlash be insufficient, then the reverse procedure must be adopted.

(**xxvii**) A tooth marking test should now be carried out, and to enable this to be done a few teeth should be painted with a suitable marking compound. The pinion should be rotated backwards and forwards by the driving flange, over the marked teeth on the crown wheel, and the markings compared with the diagram (Fig. 33), and the instructions on this diagram regarded.

HYPOID CROWN WHEEL TOOTH MARKINGS

DRIVE SIDE   *OVERRUN SIDE*

 CORRECT MARKINGS ON GEAR

 PINION CONE TOO CLOSE

 PINION CONE TOO WIDE

Fig. 33   **Crown wheel tooth markings.**

(**xxviii**) A new axle cover packing is fitted, together with the cover itself, and the latter secured with the eight setscrews.

(**xxix**) Drive the wheel bearings on to their respective axle shaft (Fig. 9), and assemble to the axle unit.

(**xxx**) The grease seals should now be tapped into the bearing housings (Fig. 8), and the assemblies fitted to each axle sleeve, followed by the brake backing plate and shoe assembly.

(**xxxi**) The four bolts are fitted through each bearing housing and brake backing plate, ensuring that both these items assume their appropriate relation with the axle sleeve, the nuts are screwed into position and firmly tightened.

(**xxxii**) The hubs are next fitted by means of a special tool or press (Fig. 34), and secured by the splined hub tightening cones, conical washers and hub securing nuts. A substantial spanner will be required to tighten the castellated securing nut. (A tightening torque of 110—125 lbs. ft. is specified. After axle No. TS.8039 the torque was increased to 125—145 lbs. ft. when a nut of a different material was introduced.) Having thoroughly tightened up this nut, the hole in the axle shaft is lined up with one of the slots in the castellated nut and the split pin is fitted.

(**xxxiii**) The brake drum is next fitted to each hub and secured thereto by means of the two countersunk grub screws.

(**xxxiv**) Fit wheel securing cones (wire wheel hubs only). Fig. 35.

Fig. 34   **The replacing of the disc wheel type hub utilising the Churchill Hub Replacing Tool No. S125.**

12. SERVICE DIAGNOSIS

Rear axle noise is usually apparent as a hum in moderate cases or as a growl in very severe cases.

Noises from the rear wheel bearings, propeller shaft bearing or tyres is often diagnosed as rear axle troubles.

Always ascertain that the noise attributed to the rear axle does actually emanate from that unit before dismantling parts.

REAR AXLE

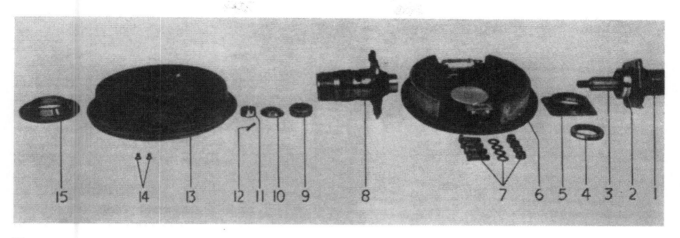

Fig. 35   " Knock on " type hub.

1. Axle casing.
2. Hub bearing.
3. Axle shaft.
4. Oil seal.

5. Seal and Bearing Housing.
6. Brake assembly.
7. Fixing bolts for brake backing plate and seal / bearing housing.

8. Hub "knock on" type.
9. Splined taper collar.
10. Washer.
11. Castellated nut.

12. Split pin.
13. Brake drum.
14. Counter sunk screws.
15. Taper collar.

CAUSE

1. Axle Noise

(a) Inadequate or improper lubrication.

(b) Teeth broken off gears.
(c) Contact of crown wheel and pinion not correctly adjusted.

2. Lubricant Leakage

(a) Leakage in general.

(b) Leakage at hub.

(c) Leakage at pinion head.

3. Axle Knock

(a) Splines on axle shafts or in differential gears badly worn.
(b) Splines on hub shell or centre of wire wheel badly worn.
(c) Incorrect shimming of planet gears in differential unit.

REMEDY

(a) Drain, flush casing out with flushing oil and replenish with correct grade of oil. See "General Data" Section A.
(b) Replace damaged parts.
(c) Noise during coasting; move the pinion away from crown wheel. Noise during driving; move the pinion toward the crown wheel. Do not move the pinion more than .004″ when making these adjustments.

(a) Reduce level of oil if overfull. Clean out breather.

(b) Clean out breather. Renew oil seal if leakage persists.

(c) Clean out breather. Renew oil seal if leakage persists.

(a) Replace worn parts.

(b) Replace worn parts.

(c) Replace present ones in use with thicker ones.

17

Service Instruction Manual

FRONT SUSPENSION

AND

STEERING

SECTION G

FRONT SUSPENSION AND STEERING

FRONT SUSPENSION

INDEX

STEERING

INDEX

TELESCOPIC STEERING UNIT

INDEX

FRONT SUSPENSION AND STEERING

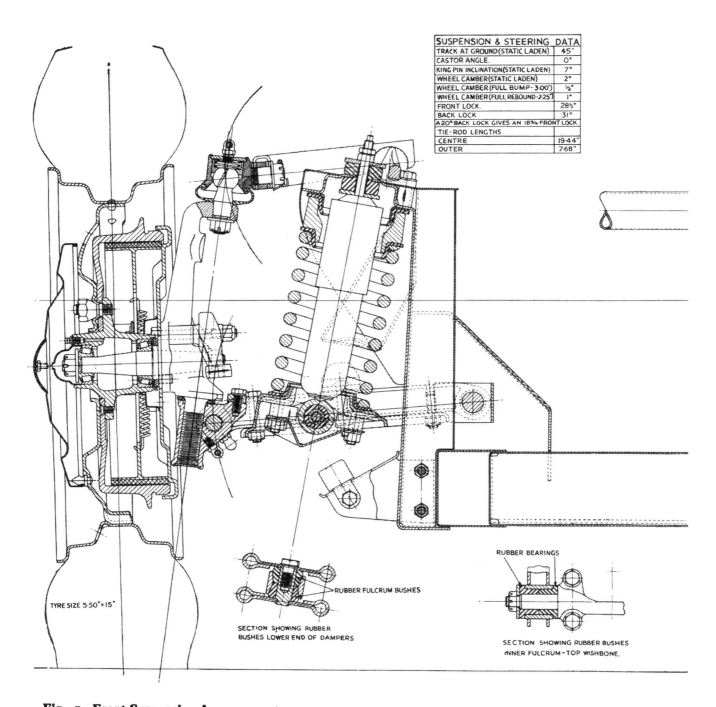

SUSPENSION & STEERING DATA	
TRACK AT GROUND (STATIC LADEN)	45"
CASTOR ANGLE.	0°
KING PIN INCLINATION (STATIC LADEN)	7°
WHEEL CAMBER (STATIC LADEN)	2°
WHEEL CAMBER (FULL BUMP -3·00")	½°
WHEEL CAMBER (FULL REBOUND -2·25")	1°
FRONT LOCK.	28½°
BACK LOCK.	31°
A 20° BACK LOCK GIVES AN 18¾ FRONT LOCK.	
TIE-ROD LENGTHS	
CENTRE	19·44"
OUTER	7·68"

TYRE SIZE 5·50"×15"

RUBBER FULCRUM BUSHES

SECTION SHOWING RUBBER
BUSHES LOWER END OF DAMPERS

RUBBER BEARINGS

SECTION SHOWING RUBBER BUSHES
INNER FULCRUM-TOP WISHBONE.

Fig. I   Front Suspension Arrangement.

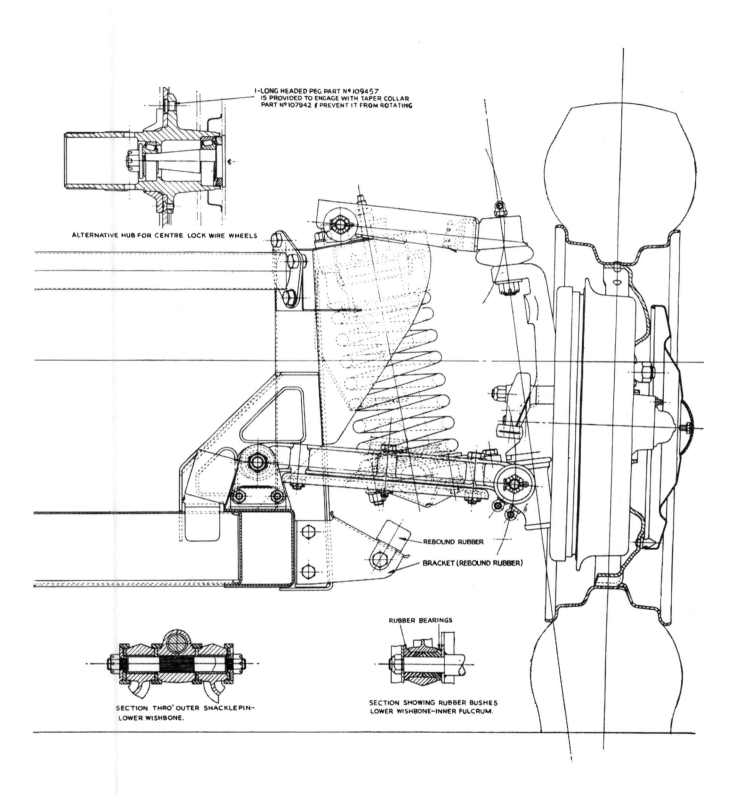

I-LONG HEADED PEG PART Nº 109457
IS PROVIDED TO ENGAGE WITH TAPER COLLAR
PART Nº 107942 & PREVENT IT FROM ROTATING

ALTERNATIVE HUB FOR CENTRE LOCK WIRE WHEELS

REBOUND RUBBER

BRACKET (REBOUND RUBBER)

RUBBER BEARINGS

SECTION THRO' OUTER SHACKLEPIN-
LOWER WISHBONE.

SECTION SHOWING RUBBER BUSHES
LOWER WISHBONE-INNER FULCRUM.

2

FRONT SUSPENSION AND STEERING

ILLUSTRATIONS

FRONT SUSPENSION AND STEERING

1. FRONT SUSPENSION DATA

Track at Ground (Static Laden)	45″
Castor Angle	Nil
King Pin Inclination (Static Laden)	7°
Wheel Camber (Static Laden)	2°
Wheel Camber (Full Bump 3″)	½°
Wheel Camber (Full Rebound 2.25″)	1°
Turning Circle	32′
Back Lock	31°
Front Lock	28.5°
A 20° Back Lock gives an 18.75° Front Lock.	
Front Wheel Alignment Parallel to ⅛″ toe in	
Length of Centre Tie Rod	19.44″
Length of Outer Tie Rod	7.68″
End Float of Lower Outer Shackle Pin Assembly	.004″ to .012″

2. DESCRIPTION (Fig. 1)

The two front suspension units are of wishbone construction. Road shocks are absorbed by low periodicity coil springs, each of these springs are controlled by a double acting telescopic shock absorber fitted inside the coil spring.

The upper wishbones are rubber bushed at their inner ends to a fulcrum pin which is attached to the spring housing, they are shaped to form a "U" and the outer ends are interlaced to accommodate a distance piece and are secured together by the screwed shank of a ball joint. This joint is fitted to, and provides the axial movement for, the upper end of the vertical link. The inner ends of the lower wishbone arms are rubber bushed on each side and are attached to the fulcrum pin mounted on the upperside of the chassis frame. The fulcrum is steadied at its extremities by two support brackets.

The outer ends of the wishbone arms, bushed with a Clevite bearing, are mounted on either end of a shackle pin. The shackle pin is splined centrally to fit transversely into the manganese bronze trunnion which is threaded to accommodate the lower end of the vertical link.

Each bushed end of the wishbone arms is located sideways on the shackle pin by means of a white metal covered steel thrust washer, bearing on the screwed trunnion on the inside and on the outer side against a steel washer which is secured by a split pinned castellated nut. During production the outer lower ends of the wishbone arms are assembled to the shackle pin to give an end float of .004″ to .012″. The need for adjustment should only occur when the front suspension units have been disturbed. Road dirt and weather are excluded from the grease lubricated bearings by special oil resisting rubber seals.

The screwed trunnion at the lower end and the ball joint at the upper end of the vertical link provide the bearings for the pivoting of the road wheels. Road dirt and weather are excluded from these bearings by a rubber gaiter interposed between the vertical link and the ball joint assembly at its upper end, at the lower end a circular rubber seal is fitted between the trunnion and the link. The thread of the trunnion is sealed off by a disc let into the lower end of the threaded bore. The steering lock stop consists of an eccentric roller bolted to the upper side of the trunnion and abuts against a machined face on the vertical link.

The vertical link, which couples the upper and lower wishbone arms as previously described, is a carbon steel stamping and carries the stub axle shaft, the brake backing plate and the steering lever.

The stub axle is of manganese molybdenum steel, which is mounted as a press taper fit in the vertical link, is secured by a split pin locked castellated nut.

The brake backing plate, with the brake shoes and hydraulic wheel cylinders attached, is secured to a machined flange on the vertical link by two setscrews with a lock plate at the lower two points and two bolts of unequal length at the upper two points. The longer of these bolts passes through the front bore of the brake plate, the vertical link, a distance piece and the steering lever and is secured by a nyloc nut; the shorter bolt is similarly secured and utilises the lower bore.

The front hub is mounted on a pair of opposed taper roller bearings carried on the stub axle shaft. The inner bearing abuts against a projecting shoulder on the vertical link and its outer ring against a flange machined in the hub. The outer ring of the

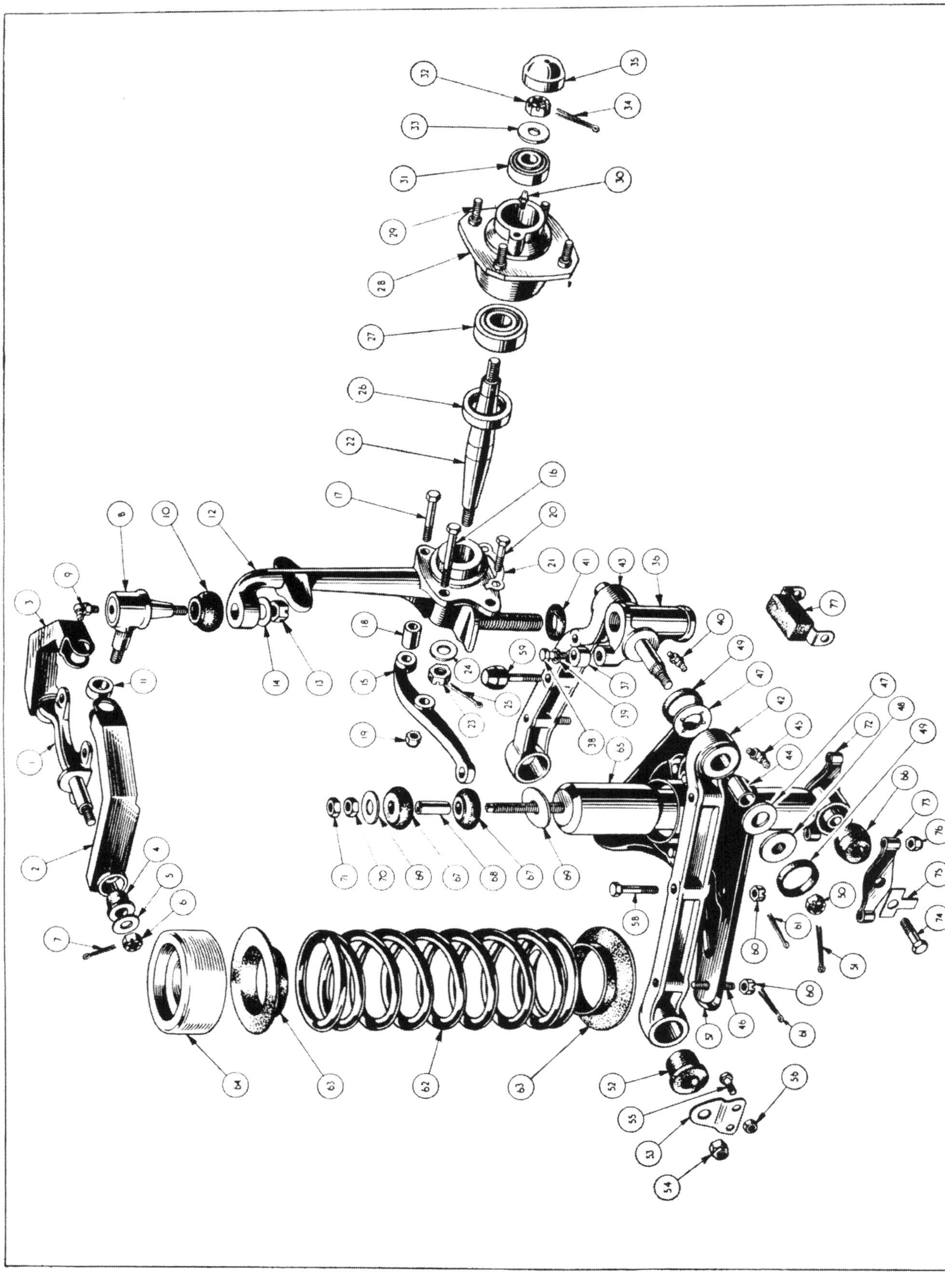

Exploded details of L.H. Front Suspension Unit.

Fig. 2

4

NOTATION FOR FIGURE 2

Ref. No.	Description	Ref. No.	Description
1.	Inner Upper Fulcrum Pin.	41.	Oil Seal.
2.	L.H. Front Upper Wishbone Arm.	42.	L.H. Front Lower Wishbone Arm Assembly.
3.	R.H. Front Upper Wishbone Arm.	43.	R.H. Front Lower Wishbone Arm Assembly.
4.	Rubber Bush.	44.	Bush for Wishbone Arm.
5.	Plain Washer.	45.	Grease Nipple.
6.	Castellated Nut.	46.	Spring Pan Studs
7.	Split Pin.	47.	Thrust Washer.
8.	Upper Wishbone Ball Joint Assembly.	48.	Lock Washer.
9.	Grease Nipple.	49.	Grease Seal.
10.	Rubber Gaiter.	50.	Castellated Nut.
11.	Upper Wishbone Distance Piece.	51.	Split Pin.
12.	Vertical Link.	52.	Rubber Bush.
13.	Castellated Nut.	53.	Support Bracket.
14.	Plain Washer.	54.	Nyloc Nut.
15.	Steering Lever.	55.	Bolt.
16.	Bolt.	56.	Nut.
17.	Bolt.	57.	Lower Spring Pan Assembly.
18.	Steering Lever Distance Piece.	58.	Bolt.
19.	Nyloc Nut.	59.	Bump Rubber.
20.	Setscrew.	60.	Castellated Nut.
21.	Locking Plate.	61.	Cotter Pin.
22.	Stub Axle.	62.	Front Road Spring.
23.	Castellated Nut.	63.	Rubber Washer.
24.	Plain Washer.	64.	Packing Piece.
25.	Split Pin.	65.	Shock Absorber.
26.	Oil Seal.	66.	Lower Rubber Mounting.
27.	Front Hub Inner Bearing.	67.	Upper Rubber Mounting.
28.	Front Hub.	68.	Metal Sleeve.
29.	Wheel Stud.	69.	Washer.
30.	Grease Nipple, fitted up to Commission No. TS.5348.	70.	Nut.
		71.	Lock Nut.
31.	Front Hub Outer Bearing.	72.	Shock Absorber Bracket and Fulcrum Pin.
32.	Castellated Nut.		
33.	"D" Washer under nut.	73.	Shock Absorber Bracket.
34.	Split Pin.	74.	Setscrew.
35.	Grease Retaining Cap.	75.	Tab Washer.
36.	Bottom Trunnion.	76.	Nut.
37.	Steering Lock Stop.	77.	Rebound Rubber.
38.	Bolt for Steering Lock Stop.		
39.	Spring Washer.		
40.	Grease Nipple.		

outer bearings bears against the flange machined in the hub and the inner cone of the race against a "D" washer, all are secured to the stub axle by a castellated nut and split pin. These bearings are adjusted by the castellated securing nut but are not pre-loaded.

Provision is made against the loss of grease by fitting a felt washer between the vertical link and inner bearing.

3. MAINTENANCE

The maintenance necessary is largely confined to periodical greasing (see Lubrication Chart in General Data Section "A".

The hub bearings are not pre-loaded and it will be necessary to ensure this condition is attained when carrying out adjustments (see page 7).

As a precautionary measure it is most desirable to check that an end float of .004″ to .012″ in the lower outer wishbone arm attachment to the shackle pin is maintained. Each arm is adjusted independently. Apart from damage at this point, tightness at this point can appreciably affect the ride of the car (see page 12, para. xii).

Front wheel alignment of parallel to $\frac{1}{8}$″ toe in should be checked if the front wheel alignment is in doubt (see below).

4. FRONT WHEEL ALIGNMENT

The track should be between parallel and $\frac{1}{8}$″ toe in.

The outer tie rods are adjustable for length and usually to give the correct track the distance between the centres of the ball joint assembles will be 7.68″.

If the wheel alignment is in doubt and a check is to be made it will be necessary to satisfy the following initial requirements :

(a) Tyre pressures are correct for all tyres.

(b) The amount of wear on both front tyres must be the same.

(c) The front wheels are true and in balance.

(d) The checking floor must be level.

(e) The car is in the static laden condition.

5. TO ADJUST FRONT WHEEL ALIGNMENT

(a) With the car satisfying the initial requirements, set the front road wheels in the straight ahead position and push the car forward a short distance.

(b) Check the alignment of the wheels with a Dunlop Optical Gauge or similar instrument.

(c) If only a fractional correction is necessary it can be made on the outer tie-rod on the opposite side to the steering box.

(d) To carry out this adjustment it is first necessary to loosen the two lock nuts and turn the tube to shorten or lengthen the tie-rod assembly. Lock the tube by the two nuts and move the car forward half a revolution of a wheel and check, and make a further adjustment if necessary.

(e) If an appreciable amount of maladjustment has to be corrected, check first the length of the outer tie-rods. Should these lengths be equal make the necessary correction to both. When they are found to be of unequal length first correct the rod nearest the steering box to 7.68″ and then make any adjustment to the further one. After making such adjustments it is a wise precaution to measure the length and if found to differ greatly from 7.68″ the front suspension should be checked for accidental damage.

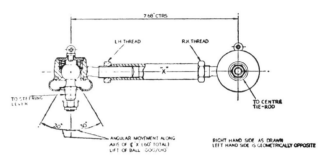

Fig. 3   **Outer Tie Rod Assembly.**

6. STEERING LOCK STOPS

The steering lock stop consists of an eccentric roller mounted on each bottom trunnion by means of a setscrew and lock washer.

It is most important that the steering lock stops come into action before the conical peg of the rocker shaft follower reaches the end of its cam path. This movement is not more than 33° either side of the mid point of the cam and will allow the steering wheel to travel approximately $2\frac{1}{4}$ turns from lock to lock.

The correct adjustment of the lock stops should allow a "Back lock" of 31° and a "Front lock" of $28\frac{1}{2}°$.

When checking this adjustment it is necessary to satisfy the following initial requirements.

(a) The tyre pressures must be correct for all four tyres.

(b) The testing ground must be flat.

(c) Car must be in the static laden condition.

7. TO SET STEERING LOCK STOPS

(a) Select a space of level ground and run the car gently forward so that the front wheels run on to the Churchill Turning measure and the back wheels on to blocks as high as the Churchill gauge (Fig. 4).

This will ensure that the car maintains its level.

(b) Measure the wheel movement from the straight ahead position.

(c) Adjust the eccentric roller by first loosening the setscrew and then turn the roller itself.

(d) When the correct degree of adjustment is attained, tighten down the setscrew so that the roller will remain in contact with the vertical link.

Fig. 4 Showing Use of Wheel Turning Measure for setting Steering Lock—V.L. Churchill Turning Measure Tool No. 121U.

NOTE: If it is impossible to obtain the correct lock positions by adjustment of the steering lock stop, this condition will indicate either a damaged steering drop arm, steering lever, or in rare cases, a fault in the steering unit. Where such difficulties do arise steps must be taken to diagnose the cause and necessary replacements fitted.

8. TO REMOVE FRONT HUB AND STUB AXLE

(a) Jack up the front of the car, remove nave plate and road wheel.

(b) Remove grease retaining cap and grease nipple from end of hub. Grease nipples were discontinued after Commission No. TS.5348.

(c) Withdraw split pin and remove castellated nut and washer from end of stub axle.

(d) Remove hub, utilising Churchill Hub Removing Tool No. M.86 or S.132.

(e) The outer hub bearing can be removed when the hub is released from the hub remover.

(f) Remove the four nuts, spring washers and bolts securing the hub grease catcher to the brake backing plate.

(g) Remove the inner wheel bearing from the stub axle, followed by the grease seal.

(h) The stub axle can be removed from the vertical link if so desired by the removal of the split pin, castellated nut and plain washer from the inner side of the vertical link.

9. TO REPLACE FRONT HUB AND STUB AXLE

(a) Fit the stub axle to the vertical link and secure with the plain washer, castellated nut locked by a split pin on the inner side of the vertical link.

(b) Seat the grease seal on its spigot of the vertical link with the felt pad towards the centre of the car, followed by the inner wheel bearing.

(c) Place the hub grease catcher in position in such a manner that the shaped end of the pressing is below the vent hole in the brake backing plate. Secure grease catcher to backing plate with four screws, spring washers and nuts.

(d) Fit the hub and outer bearing followed by the "D" aperture washer and attach castellated nut.

Adjustment of the Front Hubs

These front wheel bearings should **not** be pre-loaded.

(e) The castellated nut should be tightened to a torque loading of 10lbs. ft. and then slackened off 1½ to 2 flats according to the position of the split pin hole.

The hub bearings are now considered to be correctly adjusted and the castellated nut can be locked with the split pin.

(**f**) Fit the grease retaining cap and grease nipple to hub, and grease hub.

(**g**) Replace road wheel and nave plate. Remove lifting jack from under front of car.

10. TO REMOVE FRONT SHOCK ABSORBER

(**a**) Jack up the car, place supporting stands under the chassis frame and remove lifting jack. Remove road wheel.

(**b**) Partially compress the front road spring by placing a small lifting jack under the spring pan.

(**c**) Remove the lock nut and nut from upper end of shock absorber, followed by a plain washer and upper rubber mounting.

(**d**) Detach the rebound rubber and its bracket from the side of the chassis frame after removing the nuts, lock washers and two long bolts.

(**e**) Remove the lifting jack from below the spring pan.

(**f**) Remove the four nuts and lock washers from the underneath and centre of the spring pan. After withdrawing the rebound rubber abutment plate the shock absorber can be withdrawn through the spring plate.

(**g**) After removing the shock absorber from the car, its lower attachment brackets can be removed. Lift the tabs of the locking plate and remove the setscrew followed by one bracket and a rubber bush.

(**h**) The second bracket is removed from the shock absorber together with the rubber bush, the latter can be withdrawn from the fulcrum pin of the bracket assembly.

11. TO FIT SHOCK ABSORBER

(**a**) Examine all rubber bushes to ascertain that they are in good order. Also ensure that the fulcrum pin is securely welded to the shock absorber attachment bracket.

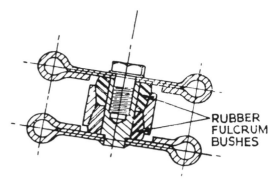

SECTION SHOWING RUBBER BUSHES, LOWER END OF DAMPERS.

Fig. 5   Section showing Rubber Bushes at lower end of Dampers.

(**b**) Press a rubber bush on to the fulcrum pin attachment bracket and feed this assembly, bush first, into the eye of the shock absorber. Press a second rubber bush on to the protruding fulcrum pin.

(**c**) Position second attachment bracket with the tab washer and secure with the setscrew. Turn over tab of washer.

(**d**) Place a large plain washer in position on the upper end of the shock absorber followed by a rubber mounting (spigot uppermost) with the metal sleeve in its centre.

(**e**) Feed the shock absorber assembly through the spring pan in such a manner that the two attachment brackets locate on the studs of the spring pan assembly and at the same time the upper attachment will pass through the spring abutment on the chassis frame. It may be necessary to compress the road spring by placing a jack under the lower wishbone assembly.

(**f**) Attach the second rubber mounting (spigot downwards) to the upper end of the damper which is protruding through the chassis frame, threading it on to the metal sleeve and followed by the plain washer and securing nut.

(**g**) Tighten this nut sufficiently to nip the plain washers and metal sleeve and lock with a second nut.

(**h**) Place the rebound rubber abutment plate in position on the lower attachment studs (welded to the spring pan) with the apex of the wedge pointing towards the centre of the car. Secure with nuts and lock washers.

(**i**) Utilising two long bolts, nuts and lock washers secure the rebound rubber and its bracket to the chassis frame.

(**j**) Remove the lifting jack from under the lower wishbones and replace the road wheel.

(**k**) Jack up front of car to remove support stands, finally remove jack.

Fig. 6 Front Road Spring being removed, utilising the Churchill Tool No. M.50.

12. TO REMOVE FRONT ROAD SPRING

(**a**) Remove front shock absorber as described on page 8.

(**b**) Withdraw the split pins from the castellated nuts on the underside of the lower wishbones. Remove the centre nut and bolt from the front wishbone arm and the bump rubber assembly from the rear wishbone arm. Feed two guide pins into the vacant holes.

(**c**) Place a small lifting jack under the spring pan, with a suitable packing between jack and pan to prevent damage to the shock absorber attachment studs on the latter.

(**d**) Remove the four remaining nuts securing the spring pan to the wishbone arms and lower jack, easing the guide pins through the wishbone arms.

(**e**) The spring can be withdrawn from its upper abutment together with rubber washers and distance piece.

An alternative method is to utilise the Churchill Tool, No. M50 in the following manner :—

(**a**) Carry out operation (**a**) and (**b**) as previously described.

(**b**) Remove the fly nut, bearing and plate from the threaded rod of the Churchill Tool followed by the "C" washer.

(**c**) Feed the rod, notched end first, through the spring pan and upper shock absorber abutment, to the protruding end fit the "C" washer.

(**d**) Feed the plate on to the threaded portion of the rod protruding from the spring pan in such a manner that the bearing seat is downwards, ensure too that the holes in the block locate on the studs of the spring pan.

(**e**) Feed bearing on to threaded rod followed by the fly nut, tighten to compress spring a small amount.

(**f**) Remove the four remaining nuts securing the spring pan to the wishbone arms.

(**g**) By slowly unscrewing the fly nut the spring pan can be lowered down the guide pins.

(**h**) When all tension is released from the road spring the guide pins and the "C" washer can be removed from the upper end of the shaft.

(**i**) Withdraw the Churchill Tool from the suspension unit together with the spring pan, spring, rubber washers and distance piece.

13. TO FIT ROAD SPRING

(**a**) Attach the rod of the Churchill Tool No. M50 to the spring abutment bracket of the front suspension unit and fit the guide pins through the centre holes of the lower wishbone arms.

9

(b) Assemble the alloy distance piece (spigot downward) on the road spring with a rubber washer interposed between, and position a second rubber washer on the spring's lower extremity.

(c) The spring and distance piece assembly is offered up to the front suspension unit followed by the spring pan, the latter located on the guide pins.

(d) Fit the plate to the threaded rod of the Churchill Tool in such a manner that the bearing will seat in its recess and the studs of the spring pan in their recesses. Follow with the bearing and fly nut.

(e) The fly nut of the tool is turned to compress the spring. Ensure that, when the spring pan closes to the wishbone arms that it is located on the attachment studs at the inner ends of the wishbone. Secure and lock washers and castellated nuts and fit two bolts with castellated nuts and lock washers at the trunnion end of the wishbone arm.

(f) When the spring pan is secured to the wishbone arms the Churchill Tool can be removed and the guide pins withdrawn from the wishbone arm.

(g) The spring pan is finally secured to the wishbone arms by a nut, bolt and lock washer at the front arm and a bump rubber assembly at the rear arm.

Lock all six nuts with split pins.

(h) The shock absorber can now be fitted as described on page 8.

14. TO REMOVE AND DISMANTLE FRONT SUSPENSION UNIT

Before dismantling the units, suitably mark the components so that they can be returned to their relative positions.

Carry out instructions as detailed for "To Remove Front Hub and Stub Axle," page 7, and "To remove Front Road Spring," page 9, then proceed as follows :—

(a) Drain the hydraulic system and disconnect the flexible hose as described in Brakes, Section "R." Remove the grease catcher by removing four nuts and bolts. Release the tabs of the locking plates and withdraw the lower

two of the four bolts securing the brake backing plate to the vertical link, followed by the upper two bolts. These bolts pass through the vertical link and distance pieces and thence through the steering lever, on the withdrawal of these bolts it will be necessary to hold the steering lever and collect the bushes. Alternately the brake plate can be removed from the vertical link without draining the system. (Fig. 7).

(b) Remove the nyloc nuts from the ends of the lower wishbone fulcrum pin, followed by the nuts, bolts and lock washers securing the fulcrum pin support brackets to the chassis frame. The support brackets can now be removed.

(c) Remove the split pins from the outer ends of the lower shackle pins. Remove the castellated nuts, grease seals and washers from both ends of the shackle pin.

(d) The wishbone arms can now be removed and the thrust washer and grease seal withdrawn from the shackle pin.

Fig. 7 The Front Suspension Unit partially dismantled.

(e) Remove the two bolts, nuts, plain and locking washers, followed by the two setscrews and spring washers, from the upper fulcrum pin.

(f) The front suspension unit can now be lifted away from the car.

(g) Withdraw the split pin from the castellated nut securing the ball joint assembly to the upper wishbone arm. Remove the castellated nut and withdraw the ball joint assembly from the wishbone arms, collecting the distance piece as the ball joint is moved.

(h) Withdraw the split pin and remove the nut and plain washer securing the ball joint assembly to the vertical link and withdraw ball joint.

(i) Withdraw the split pins from the castellated nuts at the outer ends of the upper inner fulcrum pin. Remove the large diameter plain washers and the outer rubber bushes.

(j) The wishbone arms can now be removed and the second rubber bush withdrawn from the fulcrum pin.

(k) Remove the steering stop screw from the lower end of the vertical link and detach the bottom trunnion assembly from the vertical link and collect the oil seal situated between the vertical link and the trunnion assembly.

15. TO ASSEMBLE AND REPLACE FRONT SUSPENSION UNIT

Assembly is made with strict regard to the markings on certain parts to ensure that they are returned to the same relative position.

(i) Fit a rubber bush to each end of the upper fulcrum pin.

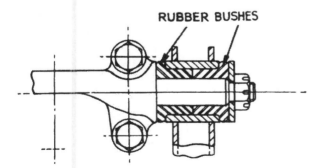

RUBBER BUSHES

SECTION SHOWING RUBBER BUSHES.
INNER FULCRUM-TOP WISHBONE.

Fig. 8 A section showing the rubber bushes of the Upper Wishbone Inner Fulcrum.

(ii) Feed the fulcrum pin into the upper wishbone arm, press the second rubber bush into the wishbone and fit the large plain washer followed by the castellated nut. This nut should be left loose at this juncture.

(iii) While similarly fitting the second wishbone arm ascertain that the other ends of the arm are positioned correctly to receive the ball pin assembly and distance piece. With the ball pin assembly toward the operator the wishbone flange on the right overlaps the one on the left. This applies to both left and right suspension units.

(iv) Feed through the upper attachment of the ball joint assembly with the distance piece between the wishbone arms and secure with the plain washer and castellated nut locked by the split pin. Tighten castellated nuts of inner upper fulcrum pins and lock with split pins.

(v) Fit the ball pin taper into the vertical link with the rubber gaiter in position and secure with the plain washer and castellated nut. Fit split pin in nut.

(vi) Offer up the inner upper fulcrum pin to the chassis frame and secure by bolts with a plain washer under its head and a lock washer with the nut at the points near the centre line of the car. Set-screws and lock washers are used for the attachment points nearer the ball joint assembly.

(vii) Ascertain that the shackle pin of the bottom trunnion assembly is mounted centrally. This pin is a press fit in the body of the casting and is prevented from turning by the imbedding of the splines, it can be centralised by the use of a press or gentle tapping with a copper faced mallet.

(viii) Fit the rubber sealing ring to the lower end of the vertical link followed by the bottom trunnion assembly, which is a screw fit on the vertical link. The trunnion is screwed home and then turned back approximately one turn so that the shackle pin lies parallel to the fore and aft line of the car but between the base of the vertical link and the chassis frame.

(ix) Feed the locking washer and steering lock stop bush on to the steering stop securing bolt and attach to the bottom trunnion assembly. The bolt is left finger tight at this juncture.

(x) Fit two rubber bushes to the inner lower fulcrum pin situated on the upper face of the chassis frame, one to each side.

(xi) Fit two thrust washers to the shackle pin, one to each side, followed by the grease seal.

(xii) The lower wishbone arms are now fitted over the rubber bushes on the inner fulcrum pin and on to the shackle pin simultaneously. Fit a second pair of rubber bushes on to the inner fulcrum pin (and into the lower wishbone arm) followed by the support bracket, the two holes of which are lowermost. Secure with the nyloc nut but do not fully tighten at this juncture.

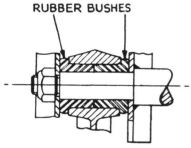

RUBBER BUSHES

SECTION SHOWING RUBBER BUSHES
LOWER WISHBONE—INNER FULCRUM.

Fig. 9 Section showing Rubber Bushes at Lower Wishbone—Inner Fulcrum.

(xiii) Secure the support brackets to the brackets welded to the chassis frame utilising bolts, nuts with lock washer. Tighten the nyloc nuts of the inner lower fulcrum pins until they are solid.

(xiv) Fit to both ends of the shackle pin at the outer end of each wishbone arm, a thrust washer followed by a special lock washer (collar inwards) followed by the rubber grease seals. These lock washers are prevented from rotating by self cutting splines. Feed on the castel-lated nuts to the ends of the shackle pin and obtain the necessary end float before locking with the split pin.

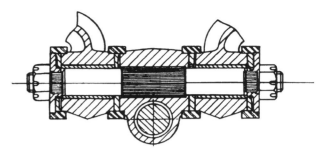

SECTION THRO' OUTER SHACKLE PIN—
LOWER WISHBONE.

Fig. 10 A section through Outer Shackle-pin and lower wishbone bearings. End float in these bearings must be .004″ to .012″.

(xv) It is essential to have .004″ to .012″ end float for the outer boss of each lower wishbone arm. As it is not possible to ascertain the end float by the usual method owing to the presence of the rubber grease seals, the following procedure is suggested.

(a) Equal tightening should be applied to the two castellated nuts and continued until the assembly is solid.

(b) The nuts should then be turned back 1½—2 flats according to the position of the split pin hole and then split pinned.

(c) The wishbone arms should then be lightly tapped outwards to displace the lock washers (now a splined fit to the shackle pin) and this should be carried out alternately on each arm to avoid altering the relationship of the shackle pin and trunnion.

(d) This method will give the recommended end float but as a final precaution the assembly should be checked for freedom of movement over its full range of operation before fitting the road spring.

Apart from damage at this point, tightness will affect the ride of the car.

(xvi) Attach the rod of the Churchill Tool No. M.50 to the spring abutment bracket and the guide rods through the centre of the lower wishbone arms.

(xvii) Assemble the alloy distance piece (spigot downward) on the road spring with a rubber washer interposed, fit a second rubber washer to the lower extremity of the road spring.

(xviii) The spring and distance piece assembly is offered up to the front suspension unit followed by the spring pan, the latter being located on the guide pins. It will be found that the rod of the Churchill Tool No. M.50 protrudes downward from the unit. Fit the plate to this rod in such a manner that the clamp bearing will seat in the recess and the studs of the spring pan fit into their recesses.

(xix) The fly nut of the tool can now be turned to compress the spring. Ensure that, as the spring pan closes to the wishbone, it is located on the attachment studs. Attach the lock washers and castellated nuts to the studs and fit the two bolts, lock washers and castellated nuts adjacent to the bottom trunnion assembly.

(xx) When the spring pan is secured to the wishbone arms the Churchill Tool can be removed.

(xxi) Remove the guide pins from the centre holes and fit the bump rubber assembly to the rear wishbone arm and secure with a lock washer and castellated nut. Fit bolt, lock washer and castellated nut to vacant hole in front wishbone arm. Lock all six nuts with split pins.

(xxii) Fit the shock absorber as described on page 8.

(xxiii) Ensuring that the taper bore of the vertical link and the taper of the stub axle are perfectly clean, feed axle into link and secure with plain washer, castellated nut and lock with a split pin.

(xxiv) Place the brake backing plate in position on the vertical link and secure by the lower bolt holes first, utilising two short setscrews and a locking plate. Through the upper holes of the brake backing plate feed the longer of the two remaining bolts, on to the shank of these bolts protruding inwards through the plate and vertical link feed a distance piece (one to each bolt). Selecting the correct steering lever, it must point forward and downward when fitted, fit this also on the protruding bolts and secure with two nyloc nuts. Finally tighten the lower pair of setscrews and turn up tabs of locking plate.

(xxv) Check that the length of outer tie-rod is correct and then connect the outer tie-rod to the steering arm and secure with the nyloc nut with plain washer.

(xxvi) Connect the flexible hose to the hydraulic line as described in "Brakes Section R".

(xxvii) Fit the hub bearings and hub as described on page 7.

(xxviii) Bleed the hydraulic system if the system has been drained and adjust brakes.

(xxix) Fit road wheels, nave plate and remove jacks.

(xxx) Check front wheel alignment as described on page 6.

(xxxi) Set the steering lock stop (see page 6).

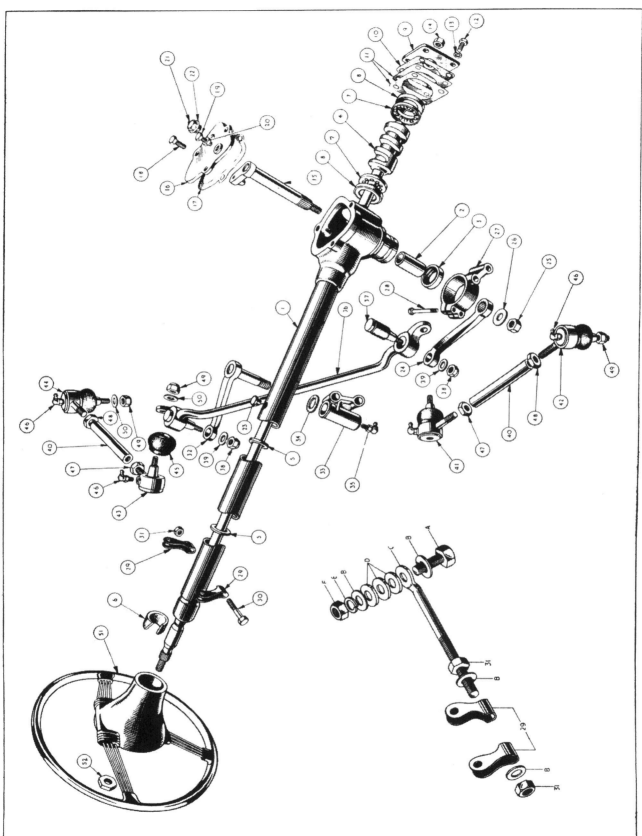

Fig. II   Exploded view of Steering Details.

FRONT SUSPENSION AND STEERING
STEERING

NOTATION FOR FIGURE 11			
Ref. No.	Description	Ref. No.	Description
1.	Outer Tube and Box Assembly.	30.	Bolt.
2.	Rocker Arm Bush.	31.	Nut.
3.	Rocker Arm Oil Seal.	32.	Idler Lever
4.	Inner Column and Cam.	33.	Idler Bracket.
5.	Rubber Ring.	34.	Oil Seal.
6.	Felt Bush.	35.	Grease Nipple.
7.	Inner Column Ball Cage.	36.	Centre Tie-rod.
8.	Ball Cage Race.	37.	Silentbloc Bush and Fulcrum Pin.
9.	End Cover.	38.	Nyloc Nut.
10.	Joint Washer.	39.	Plain Washer.
11.	Adjusting Shims.	40.	Tie-rod.
12.	Bolt.	41.	R.H. Inner End Assembly.
13.	Lock Washer.	42.	R.H. Outer End Assembly.
14.	End Plate Gland Nut.	43.	L.H. Inner End Assembly.
15.	Rocker Shaft Assembly.	44.	L.H. Outer End Assembly.
16.	Top Cover.	45.	Rubber Gaiter.
17.	Joint Washer.	46.	Grease Nipple.
18.	Bolt.	47.	R.H. Threaded Lock Nut.
19.	Rocker Shaft Adjusting Bolt.	48.	L.H. Threaded Lock Nut.
20.	Lock Nut.	49.	Nyloc Nut.
21.	Oil Filler.	50.	Plain Washer.
22.	Washer.	51.	Steering Wheel.
23.	Rubber Plug.	52.	Steering Wheel Nut.
24.	Drop Arm.	A.	Bolt
25.	Nut.	B.	Plain Washer
26.	Lock Washer.	C.	Tie Rod
27.	Trunnion Bracket.	D.	Thick Washers
28.	Bolt.	E.	Lock Washer
29.	Steering Column Clamp.	F.	Nut

A. Bolt
B. Plain Washer
C. Tie Rod
D. Thick Washers
E. Lock Washer
F. Nut
} Fitted in place of 30 after Comm. No. TS. 1390. See page 20.

1. TYPE AND DESCRIPTION

The steering gear is of the cam and lever type with a ratio of 12 to 1. The rocker shaft travel should be limited to 33° either side of the mid point of the cam by the steering lock stops and this will allow the steering wheel to travel approximately 2¼ turns from lock to lock. The cam takes the form of a spiral, whilst the lever carries a conical shaped peg which engages in this cam.

As the conical peg does not reach the bottom of the spiral cam the depth of engagement can be adjusted. This is effected by a hardened steel setscrew mounted on the top cover, the screw when turned clockwise contacts the lever's upper face and holds the conical peg in engagement with the cam.

The steering gear is a self contained and oil tight unit. The cam attached permanently to the inner column which in turn is mounted on caged ball bearings immediately above and below the cam, with a graphite impregnated bearing at its other end.

The lever, to which the conical shaped peg is attached, is an integral part of the rocker shaft assembly and the latter is mounted in a plain bearing, the bore has an oil seal fitted at its lower extremity. The shaft which protrudes through the case is splined to receive the drop arm.

The stator tube which carries the control wires of the electric horn and flashing indicators is held in position by the bottom cover plate, a gland nut and an olive, the latter also provides an oil tight seal.

The unit is attached to the chassis frame by a trunnion bracket at its lower end and braced in the body of the car to the facia panel.

2. MAINTENANCE

An oil filler is provided in the form of a rubber plug, which is located on the steering column at approximately 12″ from the steering box.

A high pressure oil should be used for replenishment. (See Lubrication Chart for Recommended Lubricants.)

The felt bush in the top of the column outer tube is graphite impregnated and should, therefore, require no additional lubrication. If owing to extreme climatic conditions a "squeak" should develop in the bush, extra lubrication should be by colodial graphite. Oil should not be used since it tends to make the bearing "sticky."

An occasional check for tightness should be made to the steering drop arm, the ball joints and also the steering box securing bolts.

Adjustment of the steering box can be affected in two ways, firstly by shims interposed between the steering box and its end cover, and secondly by a setscrew mounted in the top cover.

3. ADJUSTMENT OF STEERING BOX

Means of adjustment to take up wear is provided at two points, both of which are accessible with the steering column in position.

The **FIRST** means of adjustment is made by adding to, or taking from, the shim pack located between the end cover and the steering box. (See Note).

The thickness of the shim pack controls the amount of "float," or pre-load, of the inner column.

While a slight amount of pre-load is permissible, in no circumstances must there be any end float.

The second means of adjustment is by a hardened setscrew and locknut, situated on the top cover plate.

This screw controls the amount of lift in the rocker shaft and is adjusted with the rocker shaft in the centre of the box, that is, the straight ahead position.

The cam gear, which is integral with the inner column, is similar in shape to a spiral cam, having a greater diameter at its centre than at its extremities.

When adjusting the rocker shaft it will be noticed that at the extremities of the arc through which the rocker shaft moves, a certain amount of lift can be felt, and as the shaft moves to the centre, the amount of lift is progressively reduced.

The correct adjustment of the rocker shaft is when on turning the steering wheel from lock to lock, a very slight resistance is felt at the centre of the travel.

The point of resistance should correspond with the straight ahead position of the steering.

NOTE : The adjustment of the rocker shaft should only be made after ensuring that **NO** end float exists in the inner column.

4. TO REMOVE CONTROL HEAD FROM STEERING WHEEL

(a) Disconnect the horn and flasher control wires at the "snap connectors" situated on the wing valance. Suitably identify these wires for subsequent reconnection if the colouring is not distinguishable.

(b) Slacken off the gland nut which secures the stator tube to the end cover of the steering box.

(c) Slacken the three grub screws which are situated radially in the steering wheel hub.

(d) Withdraw the control head and stator tube from the steering column.

(e) The stator tube can now be withdrawn from the control head. These components are a slide fit just below the control head.

5. TO FIT CONTROL HEAD AND STATOR TUBE TO THE STEERING WHEEL

(a) Place the steering wheel in the straight ahead position. This position can be checked by inspecting the alignment of all four wheels.

(b) Feed the stator tube, with the anti-rattle springs in position, into the inner column of the steering unit with the tube slot uppermost and at the 12 o'clock position. Allow approximately 1 inch of tube to protrude from the end cover of the steering box.

(c) Fit the brass olive to the protruding stator tube and secure with the gland nut. Loosen nut back one turn, this is retightened in a later operation.

(d) Feed the wires from the short tube of the control head into and through the stator tube now in the steering unit. With the flasher control lever of the head at 12 o'clock ensure that the vertical lever of the stator tube plate is at the 6 o'clock position. Failure to observe this point will mean that the flashing indicators will not cancel correctly.

(e) Secure the control head in the boss of the steering wheel by tightening the three grub screws situated radially in the steering wheel hub. Do not move the steering wheel during this operation.

(f) Tighten the gland nut to secure the stator tube to the steering box end cover and reconnect wires according to the colours or identification marks.

6. TO REMOVE STEERING WHEEL

(a) First remove the stator tube and control head as described on page 16.

(b) Remove the steering wheel securing nut. If it is so desired the wheel and the top of the inner column can be "centre popped" for identification and simplified replacement.

Fig. 12 Removing the Steering Wheel, utilising the Churchill Tool No. 20SM.3600.

(c) Utilising the Churchill steering wheel remover Tool No. 20SM.3600 remove the wheel (Fig. 12).

7. TO FIT STEERING WHEEL

(a) Place the car on level ground and set the wheels in the straight ahead position.

(b) Feed the steering wheel on to the inner column of the steering unit in such a manner that the two horizontal spokes lie across the fore and aft axis of the car. If on dismantling the column and wheel previously the components have been "pop marked" it is merely necessary to align the "pops."

(c) Fit the securing nut and tighten down.

(d) Fit stator tube and control head. (See page 16.)

8. TO REMOVE STEERING UNIT

(a) Disconnect battery lead and jack up front of car. Place stands securely under frame and remove jacks.

(b) Remove front bumper and front apron as described in "Body Section N."

(c) Remove the road wheel nearest to the steering column.

(d) Using a suitable lever remove the centre tie-rod from the drop arm of the steering unit.

(e) Remove the control head from the centre of the steering wheel as described on page 16.

(f) Remove the steering wheel as described on this page.

(g) Loosen the clamp securing the column to the facia panel by slackening off the two nuts on the lower support stay (this is a nut and bolt on early production cars) (Fig. 11) and the two nuts securing the clamps to the anchor bracket. (See page 20.)

(h) Remove the clip from the rubber draught excluder.

(i) Withdraw the two bolts securing the steering unit trunnion bracket to the chassis frame.

(j) The steering unit may be drawn forward and downward through the draught excluder.

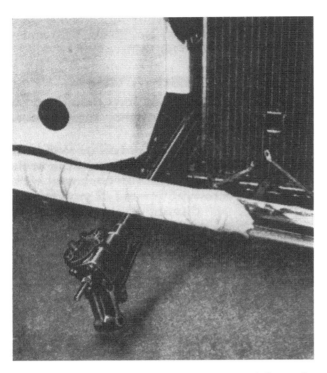

Fig. 13 **The Steering Unit being removed from the front of the car. For the purpose of this illustration the bumper has not been removed. Note the wrapping on the bumper bar to prevent wing damage.**

(k) After the removal of the steering unit the drop arm can be detached from the rocker shaft, utilising a suitable puller (Churchill Tool No. M.91) when the securing nut and lock plate have been first removed.

(l) Slacken off the two pinch bolts securing the trunnion bracket and withdraw it from the steering unit.

9. TO FIT STEERING UNIT

(a) Adjust the end float of the inner column and the rocker shaft for depth of engagement (see page 16).

(b) Fit the trunnion bracket so that the chassis mounting points are forward. Do not fully tighten these two bolts at this juncture.

(c) Attach the drop arm to the splined end of the rocker shaft in such a manner that the scribe lines on these components align and appear to be continuous. Position lock plate and tighten securing nut, lock this nut with the plate by turning its edge over the "flat" machined on the drop arm and another part of the lock plate over the nut.

(d) Place screw clip on draught excluder and feed the column of the steering unit upwards from the front of the car, through the draught excluder and clip and under the facia panel. Position the trunnion bracket in the chassis bracket and attach with two bolts and lock washers, the longer bolt also accommodates the stiffening bracket for the bumper and is fitted to the lowermost hole, the shorter of the two bolts utilises the upper hole. Leave both bolts loose at this juncture.

(e) Secure the column to the mounting bracket under the facia panel by tightening the two nuts on the lower support stay (this was a nut and bolt on early production cars) and the nuts securing the clamps to the anchor bracket. (See page 20).

(f) Tighten the two bolts securing the trunnion bracket to the chassis frame and finally the two bolts of the trunnion bracket to the steering unit.

(g) Fit the centre tie-rod to the drop arm and secure with the nyloc nut and plain washer.

(h) Tighten the draught excluder clip.

(i) Fit the steering wheel as described on page 17.

(j) Fit the control head and stator tube (see page 16).

(k) Fill steering box with high pressure oil recommended in "General Data Section A."

(l) Fit front apron and front bumper as described in "Body Section N."

(m) Replace road wheel, jack up car to remove stands and lower car to ground. Reconnect battery.

10. TO DISMANTLE STEERING UNIT

(a) Remove nut and lock plate and utilising a suitable puller (Churchill Tool No. M.91) remove the drop arm. On no account must the drop arm be removed by hammer blows as this may seriously damage the conical pin on the rocker shaft and also the cam of the centre column.

(b) Slacken off the two pinch bolts attaching the trunnion bracket to the body of the rocker shaft housing, and remove bracket.

(c) Remove cover and joint washer after withdrawing the setscrews of the steering box cover. Allow the oil to drain away.

(d) Withdraw the rocker shaft whilst protecting the rocker shaft oil seal with a thin cylinder of shim steel.

(e) Remove the setscrews and lock washers securing the end cover to the steering box, followed by the shims and joint washer.

(f) The lower bearing race and ball cage can now be removed allowing the cam to be withdrawn, together with the upper ball cage and rubber rings attached to the inner column.

(g) The split felt bush situated in the top of the outer case can now be withdrawn.

(h) The upper bearing race can be drifted out from the steering box.

(i) Drift out the bearing bush and oil seal of the rocker shaft.

11. TO ASSEMBLE STEERING UNIT

(a) Feed the rocker shaft bearing bush into the outer column and box assembly and press into position.

(b) Slide the trunnion bracket into position on the rocker shaft housing. The chassis mounting points should point forward and downward. The two bolts should be tightened just sufficiently to keep the bracket in position at this juncture.

(c) Fit the upper ball race to the steering box. Feed the inner column with the rubber rings and ball cage in position into the box.

(d) Place the second ball cage in position on the lower bearing face of the cam followed by the race.

(e) Locate a fresh joint washer together with the old shim pack on the end cover and fit to the steering box, utilising four bolts and lock washers.

(f) Check for end float. See "Adjustment of Steering Box," page 16. All float **must** be eliminated but a small amount of pre-loading is permitted. End float is adjusted by the removal or addition of shims interposed between the steer-

ing box and the end cover. Their removal decreases the end float whilst the addition of these shims increases the end float.

(g) Press the oil seal into the lower extremities of the rocker shaft body.

(h) Feed the rocker shaft into its bore through the top of the steering box and allow the conical pin to settle in the groove of the cam.

Whilst this shaft is being fitted it is essential that the oil seal lip is protected from damage, otherwise oil leaks will result.

(i) Withdraw the adjusting screw in the top cover to ensure that its shank does not bear down on to the rocker shaft lever when the cover is secured to the unit. Secure cover with three setscrews and lock washers, utilising a new joint washer.

(j) Ensure that the mounting bracket is in position as described in operation (b), for this cannot be fitted when the drop arm is attached to the rocker shaft. Position the drop arm on the splined rocker shaft so that the scribe lines align; secure with nut and lock plate, the edge of the latter is turned up to secure nut and drop arm.

(k) Having removed **all** end float as described in operation (f) adjust the depth of engagement of the rocker shaft and the cam by means of the screw mounted in the top cover. The screw is turned clockwise to increase the depth of engagement or anti-clockwise to reduce the depth. The engagement is said to be correct when slight resistance is felt when the rocker shaft is in the straight ahead position.

(l) Fit the graphite impregnated bush to the upper end of the outer column. The steering wheel securing nut is loosely attached to the inner column for safe keeping.

12. REMOVAL AND REPLACEMENT OF DROP ARM

It should be noted **that it is not possible to remove the drop arm of the steering unit without first removing the unit from the car.** This sequence is covered under "To remove Steering Unit," page 17.

The drop arm must only be removed by a special puller, Tool No. M.91 is recommended, a hammer must not be used since any blow would be transferred to the hardened conical pin in the rocker shaft lever which would in turn indent the cam gear and damage the unit.

The drop arm should only be replaced when the trunnion bracket is in position on the rocker shaft housing. The arm is set in such a manner that it will point rearwards and downwards and the scribe line on the end of the rocker shaft will align with that on the drop arm and appear to be continuous.

Should there be an absence of scribe lines on these components the rocker shaft must be set in the straight ahead position and the drop arm fitted so that it is offset 3° to the left of a line passing through the centre of the rocker shaft parallel to the centre line of the column (see Fig. 14).

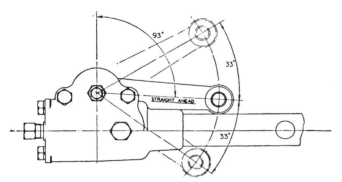

Fig. 14 Diagrammatic view of angular position of the Drop Arm.

13. TO REMOVE IDLER UNIT

(a) Jack up the car and place stands securely under the chassis frame, remove the jacks and remove the road wheel nearest to the idler unit.

(b) Remove nyloc nut and plain washer and utilising a suitable lever disconnect the centre rod from the idler lever.

(c) Remove the two bolts from the chassis frame brackets, lift out idler unit.

(d) The idler unit can be further dismantled by unscrewing the lever and fulcrum assembly from its bracket

body. The oil seal can now be removed from the base of the fulcrum pin.

14. TO FIT IDLER UNIT

(a) Ensure that the lever and fulcrum pin have full movement, this is allowed by screwing the pin into its housing and unscrewing one full turn ; ensure also that the grease seal is in good condition and that the unit is fully greased.

(b) Offer up the unit to its bracket welded to the chassis frame and secure with two bolts and lock washers.

(c) Attach centre tie-rod to the idler lever and secure with nyloc nut and plain washer.

(d) Fit road wheel, jack up car, remove stands and lower car to ground.

15. STEERING COLUMN BRACING

To provide greater steering column stability, the nut and bolt fixing for the column attachment clamps at the facia panel were replaced by a tie-rod. This tie-rod is attached at its inner end to the facia-battery box stay and grips the column clamps at its outer end by two nuts and plain washers. Cars with Commision No. TS. 1390 onwards are fitted with this tie-rod.

The rod is attached to the facia stay by a $1\frac{1}{8}$" long bolt. The bolt with a thin plain washer under its head is fed through the eye of the tie-rod with the off set uppermost, three thick plain washers are now fitted to the bolt. This assembly is offered up to the underside of the facia stay and held in position by a nut with a plain and lock washer.

An additional support bracket, clamped to the steering column by two nuts and bolts and to the front suspension unit by a third nut and bolt, was introduced at Commission No. TS. 5777. This bracket is situated between the front suspension unit and the steering box. To remove the column it will be necessary to loosen the two clamping bolts and re-tightening them on replacement of the column.

16. TELESCOPIC (ADJUSTABLE) STEERING UNIT

(a) Description (Fig. 15)

This unit is very similar to the normal equipment apart from three main features :—

(b) Steering Unit

(i) The inner column is of similar length, but its steering wheel attachment splines are of a much greater length.

(ii) The outer column is shorter than the normal equipment to allow the increased length of the inner column splines to be utilised.

(iii) The distance of the steering wheel from the driver can be increased by 2½" inches.

(c) Steering Wheel

The steering wheel is the three equi-distance spoke type and is a slide fit on the splines of the inner column, it is held at its maximum point of extension by a circlip fitted in an annular recess machined at the top of the splines. (See Fig. 16.)

The lower length of splines, between the underside of the steering wheel and the top of the outer column is covered by a telescopic metal shroud.

This metal shroud is supported at its smaller (bottom) end by a spigotted bakelite washer and positioned at its upper end under the steering wheel locking sleeve by a plated steel cup washer.

The steering wheel hub consists of a steel internally splined sleeve as its centre, with a cast aluminium surround. The lower end extruding portion of the steel insert is split, threaded and is provided with an externally tapered flange to accommodate aluminium steel lined locking sleeve.

An internal taper, corresponding to that on the lower extension of the steering wheel hub, is machined at the bottom of the locking sleeve bore. When the locking sleeve is screwed to the hub insert, a chuck action is developed, thus locking the steering wheel to the external splines on the inner column.

Fig. 15 The Telescopic (adjustable) Steering Unit.

A   Spigotted Bakelite Washer.
B   Metal Telescopic Shroud.
C   Plated Steel Cup Washer.
D   Locking Sleeve
E   Telescopic Steering Wheel
F   Flasher Control
G   Control Head

The length of these splines permit the range of adjustment, and the circlip mounted in its annular groove limits the upwards movement. The telescopic metal shroud covers and protects the splines at all points of adjustment.

(d) The Control Head

The control head mounted in the steering wheel centre is similar to the normal equipment with the exception of the stator tube. This consists of a short tube with indents at its lower end to form a key, and a longer tube with a slot at its upper end. The two tubes telescope together, the indents engaging with the slot provided.

The purpose of this key and slot is two fold, firstly to prevent rotation with the steering wheel and secondly to provide telescopic action as the steering wheel is adjusted on its splines.

2J

17. TO FIT THE TELESCOPIC (ADJUSTABLE) STEERING UNIT AND STEERING WHEEL

(a) With the exception of the steering wheel the fitting of this unit does not differ from that of the normal equipment. Follow the sequence given in "To Fit Steering Unit" (see page 18, operations (a) to (h)).

(b) Ensuring that the car is on level ground and the road wheels are aligned in the straight ahead position, thread the bakelite washer, spigot uppermost, over the splines of the inner column and locate it on the top of the outer column. Slightly grease the splines.

(c) Fit the telescopic metal shroud on to the steering column placing the smaller diameter downwards to engage the spigot of the bakelite washer. The large diameter end of the metal shroud fits into the metal cupped washer, the plane side of which abuts against the locking sleeve.

(d) With the three spokes of the steering wheel forming a "Y" and ensuring that the locking sleeve is loosened, position the wheel on the splines of the inner column so that the lowermost spoke is pointing vertically downwards.

(e) Push the wheel down to its fullest extent and tighten locking sleeve. This will uncover an annular groove in the upper end of the inner column. The circlip can now be fitted (Fig. 16).

(f) Fit the control head as described in "To fit Control Head to Telescopic Steering Wheel" (page 23).

(g) The work can be completed as described in "To fit Steering Unit" (page 18, operations (k) and (l) inclusive).

18. TO REMOVE TELESCOPIC (ADJUSTABLE) STEERING WHEEL AND STEERING UNIT

(a) Proceed as described under "To remove Steering Unit" (page 17, operations (a) to (d) inclusive).

(b) Remove the control head and stator tube as described under "To remove Control Head" (on this page).

Fig. 16 **The circlip in position on the inner column of the steering unit.**

(c) Loosen the clamping nut of the steering wheel hub and lower the wheel to its fullest extent. The hub and inner column may be "pop marked" for simplified replacement.

(d) Remove the circlip from its annular groove situated at the top of the inner column.

(e) Loosen the hub clamp to allow the steering wheel to be drawn from its column and at the same time hold the metal shroud assembly.

(f) Remove the cupped washer from the top of the metal shroud, followed by the shroud and bakelite washer from the top of the outer column.

(g) Proceed with operation (g) and onwards as detailed in "To remove Steering Unit" (page 17).

19. TO REMOVE CONTROL HEAD FROM CENTRE OF TELESCOPIC STEERING WHEEL

The sequence for removal is similar to that of the normal equipment other than the stator tube need not be released by loosening the gland nut and olive at the end cover of the steering box.

20. TO FIT CONTROL HEAD AND STATOR TUBE TO TELESCOPIC STEERING WHEEL

The procedure is the same as fitting the normal equipment but it may be considered necessary to apply a smear of grease to the upper (slotted) end of the stator tube to ensure freedom of movement. It must be pointed out that over greasing at this point may lead to corrosion of the rubber insulation of the electrical harness and cause short circuiting.

The electrical harness protruding from the stator tube must be free to allow a portion to be drawn into the tube when the steering wheel is adjusted to a higher position.

21. STEERING STIFFNESS

If after greasing all points of the steering, stiffness persists, the following procedure is recommended.

(a) Jack up the front of the car and turn the steering wheel from lock to lock. A very slight resistance should be felt when the steering is almost in the straight ahead position. If this stiffness is appreciable and extends to a distance either side of the straight ahead position, the rocker shaft adjusting screw situated in the steering box top cover is bearing too heavily on the lever head of the shaft. The screw should be unlocked and slackened off by a fraction of a turn and then relocked. Should this fail to improve the condition further investigation must be carried out.

(b) Loosen off completely the nuts of the steering column tie situated under the facia panel, followed by the two nuts securing the clamps to the anchor bracket. If the column moves more than ¼″ from its clamped position, reposition by slackening the bolts securing the steering box to its mounting bracket and the mounting bracket to the chassis frame.

(c) Move the steering unit to its correct position. Secure the mounting bracket to the chassis and the steering unit to the mounting bracket.

(d) The clamp attachments to the anchor bracket should be made finger tight and the two clamps brought together round the steering column in such a manner that the column is not displaced. Tighten the jam nuts up to the clamps and finally tighten the nuts of the clamp to anchor bracket attachment.

(e) If stiffness still persists remove the centre tie-rod from the drop arm by removing the nyloc nut and plain washer and so isolate the steering unit from the suspension unit. Check the inner column for pre-load by loosening the four bolts attaching the end cover from the steering box. Should the movement of the steering wheel become easier shims must be placed between box and end cover.

Remove the control head and steering (described on pages 16 or 20) followed by the felt bearing situated at the top of the column. Check the inner column relative to the outer column, if column appears to be displaced, it can be assumed that the inner column is bent and must be replaced.

(f) If the stiffness is traced to the ball joint assemblies, isolate the joint by removing the outer tie-rods from the steering levers. The offending ball joint can now be located and corrected.

(g) Should no stiffness be traced, the car must be jacked up and the upper and lower bearings of the vertical link examined.

ASSESSMENT OF ACCIDENTAL DAMAGE

The following illustrations are necessary for the assessment of accidental damage.

It is suggested that the suspect components are removed from the car as described in this Section, cleaned and laid on a surface plate for measuring.

The measurements taken should be compared with those shown in the appropriate illustration and a decision made as to its condition.

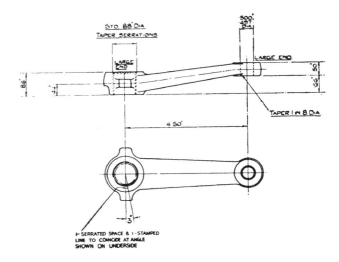

Fig. 18 The Steering Drop Arm R.H.S. L.H.S. is symmetrical but opposite handed.

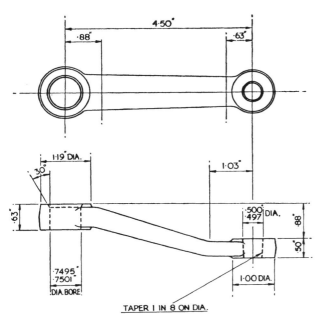

Fig. 17 The Idler Lever R.H.S. and L.H.S. are identical.

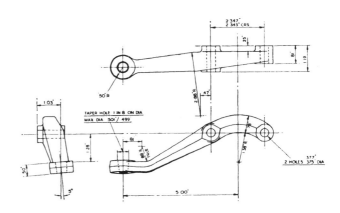

Fig. 19 The Steering Lever R.H. L.H. is symmetrical but opposite handed.

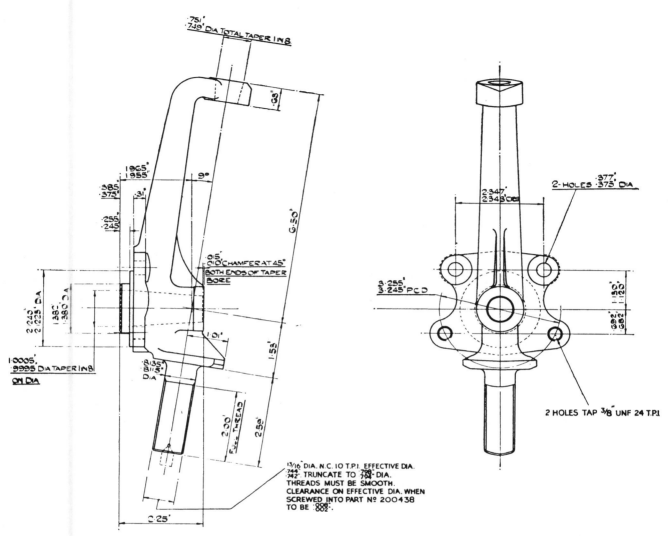

Fig. 20

The Vertical Link.

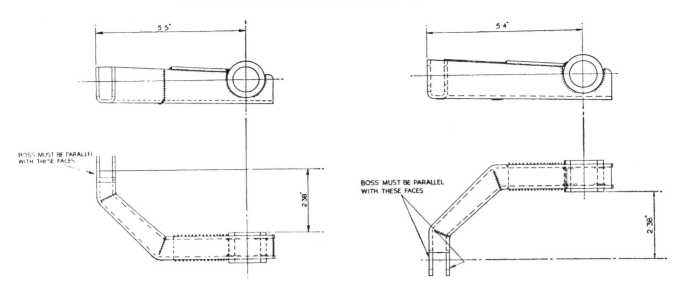

Fig. 21 R.H. front and L.H. rear upper wishbone.   Fig. 22 L.H. front and R.H. rear upper wishbone.

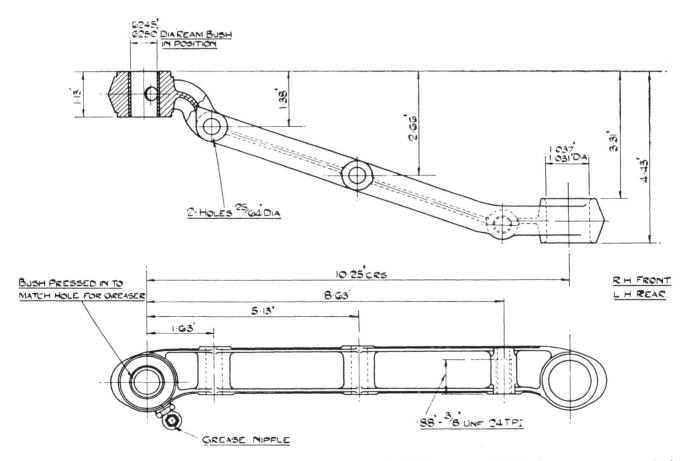

Fig. 23   The R.H. front and L.H. rear Lower Wishbone. The R.H. rear and L.H. front are symmetrical but opposite handed.

Service Instruction Manual

ROAD SPRINGS

AND

SHOCK ABSORBERS

SECTION H

ROAD SPRINGS AND SHOCK ABSORBERS

INDEX

ILLUSTRATIONS

ROAD SPRINGS AND SHOCK ABSORBERS

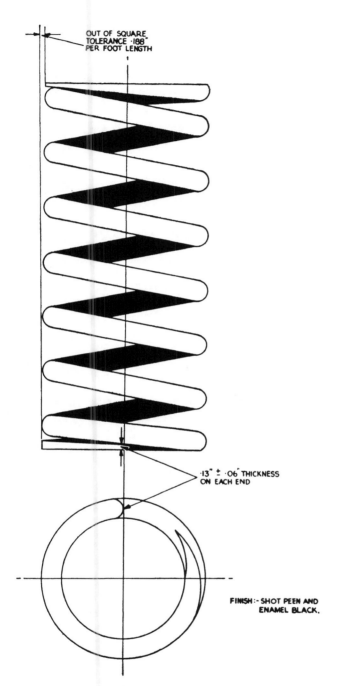

OUT OF SQUARE
TOLERANCE ·188"
PER FOOT LENGTH

·13" ± ·06" THICKNESS
ON EACH END

FINISH:- SHOT PEEN AND
ENAMEL BLACK.

SPRING DATA				
		COMPETITION		
ITEM	DIMENSION	ITEM	DIMENSION	
WIRE DIA.	·50" ± ·002"	WIRE DIA.	·52" ± ·002"	
NUMBER OF WORKING COILS	6¾	NUMBER OF WORKING COILS	6½	
MEAN DIA. OF COILS	3·5" ± ·020"/·010"	MEAN DIA. OF COILS	3·5" ± ·020"/·010"	
RATE *	310 LB/IN.	RATE *	380 LB/IN.	
FREE LENGTH	9·75" (APPROX)	FREE LENGTH	9·19" (APPROX)	
FITTED LENGTH	6·75" ± ³/₃₂"	FITTED LENGTH	6·75" ± ³/₃₂"	
SOLID LENGTH	4·13" 4·26" MAX	STATIC DEFLECTION	2·44"	
STRESS AT SOLID	151,000 LB/□"	FITTED LOAD *	925 LB	
HAND OF HELIX	RIGHT	SOLID LENGTH	4·16" 4·29 MAX	
FITTED LOAD *	925 LB.	STRESS AT SOLID	148,000 LB/□"	
STATIC DEFLECTION	3·0" APPROX.	HAND OF HELIX	LEFT	
MATERIAL	SILICO MANGANESE GROUND BAR	MATERIAL	SILICO MANGANESE GROUND BAR	

GENERAL DATA			
ITEM	DIMENSION	ITEM	DIMENSION
MAX. WHEEL BUMP	3·0"	MAX. WHEEL BUMP	3·0"
MAX. WHEEL REBOUND	2·25"	MAX. WHEEL REBOUND	2·25"
SPRUNG LOAD AT WHEEL	540 LB.	SPRUNG LOAD AT WHEEL	540 LB.
LEVERAGE	·585"	LEVERAGE	·585"
STATIC WHEEL DEFLECTION	5·1"	STATIC WHEEL DEFLECTION	4·15"
WHEEL RATE	106 LB/IN.	WHEEL RATE	130 LB/IN
PERIODICITY	83 OSC/MIN.	PERIODICITY	92 OSC/MIN
WAHL FACTOR	1·213	WAHL FACTOR	1·223
STATIC STRESS	80,000 LB/□"	STATIC STRESS	71,600 LB/□"
STRESS AT MAX. BUMP	140,800 LB/□"	STRESS AT MAX. BUMP	138,000 LB/□"
LADEN HEIGHT AT WHEEL	± ⅛"	LADEN HEIGHT AT WHEEL	± ⅛"
CLEARANCE FROM CHOC-A-BLOC AT MAX. WHEEL BUMP	·37"	CLEARANCE FROM CHOC-A-BLOC AT MAX WHEEL BUMP	·34"

* IMPORTANT

Fig. 1   Front Road Spring.   For illustration purposes only the Competition Spring is shown. This spring has a left-hand helix.

1

FRONT SPRING

1. DESCRIPTION

The low periodicity coil spring used in the front suspension of this car is illustrated in Fig. 1. This illustration also gives the data of both the normal road spring and the competition spring. This competition spring can easily be distinguished from the normal type, for it has a left-hand helix. Damping action is provided by a direct acting telescopic type shock absorber, mounted centrally through the coil spring.

2. MAINTENANCE

Very little maintenance should be required during the lifetime of the car. There is no lubrication required, and the only possible maintenance would be to replace the rubber washers, or to check the spring against the data given in Fig. 1.

3. TO REMOVE OR REPLACE

These operations are fully covered in the "Front Suspension, Section G" of this manual.

REAR ROAD SPRINGS

1. DESCRIPTION (Fig. 2)

Semi-elliptical laminated springs are used which have their location point with the axle below and forward of the centre, so that the longer end of each spring is fitted toward the rear of the car.

The forward fulcrum of the spring has a silentbloc bush and is mounted on a bolt protruding from the outer side of the chassis frame. The attachment is completed by a "D" washer and split pinned castellated nut. The rear fulcrum is a shackle assembly utilising split rubber bushes interposed between the pins, the spring or the chassis frame. The attachment is completed by nuts and lock washers situated between the spring and the chassis frame.

2. MAINTENANCE

The only lubrication required is that for the spring leaves, on no account must the rubber or silentbloc bushes be lubricated. Over lubrication of the spring leaves should be avoided. After the springs have been cleaned, brush the blades at their edges with engine oil, this will allow sufficient oil to penetrate between the leaves and provide inter-leaf lubrication.

Lubrication of spring blades is chiefly required at the ends of the leaves where one presses upon the next and where the maximum relative motion occurs.

The clips should be inspected and any looseness corrected by pinching the "ears" closer to the spring. Failure to keep these clips tight often causes "knocks" at the rear of the car.

3. TO REMOVE REAR ROAD SPRING

(a) Jack up the body at the rear of the car sufficiently to take the weight off the road spring.

(b) Remove the rear wing stay situated behind the rear wheel between the chassis and wing itself.

(c) Holding the hexagon of the shock absorber-link remove the nyloc attachment nut.

(d) Remove the two nuts and lock washers, followed by the plate of the shackle assembly at the rear end of the spring. Withdraw the plate and pin assembly and collect the rubber bushes from the spring eye and the chassis bracket.

(e) Screw a $\frac{5}{16}'' \times 24$ UNF bolt into the head of the forward fulcrum bolt to a depth of $\frac{1}{2}''$. Withdraw the split pin to remove nut and "D" washer. Utilising a lever under the head of the $\frac{5}{16}''$ UNF bolt, the fulcrum bolt can now be withdrawn from the spring and chassis frame.

(f) Supporting the spring by a small jack remove the four nyloc nuts of the two "U" bolts attaching the spring to the axle, remove the "U" bolts and the spring plate from the shock absorber link.

(g) The road spring and the supporting jack is now removed from under the car to a bench.

(h) The silentbloc bush can now be removed from the forward eye of the spring.

2

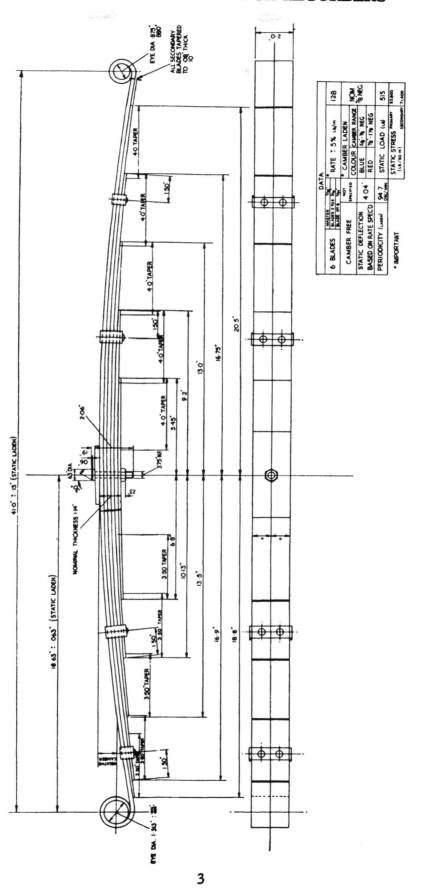

Fig. 2   Rear Road Spring.

3

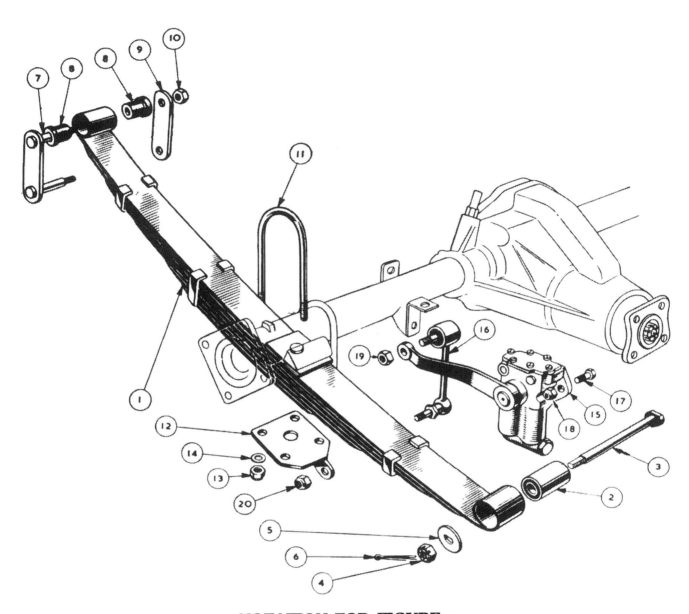

NOTATION FOR FIGURE 3.

Ref. No.	Description	Ref. No.	Description
1.	Rear Road Spring	12.	Right Hand Shock Absorber Plate Assembly.
2.	Silentbloc Bush.		
3.	Front Attachment Bolt.	13.	Nyloc Nut.
4.	Castellated Nut.	14.	Plain Washer.
5.	"D" Washer.	15.	Shock Absorber.
6.	Split Pin.	16.	Shock Absorber Link.
7.	Shackle Pin and Plate Assembly.	17.	Attachment Bolt.
8.	Rubber Bush.	18.	Attachment Nut.
9.	Shackle Plate.	19.	Nut for Link Upper Attachment.
10.	Nut.	20.	Nut for Link Lower Attachment.
11.	"U" Bolt.		

4. TO FIT REAR ROAD SPRING

(a) Press the silentbloc bush into the forward eye of the road spring and ensure that the eight split rubber bushes are in good condition.

(b) Offer up the spring, short end forward, to a position above the rear shackle bracket of the chassis frame and below the axle. Support the spring on a small jack and attach spring plate loosely to the shock absorber link.

(c) Fit the "U" bolts over the axle either side of the spring and through the spring plate, secure with four nyloc nuts.

(d) Secure shock absorber link to spring plate.

(e) Feed the front attachment bolt from inner side of the chassis frame through its support tube into the silentbloc bush of the road spring and allow the machined flat on its head to bed against its abutment on the inner side of the chassis frame. Secure the fulcrum bolt on its outer side by a "D" washer and castellated nut locked by a split pin.

(f) Fit the two rubber half bushes to the road spring rear eye—one from each side. Press a second pair of half bushes into the shackle eye on the chassis frame.

(g) Press the shackle pins of the shackle assembly through the rubber bushes and after positioning the inner shackle plate on the pin extremities, between the shackle assembly and chassis side member, fit and secure the two nuts and lock washers.

(h) Replace the rear wing stay, positioning it behind the rear wheel in the wing valance and chassis bracket provided and securing with bolts, nuts, plain and lock washers.

(i) Remove the jacks from under the body of the car.

5. REAR ROAD SPRING OVERHAUL

The better procedure to adopt when dealing with a road spring which has settled badly or where blades have broken is to fit a replacement.

The only provision the Spares Department make for these springs, other than complete replacements, is the supply of the master blade.

The spring, on being removed from the chassis, should be laid on a surface plate and measured, the measurements taken should be compared with those given in Fig. 2 and a decision made as to its condition.

6. TO DISMANTLE REAR ROAD SPRING

(a) Drift out the silentbloc bush from the forward eye of the master blade.

(b) Gripping the spring in a vice, prise open the clips sufficiently to allow the removal of the leaves.

(c) Remove the centre bolt and dismantle the spring.

(d) Clean and examine the blades for cracks or breakages. Damage is most likely to occur toward the centre hole of each blade.

(e) Examine centre bolt for damage and wear.

7. TO ASSEMBLE REAR ROAD SPRING

(a) Grease the blades with a graphite grease, particularly at the ends where one blade contacts the one above.

(b) Feed the leaves on to the centre bolt and utilising a press or vice compress the assembly sufficiently to attach the nut of the centre bolt.

(c) Tap the clips over with a hammer and an anvil so that they grip the blades firmly. Failure to ensure complete tightness will result in "knocks" when the car is in use.

(d) Press the silentbloc bush into the forward eye of the master blade and ensure that it does not become contaminated with grease.

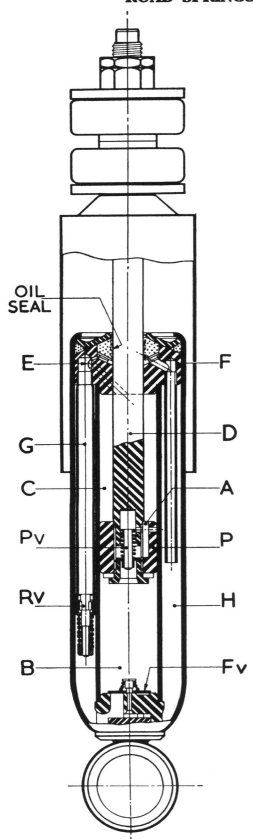

OIL
SEAL

E

F

G

D

C

A

Pv

P

Rv

H

B

Fv

Fig. 4   The Front Shock Absorber.

FRONT SHOCK ABSORBER

1. DESCRIPTION

A telescopic type shock absorber is fitted, utilising a stem fixing at the top with rubber bushes, large diameter steel washers and lock nuts. At the lower end it is first attached to a fulcrum pin bracket with rubber bushes interposed between shock absorber eye and fulcrum pin, the bracket assembly is secured to the lower side of the spring pan. The body of the shock absorber is in the centre of the coil spring.

2. MAINTENANCE

The shock absorber is a sealed unit and requires no topping up. If it is found to be unserviceable it must be replaced.

The only maintenance that can be required is the renewal of the rubber mountings. This is detailed in the "Front Suspension, Section G" under "To remove front shock absorber."

NOTATION FOR FIGURE 4

A. Port in Piston.
B. Portion of Cylinder below Piston.
C. Portion of Cylinder above Piston.
D. Piston Rod.
E. Port in Piston Rod Guide.
F. Piston Rod Guide.
Fv. Foot Valve.
G. Foam Tube.
H. Oil Reservoir.
P. Piston
Pv. Piston Valve.
Rv. Rebound Valve.

---

3. OPERATION OF THE TELESCOPIC SHOCK ABSORBER (Fig. 4)

This shock absorber operates by the one-way circulation of oil. By this method of circulation the oil moves all the time the unit is in operation thus keeping the unit cool under the most arduous conditions of service. The valve gear is simple, of robust construction, and is self cleaning.

On the bump stroke, the oil pressure opens the piston valve (Pv) against the spring load and oil passes through the ports (A) in the piston (P) from the lower to the upper portion of the cylinder (B to C). The excess oil volume equal to the displacement of the piston rod (D) passes through the ports (E) in the piston rod guide (F), down the anti-foam tube (G) and into the reservoir (H) by way of the rebound valve (Rv).

On the rebound stroke, however, the piston valve (Pv) closes and oil passes through the ports (E) in the piston rod guide (F), down the anti-foam tube (G), opens the rebound valve (Rv) against the spring load and passes into the reservoir (H). At the same time the foot valve plate (Fv) lifts and oil is recuperated to the lower part of the cylinder (B). General slow speed damping is accomplished by bleed orifices built into the valve mechanism.

The maximum load of compression (bump) is 200 lbs. and on extension (rebound) 500 lbs.

4. TO REMOVE OR REPLACE FRONT SHOCK ABSORBER

This is detailed in the "Front Suspension, Section G" under this heading.

REAR SHOCK ABSORBER

1. DESCRIPTION (Fig. 5)

The shock absorber body is attached to the brackets welded to the upper sides of the chassis frame and linked to the rear axle by an arm, splined to the shock absorber spindle, and a connecting link to a plate assembly mounted on the underside of the road spring.

The body has two equal sized cylinders accommodating steel pistons which are recipricated through short connecting rods and are coupled to the crank plate which is attached to the spindle.

When the axle moves relative to the car (this movement is allowed by the road spring) the arm is moved up or down, and as it is splined to a spindle, the latter rotates. The spindle is a splined fit in the crank plate, this plate being coupled by means of connecting rods to the pistons, in which are situated lightly loaded recuperating valves. The pressure is built up in one cylinder or the other and since the cylinders are connected by ports in the body to the valve chamber, this pressure is dependent on the valve setting.

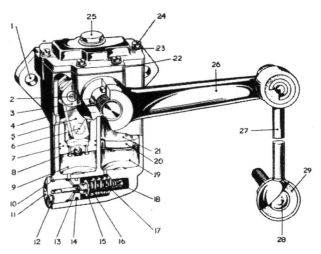

Fig. 5   **The Rear Shock Absorber.**

NOTATION FOR FIGURE 5

Ref. No.	Description
1.	Mounting Holes.
2.	Crank Pin.
3.	Crank Plate.
4.	Oil Seal.
5.	Connecting Rod.
6.	Piston Pin.
7.	Compression or Bump Piston.
8.	Recuperating Valve.
9.	Compression or Bump Cylinder.
10.	Ring Seal.
11.	Valve Screw.
12.	Valve Screw Washer.
13.	Rebound Valve.
14.	Ring Seal.
15.	Compression Valve.
16.	Compression Washer.
17.	Compression Spring.
18.	Rebound Spring.
19.	Rebound Cylinder.
20.	Rebound Piston Seal.
21.	Rebound Piston.
22.	Gasket.
23.	Shake Proof Washer.
24.	Lid Screw.
25.	Filler Plug.
26.	Arm.
27.	Connecting Link.
28.	Ball End Bolt.
29.	Rubber Cushion.

The unit is filled to the base of the filler plug boss which prevents over filling and maintains the necessary air space essential to satisfactory operation. The working mechanism is completely submerged in oil which is prevented from leaking along the spindle by means of oil seals.

2. MAINTENANCE

The damper requires very little attention but the fluid level should be checked every 15,000 miles. It should be topped up to the lower reaches of the filler boss and only with Armstrong Shock Absorber Oil No. 624 should be used, the guarantee of this particular component becomes void if any other oil is used.

Every precaution must be taken to ensure that no lubrication is given to the rubber mountings of the connecting link.

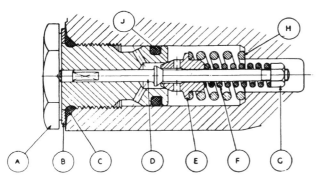

Fig. 6. Sectional view of Rebound and Compression Valve of Rear Shock Absorber.

A Valve Screw
B Valve Screw Washer
C Ring Seal
D Rebound Valve
E Compression Valve
F Rebound Valve Spring
G Rebound Valve Spring Nut
H Compression Spring
J Ring Seal

3. VALVE OPERATION

To accomplish general damping of the car springs, a small bleed is built into the valve. This operates both on compression (axle moving up) and on rebound (axle moving down). As bumps become more severe on compression, pressure builds up in the compression cylinder and blows the compression valve off its seat at a/pre-determined pressure controlled by the outer spring.

As the speed of the rebound increases, pressure is built up in the rebound cylinder and blows the rebound valve off its seat at a pre-determined pressure controlled by the inner spring.

It will be clear that by suitable selection of springs in the valve, any range from zero to a maximum rating of the shock absorber can be obtained in either direction.

4. TO REMOVE REAR SHOCK ABSORBER

(a) Jack up the rear of the car and remove the road wheel nearest to the shock absorber to be removed.

(b) Remove the nyloc nut and plain washer from the connecting rod attachment to the spring plate. It may be necessary to hold the hexagon on the inner side of the spring plate.

(c) Remove the nut and lock washer from the upper joint of the connecting link. Utilising a suitable extractor, remove the link from the shock absorber arm, this is a taper fit. Remove the connecting link from between chassis frame and spring.

(d) Remove the bolts and nyloc nuts securing the body of the shock absorber to its bracket on the chassis frame and withdraw the shock absorber and connecting link.

5. TO FIT REAR SHOCK ABSORBER

(a) Remove the connecting link from the shock absorber arm.

(b) Offer up the shock absorber to its bracket on the chassis frame in such a manner that the body faces outwards and the arm points rearwards. Secure with two bolts and nyloc nuts.

(c) With the spherical knuckle of the connecting link lowermost, offer up the link to the shock absorber arm and spring plate, the link should be positioned between the road spring and chassis frame, with the nuts away from the centre line. Holding the hexagon of the lower attachment bolt secure the link to the spring plate with a nyloc nut and plain washer.

(d) Utilising a nut and lock washer secure the connecting link to the shock absorber arm.

(e) Fit road wheel and remove jacks.

Service Instruction Manual

FRAME UNIT

SECTION J

FRAME UNIT

INDEX

ILLUSTRATIONS

FRAME UNIT

1. DESCRIPTION. Fig. 1.

A rigid structure is provided, the frame side members being formed by opposed steel pressings welded together, giving tubular type side members of rectangular section.

Welded at the front and rear ends of the side members are two tubular cross members. The front tube is 1⅛″ diameter and is supported by the steering unit and idler unit mounting brackets welded to the side members. The rear tube of 1¾″ diameter is welded between the two side members. A second tube, just forward of the rear one, protrudes through the side members and to it the rear road spring shackle brackets are welded.

The centre of the frame is braced by channel sectioned steel pressings forming a rigid cruciform structure and stiffened at its centre by heavy gauge plates. This structure carries the gearbox and handbrake mounting brackets.

At the front end but a little to the rear of the tubular cross member, the frame is braced by opposed steel pressings welded together and forming a rigid box section cross member. This member forms the lower points of attachment for the front suspension and engine mountings. It is built up to form the upper abutments for the front road springs and this upper structure is braced by a detachable tubular cross member and by supports to the two side members.

Welded in position approximately half-way along the inner side of each side member are the jacking brackets. To each cruciform member is welded an outrigger body support bracket, these brackets pass through and are supported by the side members. There are four such brackets.

The complete frame is protected from corrosion by rust proofing.

2. THE ASSESSMENT OF ACCIDENTAL DAMAGE

For this purpose reproduction drawings of the chassis frame giving the necessary dimensions are given as Fig. 1.

Even when a car has suffered only superficial damage it is possible that the frame members have been displaced which will result in the road wheels failing to track correctly and it is recommended that the frame is checked for squareness.

It is possible to check the frame dimensionally to a satisfactory degree of accuracy without first removing the body. For clarity the chassis frame only is shown in the illustrations of this section.

Details of checks for "twist," "cradling," "squareness," and "bowing," are given in this section. By carrying out these checks in the order mentioned a great deal of work is eliminated.

FRAME UNIT

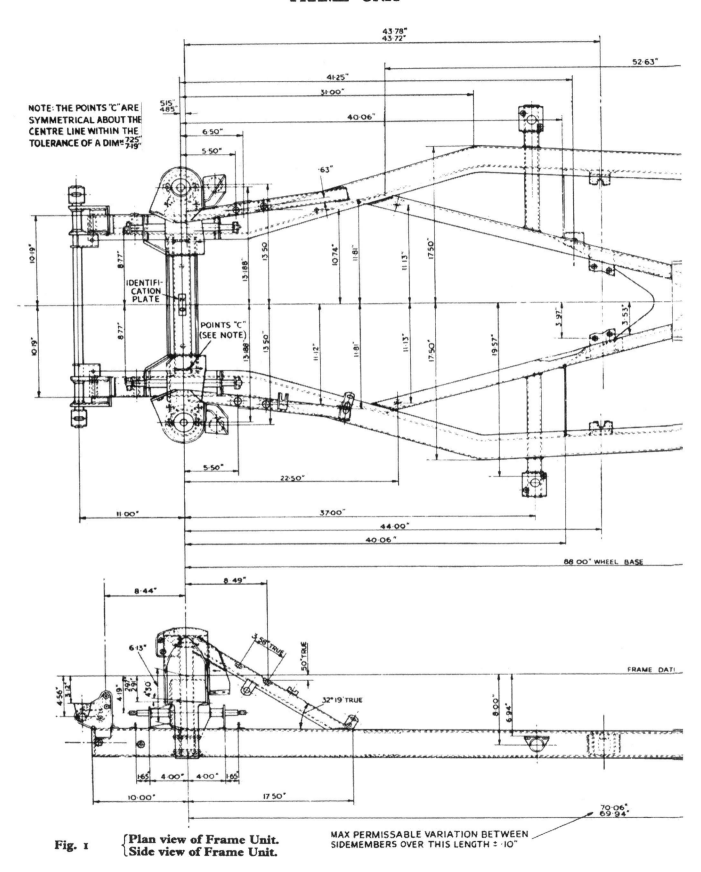

NOTE: THE POINTS "C" ARE SYMMETRICAL ABOUT THE CENTRE LINE WITHIN THE TOLERANCE OF A DIM.

IDENTIFICATION PLATE

POINTS "C" (SEE NOTE)

88.00" WHEEL BASE

FRAME DATUM

MAX PERMISSABLE VARIATION BETWEEN SIDEMEMBERS OVER THIS LENGTH ± .10"

Fig. 1 { Plan view of Frame Unit. Side view of Frame Unit.

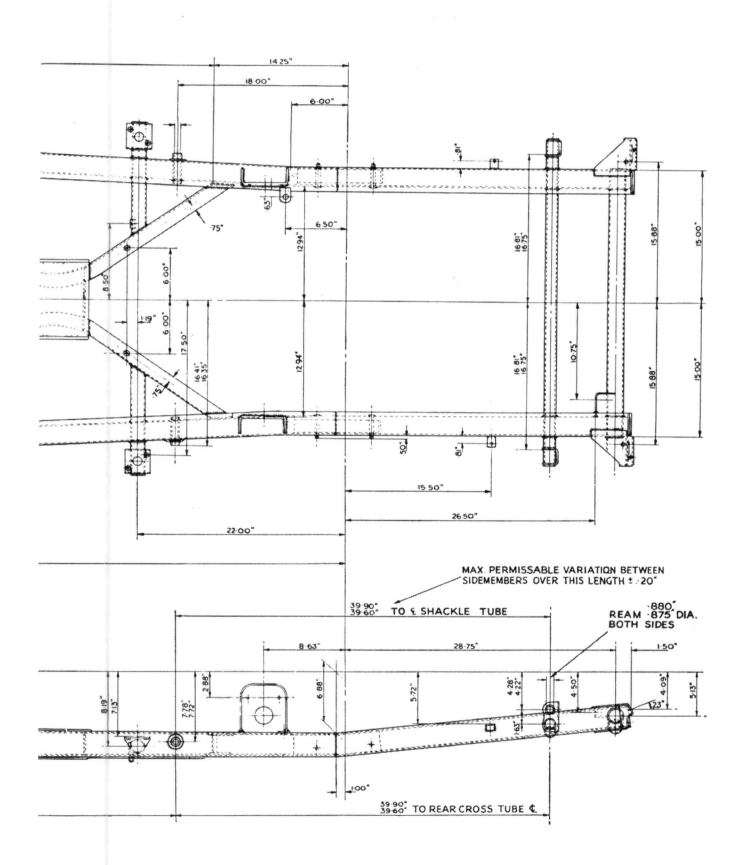

MAX. PERMISSABLE VARIATION BETWEEN
SIDEMEMBERS OVER THIS LENGTH ± ·20"

39·90"
39·60"   TO ₵ SHACKLE TUBE

·880"
REAM ·875" DIA.
BOTH SIDES

39·90"
39·60"   TO REAR CROSS TUBE ₵

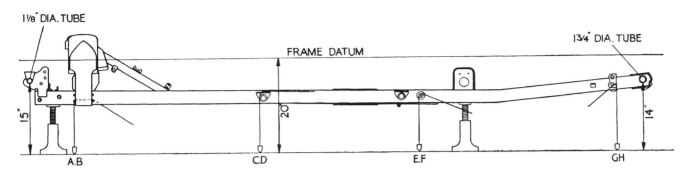

Fig. 2 The car prepared for the assessment of accidental damage, in particular the assessment of "cradling" (for clarification purposes only the chassis frame is shown).

3. **PREPARATION OF CAR (Fig. 2)**

   (a) Select a clean level floor space and jack up the car, utilising four screw jacks. It is suggested that two jacks are placed near the front box section cross member and the second two under the side members at the rear.
   Remove all four road wheels.

   (b) Adjust the rear jacks until the straight portion of the rearmost tubular cross member is 14″ from the ground, measured as close to the side member as is practical.

   (c) Adjust the two front jacks similarly until the foremost tubular cross member is 15″ from the ground, measured as close to the steering and idler mounting brackets as possible.

   (d) Remove the front rebound buffer and bracket from each side of the chassis frame by withdrawing two bolts, nuts and lock washers.

   (e) Through the lower bolt hole pass the plumb bob cord from front to rear. Mark the floor directly under the plumb bob pointer. This operation is repeated on the other side of the chassis frame and so creates points A and B (Fig. 3).

   (f) From inside the car adjacent to the front door posts raise the carpet and remove the most forward body securing bolts from the forward outrigger body supports.

   (g) Pass the plumb bob cord from below through the bolt hole. Mark the floor immediately below the plumb bob pointer. The operation is repeated on the other side and so creates points C and D (Fig. 3).

   (h) Withdraw the split pins to remove the castellated nuts and "D" washers from the rear road spring front fulcrum pins.

   (i) Pass the cord of the plumb bob over the fulcrum pin in such a manner that the bob hangs in front of the pin. Mark the floor immediately below the plumb bob pointer. This operation is repeated on the other side of the chassis and creates the points E and F (Fig. 3).

   (j) Thread the cord of the plumb bob from the rear and through the lower jig hole in the rear road spring shackle bracket. Mark the floor immediately below the plumb bob pointer. This operation, when repeated on the other side of the chassis frame, creates points G and H (Fig. 3).

4

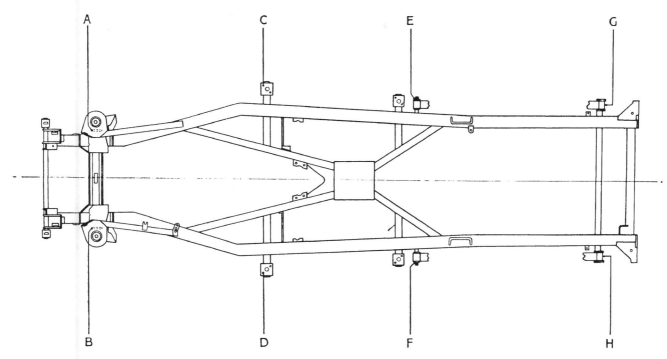

Fig. 3 Illustrating the eight points of the chassis which are generated on the floor below.

4. CHECKING THE SIDE MEMBERS FOR TWIST

If, by adjusting the screw jacks under the chassis frame as described in "Preparation of Car" page 3 operation **a—c**, it is found to be an impossibility to bring the front cross member and the straight portions of the rear cross members parallel to the ground, the frame can be considered to be "twisted."

5. CHECKING SIDE MEMBERS FOR CRADLING

(**a**) Having prepared the car as detailed in "Preparation of Car" page 3 operations **a—c**, it is now standing with the datum line parallel to the ground and this line is 20″ from the ground (Fig. 2).

(**b**) Referring to Fig. 1 it will be observed that all dimensions are given from this datum line and by simple subtraction of these dimensions from 20″ it is pos-

sible to calculate their height above the ground.

As an example, when checking the position of one of the front "out rigged" body supports, the dimension given is 6.94″ from the top of the support to the datum, therefore if we subtract 6.94″ from 20″ the result will be 13.06″ which should be the distance between the top of the support and the floor.

(**c**) Measure the height above the ground at several points and subtract the dimensions obtained from 20″. By comparing the results with the drawing dimensions, it will be possible to determine whether the frame is true.

(**d**) (**i**) When the difference is greater than the drawing, the chassis frame is "bowed" downward.

(**ii**) When the difference is less that the drawing the chassis frame is "bowed" upwards.

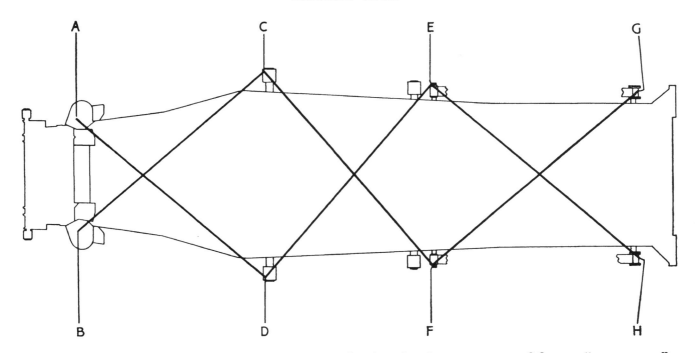

Fig. 4   Utilising the eight generated points for the assessment of frame "squareness."

6. CHECKING SIDE MEMBERS FOR SQUARENESS (Fig. 4)

(a) It is assumed that the car has been prepared and the eight points generated on the floor below. Replace the road wheel and rebound rubber bracket. The car is now moved so that the position of the markings can be examined.

(b) Utilising a suitable measure ascertain the lengths of the diagonals AD, BC, CF, DE, EH, and FG.

(c) If the chassis frame is square the length AD will equal BC, CF will equal DE and EH will equal FG.

(d) (i) When BC, DE, and FG are of greater length than AD, CF and EH respectively the left hand (BH) side member is forward of the right hand (AG) side member.

(ii) When AD, CF and EH are of greater length than BC, DE and FG respectively the right hand (AG) side member is forward of the left hand (BH) side member.

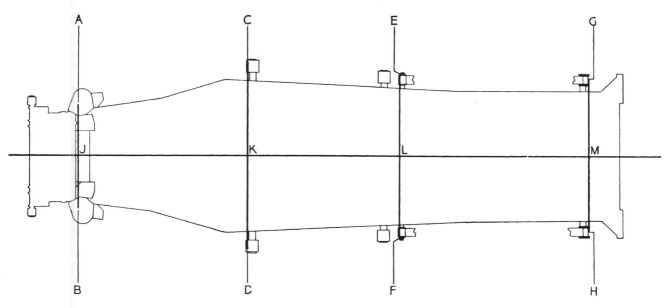

Fig. 5 Utilising the eight generated points for the assessment of "bowing."

7. CHECKING THE SIDE MEMBERS FOR BOWING (Fig. 5)

(a) Having gained access to the points generated on the floor beneath the car, join the points A to B, C to D, E to F and G to H.

(b) Accurately determine the mid-points of the lines AB, CD, EF and GH. Call these points J, K, L and M respectively.

(c) With a suitable straight edge join point J to point M.

(d) (i) If this line passed through points K and L the side members are correctly aligned.

(ii) When the points K and L lay to the right of the line JM the side members are "bowed" to the right.

(iii) When the points K and L lay to to the left of the line JM the side members are "bowed" to the left.

Service Instruction Manual

PROPELLER SHAFT

SECTION K

PROPELLER SHAFT

INDEX

ILLUSTRATIONS

PROPELLER SHAFT

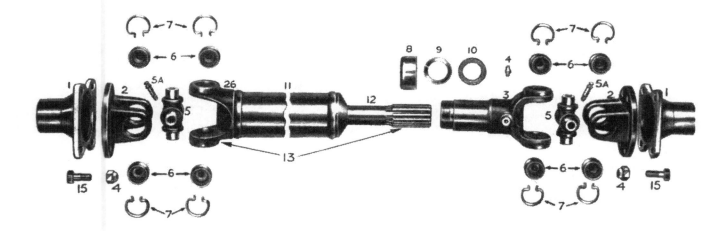

Fig. 1   Propeller   Shaft   Details.

NOTATION FOR FIG. 1			
Ref. No.	Description	Ref. No.	Description
1	Companion Flange.	8	Dust Cap.
2	Flange Yoke.	9	Steel Washer.
3	Sleeve Yoke Assembly.	10	Cork Washer.
4	Nipple for Splines.	11	Tube.
5	Spider Journal Assembly (less Nipple).	12	Splined Stub Shaft.
5A	Nipple for Journal Assembly.	13	Propeller Shaft Assembly.
6	Bearing Race Assembly.	14	Simmonds Nut.
7	Snap Ring.	15	Flange Attachment Bolts.

1. DESCRIPTION

The propeller shaft and universal joints fitted to this model are the Hardy Spicer Series 1300, the tube diameters being 2″, and the overall length of the assemblies being 2′ 4$\frac{9}{16}$″.

Details of these propeller shafts are as shown in exploded form in Fig. 1.

When the rear axle rises and falls, with the flexing of the springs, the arc of the axle's travel necessitates variations in the length of the propeller shaft which is provided for by the fitting of a sliding spline at the front end of the assembly. The splined end of the propeller shaft is shown under Notation 13 in Fig. 1.

A universal joint is supplied at each end, consisting of a central spider having four trunnions, four needle roller bearings and two yokes as can be appreciated by a study of Fig. 1.

2. LUBRICATION

Each spider is provided with an oil nipple and there is one fitted on the sleeve yoke assembly (3) to lubricate the sliding spline. After dismantling and before re-assembly, the inside splines of the sleeve yoke should be liberally smeared with oil. Each of the two journal assemblies are provided with an oil nipple which should be lubricated each 5,000 miles in accordance with the lubrication recommendation made in the summary in " General Data " Section.

If a large amount of oil exudes from the oil seals, the joint should be dismantled and new oil seals fitted.

1

3. MAINTENANCE INSTRUCTIONS
To test for wear

Wear on the thrust faces is located by testing the lift in the joint by hand.

Any circumferential movement of the shaft relative to the flange yokes indicates wear in the needle roller bearings and/or the sliding splines.

4. REMOVAL OF PROPELLER SHAFT

(a) Jack up one rear wheel clear of the ground to enable the propeller shaft to be rotated.

(b) Remove nuts from bolts at both flange yokes engaging first gear, as necessary to hold the shaft from turning when slackening nuts.

(c) Tap out bolts and remove propeller shaft assembly.

5. TO DISMANTLE PROPELLER SHAFT

Before commencing to dismantle propeller shaft see if " arrow " location marks are visible when the parts are clean. If no markings are visible, re-mark to ensure correct re-assembly.

Having unscrewed the dust cap (8, Fig. 1), pull sleeve yoke assembly (3, Fig. 1) off shaft. Clean enamel from snap rings and top of bearings races. Remove all snap rings by pinching ears together with a suitable pair of circlip pliers and subsequently prising out these with a screwdriver. If ring does not snap out of groove readily, tap end of bearing race lightly inwards to relieve the pressure against ring. Holding joint in left hand with splined

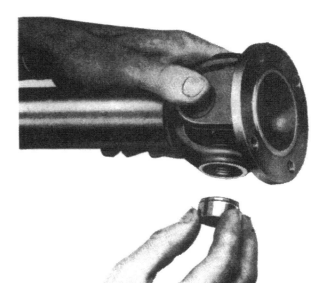

Fig. 3   Removing Bearing.

sleeve yoke lug on top, tap yoke arms lightly with a soft hammer as shown in

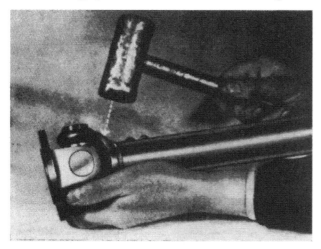

Fig. 2   Tapping Tube Yoke to release Bearing.

Fig. 4   Removing Bearing Race with Special Punch.

Fig. 2. Top bearing should begin to emerge, turn joint over and finally remove with fingers as shown in Fig. 3.

If necessary tap bearing race from inside with small diameter bar, as shown in Fig. 4, taking care not to damage the bearing race. This operation will destroy the oil seal and necessitate fitting replacement parts when re-assembling, keep joint in this position whilst removing bearing race, so as to avoid dropping the needle rollers.

Repeat the operation described in previous paragraph for opposite bearing. The splined sleeve yoke can now be removed as shown in Fig. 5.

Rest the two exposed trunnions on wood or lead blocks, then tap flange yoke with soft hammer to remove the two remaining bearing races.

Fig. 5   Removing the Yoke.

6. TO EXAMINE AND CHECK FOR WEAR

The parts most likely to show signs of wear after long usage are the bearing races and spider trunnions. Should looseness in the fit of these parts, load markings, or distortion be observed, they must be renewed complete, as no oversize journal bearing races are provided. It is essential that bearing races are a light drive fit in the yoke. In the rare event of wear having taken place in the yoke cross hole, the holes will most certainly be oval, and such yokes must be replaced.

In the case of wear of the cross holes in a fixed yoke, which is part of the tubular shaft assembly, only in cases of absolute emergency should this be replaced by welding in a new yoke. The normal pro-cedure is to replace by a complete shaft assembly. The other parts likely to show signs of wear are the splines of the sleeve yoke, or splined stub shaft. A total of .004" circumferential movement, measured on the outside diameter of the spline, should not be exceeded.

In the event of the splined stub shaft requiring renewal this must be dealt with in the same way as the fixed yoke, i.e., a replacement tubular shaft assembly fitted.

7. TO ASSEMBLE

See that the trunnion assemblies are well lubricated with one of the oils recommended. Assemble needle rollers in bearing recess, smearing the walls of the races with vaseline, or lubricant, to retain the rollers in place.

It is advisable to replace cork gaskets and gasket retainers (oil seals) on the trunnions using a tubular drift as shown in Fig. 6. The spider journal shoulders should be shellacked prior to fitting retainers to ensure a good oil seal. Ensure that the trunnions are clean and free from shellac before fitting needle rollers.

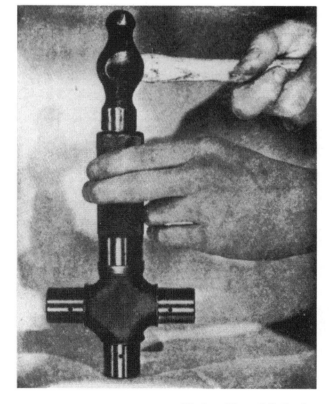

Fig. 6   Fitting New Oil Seals.

3

Insert spider in flange yoke. Then using a soft-nosed drift about $\frac{1}{32}''$ smaller in diameter than the hole in the yoke, tap the bearing into position. It is essential that bearing races are a light drive fit in the yoke holes. Repeat this operation for the other three bearings.

Refit snap rings with a suitable pair of circlip pliers, ensuring that rings engage properly with their respective grooves. If joint appears to bind after assembly, tap lightly with a soft hammer, thus relieving any pressure of the bearings on the ends of the trunnions.

WHEN REPLACING SLIDING JOINT ON SHAFT BE SURE THAT SLIDING AND FIXED YOKES ARE IN THE SAME PLANE AND ARROW MARKINGS COINCIDE. A single universal joint does not transmit uniform motion when the driving and driven shafts are out of line, but when two joints are used as in the case of a propeller shaft, and are set in correct relation the one to the other, the errors of one are corrected by the discrepancies of the other, and uniform motion is then transmitted. Hence the importance of re-engaging the splines correctly when they have been taken apart.

8. **TO FIT PROPELLER SHAFT**
Wipe companion flange and flange yoke faces clean, to ensure the pilot flange registering properly and joint faces bedding evenly all round. Insert bolts, and see that all nuts are evenly tightened all round and are securely locked. Dust cap to be screwed up by hand as far as possible. Sliding joint is always placed towards front of vehicle.

Service Instruction Manual

WHEELS AND TYRES

SECTION L

WHEELS AND TYRES

INDEX

ILLUSTRATIONS

WHEELS AND TYRES

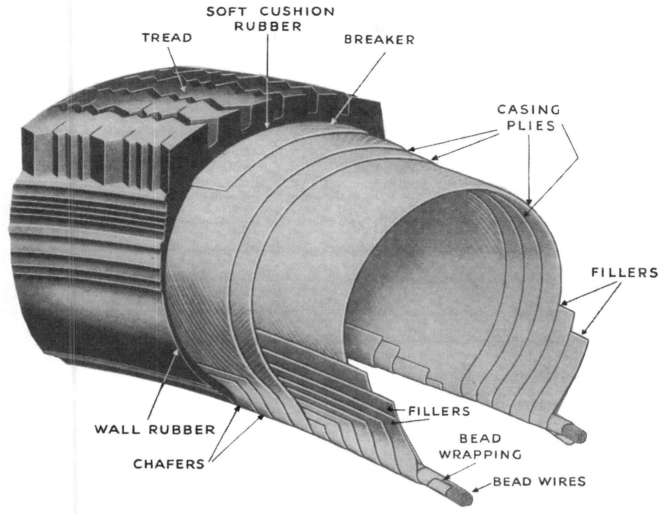

SOFT CUSHION
RUBBER

TREAD

BREAKER

CASING
PLIES

FILLERS

FILLERS

WALL RUBBER

CHAFERS

BEAD
WRAPPING

BEAD WIRES

Fig. 1   Tyre Construction.

Construction of Tyre

One of the principal functions of the tyres fitted to a car is to eliminate high frequency vibrations. They do this by virtue of the fact that the unsprung mass of each tyre—the part of the tyre in contact with the ground—is very small.

Tyres must be flexible and responsive. They must also be strong and tough to contain the air pressure, resist damage, give long mileage, transmit driving and braking forces, and at the same time provide road grip, stability, and good steering properties.

Strength and resistance to wear are achieved by building the casing from several plies of cord fabric, secured at the rim position by wire bead cores, and adding a tough rubber tread (Fig. 1).

Part of the work done in deflecting the tyres on a moving car is converted into heat within the tyres. Rubber and fabric are poor conductors and internal heat is not easily dissipated. Excessive temperature weakens the tyre structure and reduces the resistance of the tread to abrasion by the road surface.

Heat generation, comfort, stability, power consumption, rate of tread wear, steering properties and other factors affecting the performance of the tyres and car are associated with the degree of tyre deflection. All tyres are designed to run at predetermined deflections, depending upon their size and purpose.

Load and pressure schedules are published by all tyre makers and are based on the correct relationship between tyre deflection, tyre size, load carried and inflation pressure. By following

1

the recommendations, the owner will obtain the best results both from the tyres and the car.

Tyre Pressures:

Correct tyre pressures for 5.50"—15" are : Front 22 lbs.   Rear 24 lbs.

Note.—Pressures should be checked when the tyres are cold, such as after standing overnight, and not when they have attained normal running temperatures.

Pressures shown are for normal motoring when sustained high speeds are not possible.

Special Pressures for High Speed Motoring

(a) For touring at sustained speeds in excess of 85/90 m.p.h., pressure in front and rear tyres should be increased by 6 lb. per sq. in.

(b) For predominantly and regularly high speed touring of continental type, pressures in front and rear tyres should be increased by 8 lbs. per sq. in.

Tyres lose pressure, even when in sound condition, due to a chemical diffusion of the compressed air through the tube walls. The rate of loss in a sound car tyre is usually between 1 lb. and 3 lbs. per week, which may average 10% of the total initial pressure.

For this reason, and with the additional purpose of detecting slow punctures, *pressures should be checked with a tyre gauge applied to the valve not less often than once per week.*

Any unusual pressure loss should be investigated. After making sure that the valve is not leaking the tube should be removed for a water test.

Do not over-inflate, and do not reduce pressures which have increased owing to increased temperature. (See " Factors Affecting Tyre Life and Performance," page 3).

(a) **Valve Cores and Caps**

Valve cores are inexpensive and it is a wise precaution to renew them periodically

Valve caps should always be fitted, and renewed when the rubber seatings have become damaged after constant use.

(b) **Tyre Examination**

Tyres on cars submitted for servicing should be examined for :—

Inflation pressures.
Degree and regularity of tread wear.
Misalignment.
Cuts and penetrations.
Small objects embedded in the treads, such as flints and nails.
Impact bruises.
Kerb damage on walls and shoulders.
Oil and grease.
Contact with the car.

Oil and grease should be removed by using petrol sparingly. Paraffin is not sufficiently volatile and is not recommended.

If oil or grease on the tyres results from over-lubrication or defective oil seals suitable correction should be made.

(c) **Repair of Injuries**

Minor injuries confined to the tread rubber, such as from small pieces of glass or road dressing material, require no attention other than the removal of the objects. Cold filling compound or " stopping " is unnecessary in such cases.

More severe tread cuts and wall rubber damage, particularly if they penetrate to the outer ply of the fabric casing, require vulcanised repairs. The Dunlop Spot Vulcanising Unit is designed for this purpose and it is also suitable for all types of tube repairs.

Injuries which extend into or through the casing, except clean nail holes, seriously weaken the tyre. Satisfactory repair necessitates new fabric being built in and vulcanised. This requires expensive plant and should be undertaken by a tyre repair specialist or by the tyre maker.

Loose gaiters and " stick-in " fabric repair patches are not satisfactory substitutes for vulcanised repairs and should be used only as a temporary " get-you-home " measure if the tyre has any appreciable tread remaining. They can often be used successfully in tyres which are nearly worn out and which are not worth the cost of vulcanised repairs.

Clean nail holes do not necessitate cover repairs. If a nail has penetrated the cover the hole should be sealed by a tube patch attached to the inside of the casing. This will protect the tube from possible chafing at that point.

If nail holes are not clean, and particularly if frayed or fractured cords are visible inside the tyre, expert advice should be sought.

I. FACTORS AFFECTING TYRE LIFE AND PERFORMANCE

(a) Inflation Pressures

Other things being equal there is an average loss of 13% tread mileage for every 10% reduction in inflation pressure below the recommended figure. The tyre is designed so that there is minimum pattern shuffle on the road surface and a suitable distribution of load over the tyre's contact area when deflection is correct.

Moderate under-inflation causes an increased rate of tread wear although the tyre's appearance may remain normal. Severe and persistent under-inflation produces unmistakable evidence on the tread (Fig. 2). It also causes structural failure due to excessive friction and temperature within the casing (Figs. 3 and 4). Pressures which are higher than those recommended for the car reduce comfort. They may also reduce tread life due to a concentration of the load and

Fig. 3   Breaking up of Casing due to overflexing and heat generation.

Fig. 4   Running deflated destroyed this Tyre.

wear on a smaller area of tread, aggravated by increased wheel bounce on uneven road surfaces. Excessive pressures overstrain the casing cords, in addition to causing rapid wear, and the tyres are more susceptible to impact fractures and cuts.

(b) Effect of Temperature

Air expands with heating and tyre pressures increase as the tyres warm up. Pressures increase more in hot weather than in cold weather and as the result of high speed. These factors

Fig. 2   Excessive Tyre Wear due to persistent under-inflation.

are taken into account when designing the tyre and in preparing Load and Pressure Schedules.

Pressures in warm tyres should not be reduced to standard pressures for cold tyres. " Bleeding " the tyres increases their deflections and causes their temperatures to climb still higher. The tyres will also be under-inflated when they have cooled.

(c) Speed

High speed is expensive and the rate of tread wear may be twice as fast at 50 m.p.h. as at 30 m.p.h.

High speed involves :—

(i) Increased temperatures due to more deflections per minute and a faster rate of deflection and recovery. The resistance of the tread to abrasion decreases with increase in temperature.

(ii) Fierce acceleration and braking.

(iii) More tyre distortion and slip when negotiating bends and corners.

(iv) More " thrash " and " scuffing " from road surface irregularities.

(d) Braking

" Driving on the brakes " increases rate of tyre wear, apart from being generally undesirable. It is not necessary for wheels to be locked for an abnormal amount of tread rubber to be worn away.

Other braking factors not directly connected with the method of driving can affect tyre wear. Correct balance and lining clearances, and freedom from binding, are very important. Braking may vary between one wheel position and another due to oil or foreign matter on the shoes even when the brake mechanism is free and correctly balanced.

Brakes should be relined and drums reconditioned in complete sets. Tyre wear may be affected if shoes are relined with non-standard material having unsuitable characteristics or dimensions, especially if the linings differ between one wheel position and another in such a way as to upset the brake balance. Front tyres, and particularly near front tyres, are very

sensitive to any condition which adds to the severity of front braking in relation to the rear.

" Picking up " of shoe lining leading edges can cause grab and reduce tyre life. Local " pulling up " or flats on the tread pattern can often be traced to brake drum eccentricity. (Fig. 5.) The braking varies during each wheel revolution as the minor and major

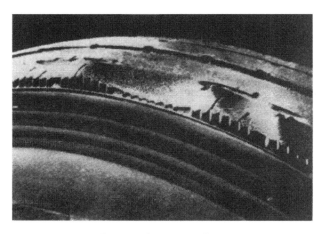

Fig. 5 Local excessive wear due to Brake Drum Eccentricity.

axes of the eccentric drum pass alternately over the shoes. Drums should be free from excessive scoring and be true when mounted on their hubs with the road wheels attached.

(e) Climatic Conditions

The rate of tread wear during a reasonably dry and warm summer can be twice as great as during an average winter.

Water is a rubber lubricant and tread abrasion is much less on wet roads than on dry roads. Also the resistance of the tread to abrasion decreases with increase in temperature. Increased abrasion on dry roads, plus increased temperatures of tyres and roads cause faster tyre wear during summer periods. For the same reasons tyre wear is faster during dry years with comparatively little rainfall than during wet years.

When a tyre is new its thickness and pattern depth are at their greatest. It follows that heat generation and pattern distortion due to flexing, cornering, driving and braking are greater than when the tyre is part worn. Higher tread mileages will usually be obtained if new tyres are fitted in the autumn or winter rather than in the spring or summer. This practice also tends to reduce the risk of road delays because tyres are more easily cut and penetrated when they are wet than when they are dry. It is therefore advantageous to have maximum tread thickness during wet seasons of the year.

(f) Road Surface

The extent to which road surfaces affect tyre mileage is not always realised.

Present day roads generally have better non-skid surfaces than formerly. This factor, combined with improved car performance, has tended to cause faster tyre wear, although developments in tread compounds and patterns have done much to offset the full effects.

Road surfaces vary widely between one part of the country and another, often due to surfacing with local material. In some areas the surface dressing is coarser or of larger "mesh" than in others. The material may be comparatively harmless rounded gravel or more abrasive crushed granite or knife edged flint. Examples of surfaces producing very slow tyre wear are smooth stone setts and wood blocks but their non-skid properties are poor.

Bends and corners are severe on tyres because a car can be steered only by misaligning its wheels relative to the direction of the car. This condition applies to the rear tyres as well as to the front tyres. The resulting tyre slip and distortion increase the rate of wear according to speed, load, road camber and other factors. (Fig. 6.)

The effect of hills, causing increased driving and braking torques with which the tyres must cope, needs no elaboration.

Road camber is a serious factor in tyre wear and the subject is discussed on page 8.

An analysis of tyre performance *must* include road conditions.

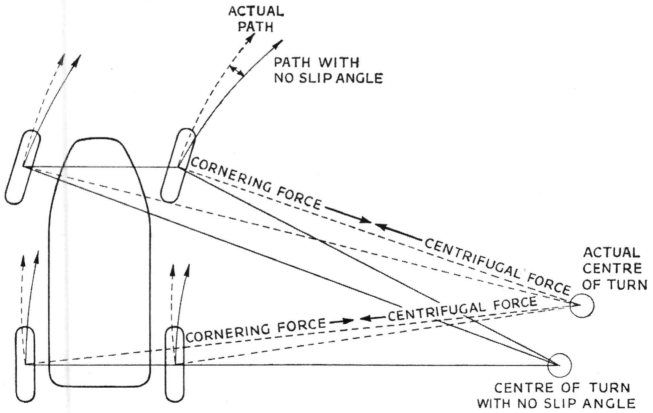

ACTUAL PATH

PATH WITH NO SLIP ANGLE

CORNERING FORCE

CENTRIFUGAL FORCE

ACTUAL CENTRE OF TURN

CORNERING FORCE

CENTRIFUGAL FORCE

CENTRE OF TURN WITH NO SLIP ANGLE

Fig. 6   Diagrammatic Illustration of Slip Angles.

5

(g) Impact Fractures

In order to provide adequate strength, resistance to wear, stability, road grip and other necessary qualities, a tyre has a certain thickness and stiffness. Excessive and sudden local distortion such as might result from striking a kerb, a large stone or brick, an up-standing manhole cover, or a deep pothole may fracture the casing cords. (Figs. 7 and 8.)

Impact fractures often puzzle the car owner because the tyre and road spring may have absorbed the impact without his being aware of anything unusual; only one or two casing cords may be fractured by the blow and the weakened tyre fails some time later; there is usually no clear evidence on the outside of the tyre unless the object has been sufficiently sharp to cut it. This damage is not associated solely with speed and care should be exercised at all times, particularly when drawing up to a kerb or parking against one.

2. SPECIAL TYPES OF IRREGULAR TREAD WEAR

(a) "Heel and toe" or "saw tooth" wear

This is the condition where one end of each pattern segment or stud is more worn than the other (Fig. 9). To some extent it is latent in any non-skid

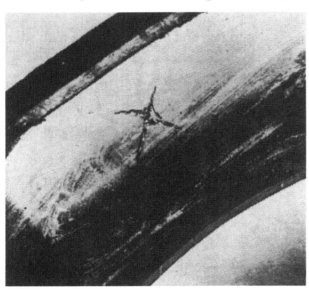

Fig. 7   Severe impact has fractured this Casing.

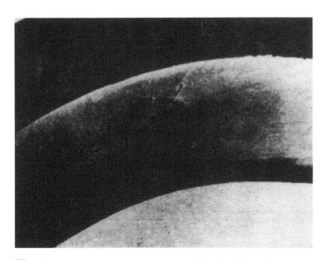

Fig. 8   A double fracture.

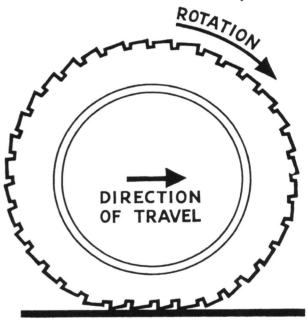

Fig. 9   "Heel and Toe" Wear.

pattern design and severe service conditions may cause it to develop.

When each successive portion of a running tyre comes under load the tread is flattened and there is limited pattern distortion and shuffle on the road surface. Additional movement is caused by braking, driving and the tyre's own rolling resistance, which acts as a constant retarding force.

On rear wheels the effects of braking and rolling resistance are offset by the effects of driving. Rear tyres usually wear evenly if they are properly maintained. Front tyres are at a disadvantage in this respect and their pattern displacement tends to be always in the same direction.

Fig. 10 illustrates the basic cause of

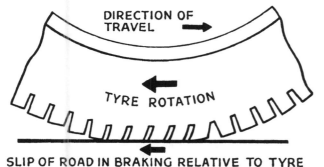

DIRECTION OF TRAVEL

TYRE ROTATION

SLIP OF ROAD IN BRAKING RELATIVE TO TYRE

Fig. 10 Showing the effect of braking and rolling resistance on Tyre Tread.

" heel and toe " wear. If the tyre is assumed to be on a locked wheel and sliding forward, the abrasive road surface may be likened to a file passing across the tread. The manner in which the flexible rubber studs will be worn is clear. There is a similar but less marked effect when the tyre is re-

Fig. 11 " Spotty " Wear due to a variety of causes.

volving but trying to " hang back " under the forces of braking and rolling resistance.

Modern tyre patterns designed for use on hard road surfaces are very stable They do not consist of separate unsupported studs or blocks such as are shown in the diagram. In normal conditions " heel and toe " wear should be absent or barely noticeable but any localised forces such as from eccentric brake drums, fierce or binding brakes, incorrect brake balance and severe front braking will usually cause this type of wear to appear amongst other evidence of these troubles. An unsuitable tyre contact area and distribution of load, resulting from road camber, wheel camber, or excessive deflection, will also produce " heel and toe " wear.

Regular interchanging of tyres will prevent or reduce irregular wear (see page 11).

(b) " Spotty " Wear

Fig. 11 shows a type of irregular wear which sometimes develops on front tyres and particularly on near front tyres. The causes are difficult to diagnose although evidence of camber wear, misalignment, under-inflation, or braking troubles may be present.

Front tyres are at a disadvantage due to their fore and aft slip and distortion being in one direction. Front tyres are connected to the car through swivelling stub axles and jointed steering linkage and they are subjected to complicated movements resulting from steering, spring deflection, braking and camber. Load transference during braking causes increased loading and pattern displacement on front tyres, and adds to the severity of front tyre operation. Unbalance of the rotating assembly may also contribute to a special form of irregular wear with one half of the tyre's circumference more worn than the other half. Unbalance alone does not cause the type of " spotty " wear illustrated but the unbalance usually becomes progressively worse as the irregular or unequal wear develops. The nature of " spotty " wear—the pattern being much worn and little

7

worn at irregular spacings round the circumference—indicates an alternating " slip-grip " phenomenon but it is seldom possible to associate its origin and development with any single cause.

It is preferable to check all points which may be contributory factors. The front tyre and wheel assemblies

Fig. 12   **Fins and Feathers due to severe misalignment.**

may then be interchanged, which will also reverse their direction of rotation, or better still the front tyres may be interchanged with the rear tyres.

Points for checking are :—

(**a**) Inflation pressures and the consistency with which the pressures are maintained.

(**b**) Brake freedom and balance, shoe settings, lining condition, drum condition and truth.

(**c**) Wheel alignment.

(**d**) Camber and similarity of camber of the front wheels.

(**e**) Play in hub bearings, king pin bearings, suspension bearings and steering joints.

(**f**) Wheel concentricity at the tyre bead seats. S.M.M. & T. tolerances provide for a radial throw not exceeding $\frac{3}{32}''$, but this may be affected by impact or other damage.

(**g**) Balance of wheel and tyre assemblies.

(**h**) Condition of road springs and shock absorbers.

Corrections which may follow a check of these points will not always effect a complete cure and it may be necessary to continue to interchange wheel positions and reverse directions of rotation at suitable intervals.

Irregular wear may be inherent in the local road conditions such as from a combination of steep camber, abrasive surfaces, and frequent hills and bends. Driving methods may also be involved. Irregular wear is likely to be more prevalent in summer than in winter, particularly on new or little worn tyres.

3.   **WHEEL ALIGNMENT AND ITS ASSOCIATION WITH ROAD CAMBER**

It is very important that correct wheel alignment should be maintained. Misalignment causes a tyre tread to be scrubbed off laterally because the natural direction of the wheel differs from that of the car.

An upstanding sharp " fin " on the edge of each pattern rib is a sure sign of misalignment and it is possible to determine from the position of the " fins " whether the wheels are toed in or toed out (Fig. 12).

" Fins " on the inside edges of the pattern ribs—nearest to the car—and particularly on the nearside tyre indicate toe in. "Fins" on the outside edges, particularly on the offside tyre, indicate toe out.

With minor misalignment the evidence is less noticeable and sharp pattern edges may be caused by road camber even when wheel alignment is correct. In such cases it is better to make sure by checking with an alignment gauge.

Road camber affects the direction of the car by imposing a side thrust and if left to follow its natural course the car will drift towards the nearside. This is instinctively corrected by steering towards the road centre.

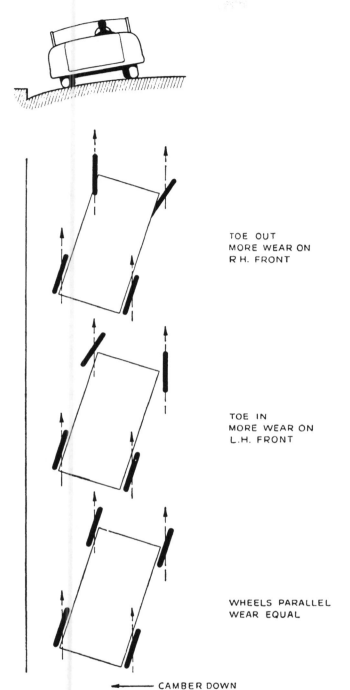

TOE OUT
MORE WEAR ON
R H. FRONT

TOE IN
MORE WEAR ON
L.H. FRONT

WHEELS PARALLEL
WEAR EQUAL

◄—— CAMBER DOWN

Fig. 13   **Exaggerated Diagram showing effect of road camber on a car's progress.**

As a result the car runs crab-wise, diagrammatically illustrated in an exaggerated form in Fig. 13. The diagram shows why nearside tyres are very sensitive to too much toe in and offside tyres to toe out. It also shows why sharp " fins " may appear on one tyre but not on the other and why the direction of misalignment can be determined by noting the position of the " fins." Severe misalignment produces clear evidence on both tyres.

The front wheels on a moving car should be parallel. Tyre wear can be affected noticeably by quite small variations from this condition. It will be noted from the diagram that even with parallel wheels the car is still out of line with its direction of movement, but there is less tendency for the wear to be concentrated on any one tyre.

The near front tyre sometimes persists in wearing faster and more unevenly than the other tyres even when the mechanical condition of the car and tyre maintenance are satisfactory. The more severe the average road camber the more marked will this tendency be. This is an additional reason for the regular interchanging of tyres.

(a) Precautions when measuring Wheel Alignment

(i) The car should have come to rest from a forward movement. This ensures as far as possible that the wheels are in their natural running positions.

(ii) It is preferable for alignment to be checked with the car laden.

(iii) With conventional base-bar tyre alignment gauges measurements in front of and behind the wheel centres should be taken at the same points on the tyres or rim flanges. This is achieved by marking the tyres where the first reading is taken and moving the car forwards approximately half a road wheel revolution before taking the second reading at the same points. With the Dunlop Optical Gauge two or three readings should be taken with the car moved forwards to different positions—180° road wheel turn for two readings and 120° for three readings. An average figure should then be calculated.

Wheels and tyres vary laterally within their manufacturing tolerances, or as the result of service, and alignment figures obtained without moving the car are unreliable.

9

4. CAMBER, CASTOR AND KING PIN INCLINATION

These angles normally require no attention unless they have been disturbed by a severe impact or abnormal wear of front end bearings. It is always advisable to check them if steering irregularities develop.

Wheel camber, usually combined with road camber, causes a wheel to try to turn in the direction of lean, due to one side of the tread attempting to make more revolutions per mile than the other side. The resulting increased tread shuffle on the road and the off centre tyre loading tend to cause rapid and one-sided wear. If wheel camber is excessive for any reason the rapid and one-sided tyre wear will be correspondingly greater. Unequal cambers introduce unbalanced forces which try to steer the car one way or the other. This must be countered by steering in the opposite direction which results in still faster tread wear. When tyre wear associated with camber results from road conditions and not from car condition little can be done except to interchange or reverse the tyres. This will prevent one-sided wear, irregular wear, and fast wear from developing to a maximum degree on any one tyre, usually the near front tyre.

Castor and king pin inclination by themselves have no direct bearing on tyre wear but their measurement is often useful for providing a general indication of the condition of the front end geometry and suspension.

5. TYRE AND WHEEL BALANCE

    (a) Static Balance

        In the interests of smooth riding, precise steering, and the avoidance of high speed " tramp " or " wheel hop," all Dunlop tyres are balance checked to predetermined limits.

        To ensure the best degree of tyre balance the covers are marked with white spots on one bead, and these indicate the lightest part of the cover. Tubes are marked on the base with black spots at the heaviest point. By fitting the tyre so that the marks on the cover bead exactly coincide with the marks on the tube, a high degree of tyre balance is achieved (Fig. 14). When using tubes which do not have the coloured spots it is usually advan-

tageous to fit the covers so that the white spots are at the valve position.

Some tyres are slightly outside standard balance limits and are corrected before issue by attaching special loaded patches to the inside of the covers at

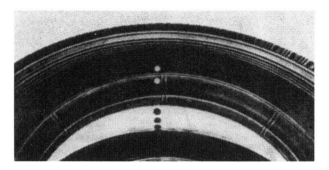

Fig. 14   **The correct relationship between Tyre and Tube.**

the crown. These patches contain no fabric, they do not affect the local stiffness of the tyre and should not be mistaken for repair patches. They are embossed " Balance Adjustment Rubber."

The original degree of balance is not necessarily maintained and it may be affected by uneven tread wear, by cover and tube repairs, by tyre removal and refitting or by wheel damage and

Fig. 15   **Dunlop Tyre Balancing Machine.**

eccentricity. The car may also become more sensitive to unbalance due to normal wear of moving parts.

If roughness or high speed steering

troubles develop, and mechanical investigation fails to disclose a possible cause, wheel and tyre balance should be suspected.

A Tyre Balancing Machine is marketed by the Dunlop Company to enable Service Stations to deal with such cases. This is shown in Fig. 15; a second, marketed by Messrs. V. L. Churchill Ltd., in Fig. 16.

(b) Dynamic Balance

Static unbalance can be measured when the tyre and wheel assembly is stationary. There is another form known as dynamic unbalance which can be detected only when the assembly is revolving.

eccentric wheels give the same effect. During rotation the offset weight distribution sets up a rotating couple which tends to steer the wheel to right and left alternately.

Dynamic unbalance of tyre and wheel assemblies can be measured on the Dunlop Tyre Balancing Machine and suitable corrections made when cars show sensitivity to this form of unbalance. Where it is clear that a damaged wheel is the primary cause of severe unbalance it is advisable for the wheel to be replaced.

6. CHANGING POSITION OF TYRES

There have been references to irregular tread wear and there may be different rates

ELECTRONIC UNIT

WHEEL BLOCK

CONNECTION FOR MAINS SUPPLY

PICK-UP

WHITE TAPE

SPINNER MOTOR

WEIGHT TOOL & WEIGHTS

Fig. 16   Churchill 120 Electronic Wheel Balance.

There may be no heavy spot—that is, there may be no natural tendency for the assembly to rotate about its centre due to gravity—but the weight may be unevenly distributed each side of the tyre centre line (Fig. 17). Laterally

of wear between one tyre and another. It has also been stated that irregular wear is confined almost entirely to front tyres and that the left-hand front tyre is likely to be more affected than the right-hand front tyre.

11

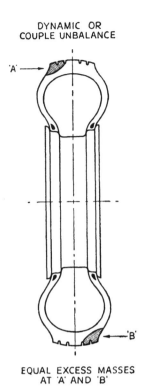

DYNAMIC OR
COUPLE UNBALANCE

'A' →

EQUAL EXCESS MASSES
AT 'A' AND 'B'

Fig. 17   **Dynamic or Couple Unbalance.**

The causes may lie in road conditions, traffic conditions, driving methods and certain features of design which are essential to the control, steering and driving of a car. Close attention to inflation pressures and the mechanical condition of the car will not always prevent irregular wear.

It is therefore recommended that front tyres be interchanged with rear tyres at least every 2,000 miles. Diagonal interchanging between left-hand front and right-hand rear and between right-hand front and left-hand rear provides the most satisfactory first change because it reverses the directions of rotation.

Subsequent interchanging of front and rear tyres should be as indicated by the appearance of the tyres, with the object of keeping the wear of all tyres even and uniform.

7. PRESSED STEEL WHEELS

S.M.M. & T. standard tolerances are—

(a) Wobble

The lateral variation measured on the vertical inside face of a flange shall not exceed $\frac{3}{32}$".

(b) Lift

On a truly mounted and revolving wheel the difference between the high and low points, measured at any location on either tyre bead seat, shall not exceed $\frac{3}{32}$".

Radial and lateral eccentricity outside these limits contribute to static and dynamic unbalance respectively. Severe radial eccentricity also imposes intermittent loading on the tyre. Static balancing does not correct this condition which can be an aggravating factor in the development of irregular wear.

A wheel which is eccentric laterally will cause the tyre to " snake " on the road but this in itself has no effect on the rate of tread wear.

At the same time undue lateral eccentricity is undesirable and it affects dynamic balance.

There is no effective method of truing eccentric pressed steel wheels economically and they should be replaced.

Wheel nuts should be free on their studs. When fitting a wheel all the nuts should be screwed up very lightly, making sure that their seatings register with the seatings in the wheel.

Fig. 18   **Wire Wheel and Hub Cap.**

Final tightening should be done progressively and alternately by short turns of opposite nuts to ensure correct seating and to avoid distortion.

Wheels with damaged or elongated stud holes, resulting from slack nuts, should be replaced.

Rim seatings and flanges in contact with the tyre beads should be free from rust and dirt.

8. WIRE WHEELS (Fig 18)

See "Front Suspension and Steering" Section also "Rear Axle" Section for special hubs.

(a) To Remove Wheels

(i) Jack up the car.

(ii) With a copper headed mallet tap the lugs of the hub cap in the direction stated thereon :—

UNDO—➤UNDO
RIGHT SIDE { Caps fitted on right-hand side of car.

UNDO◄—UNDO
LEFT SIDE { Caps fitted on left-hand side of car.

(iii) By gripping the tyre with both hands the wheel can be pulled off the hub.

(b) To Replace Wheels

(i) Lightly grease the splines of the hub, and the thread of the hub cap.

(ii) Slide wheel on to hub and secure the hub caps.

(iii) Tap the lugs of the cap with the copper headed mallet to secure the wheel.

RIGHT-HAND SIDE CAPS ARE TURNED ANTI-CLOCKWISE TO TIGHTEN. LEFT-HAND SIDE CAPS ARE TURNED CLOCKWISE TO TIGHTEN.

(iv) Remove jacks.

(c) Examination

This should be done periodically every 5,000 miles or at more frequent intervals if the car is used for competition driving or racing.

After cleaning the wheels they should be examined for faults paying particular attention to the following :—

(i) Spokes

Looseness can be corrected and damaged spokes replaced but care must be taken to ensure that the position of the rim relative to the hub shell is not disturbed (Fig. 19).

No undue load must be placed on any one spoke and all spokes must be under the same relative tension. The correct tension is that that will give a flexible but strong wheel. If the tension is too high the wheel will become rigid and loose its advantage over the disc wheel. Or, if too loose, undue strain will be placed on the spokes resulting in breakages.

This tension can be ascertained by drawing a light spanner or similar metal object across the spokes. When the spokes are correctly tensioned they will emit a "ringing note", however, if the spokes are slack the "ring" will be flat. Spoke tensioning is best carried out with the tyre and the tube removed and any protruding spoke heads filed off flush to the nipple.

Note—The building of wire wheels is a specialised trade and this Company and the wheel manufacturers advise that a wheel specialist is consulted if the condition of the wheel is in doubt.

(ii) Hub Shells. The splines should be examined for wear, this is often caused by looseness of the wheel on the axle hub. Excessive wear on these splines will mean the replacement of the hub shell. Rust caused by water entering from outside should be cleaned off and a smear of grease used to protect the interior of the shell and ease the fitting and removing of the wheel from the axle hub.

(iii) Rims

All rust should be cleaned off the exterior of the rim and the affected portion protected with enamel or similar finish. When the tyres are changed the interior of the rims can be inspected for corrosion. Particular attention must be paid to the corrosion, if it is not cleaned away the tyre will become affected.

(d) Wheel Building (See Fig. 19)

The spokes should be laced as shown in the illustration and particular attention must be paid to the positioning of the valve hole, failure to observe this point will mean that the valve stem of the inner tube will foul one or more spokes, resulting in insufficient clearance to connect an air line.

The hub shell, spokes and rim should be loosely assembled and the rim brought into true position relative to the hub, ensuring that the outside dish is maintained.

When this condition is reached the wheel should be mounted on a running hub, each pair of spokes should be carefully tensioned a small amount at a time, working from one pair ·and thence to the diametrically opposite pair. Afterwards, repeating the procedure with the opposed pairs which are located at right angles to the original pairs.

At each stage of the tensioning the truth of the wheel should be checked both for lateral (buckle) and up and down movement (gallop). Then checking any buckle or gallop by giving a slight additional or reduction of tension to the appropriate spoke or sets of spokes.

It is important that as little additional tension as possible is given when truing the wheel. The desired condition when the wheel is finally true is that each spoke should have as near as possible the same tension as its neighbour. This condition can be attained by slackening the tension of one spoke, as as well as increasing that in the opposite spoke, to position the rim correctly.

An experienced wheel builder will be able to gauge when the correct tension has been reached, either by the general feel of the spokes or by the ringing note which the spokes will give when lightly struck with a small spanner or similar metal object.

When building is complete the spoke ends should be examined to ensure that none protrude through the nipple.

Any protrusions should be filed off and the filings brushed away from the rim.

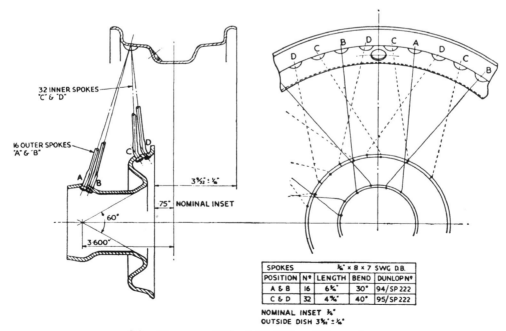

32 INNER SPOKES 'C' & 'D'

16 OUTER SPOKES 'A' & 'B'

3·75" ± ¼"

·75" NOMINAL INSET

60°

3·600"

SPOKES	¼" × 8 × 7 SWG D.B.			
POSITION	Nº	LENGTH	BEND	DUNLOP Nº
A & B	16	6¼"	30°	94/SP 222
C & D	32	4¾"	40°	95/SP 222

NOMINAL INSET ¾"
OUTSIDE DISH 3¾" ± ¼"

Fig. 19   **Wheel Building Dimensions.**

Service Instruction Manual

ELECTRICAL EQUIPMENT

SECTION M

ELECTRICAL EQUIPMENT

INDEX

ELECTRICAL EQUIPMENT

INDEX

ELECTRICAL EQUIPMENT
ILLUSTRATIONS

ELECTRICAL EQUIPMENT

BATTERIES

Models GTW7A/2, GTW9A/2, GT9A/2 and GTZ9A/2.

1. ROUTINE MAINTENANCE

Every 1,000 miles, or monthly (weekly in hot climates) examine the level of the electrolyte in the cells, and if necessary add *distilled* water to bring the level up to the top of the separators. A convenient method of adding the distilled water is by means of the Lucas Battery Filler, a device which automatically ensures that the correct level is attained. The action of resting the nozzle of the battery filler on the separators opens a valve and allows distilled water to flow into the cell, this being indicated by air bubbles rising in the filler. When the correct level has been reached air bubbles cease and the battery filler can then be withdrawn from the cell. A special non-spill nozzle prevents leakage from the filler.

Some earlier batteries incorporated correct-acid-level devices for ease of topping up. These consist of a central plastic tube with a perforated flange—one being located in each cell filler hole. The method of topping up is as follows :—

Pour distilled water into the perforated flange *not down the central tube* of the correct-acid-level device until no more water will enter the cell and the water begins to rise in the filling hole. This will happen when the electrolyte level reaches the bottom of the central tube and prevents further escape of air displaced by the topping-up water. Lift the tube slightly and allow the small amount of visible water in the filling hole to drain into the cell.

WARNING : Do not repeat these operations. The acid level will be correct and the rubber plugs can be refitted.

2. SERVICE DATA

(a) Capacity and Charging Rates

Battery	No. of Plates in each cell	Ampere-hour capacity		Volume of electrolyte required to fill one cell	Initial Charging Current (Amps.)	Normal Recharge Current (Amps.)
		at 10 hour rate	at 20 hour rate			
GTW7A/2	7	38	43	$\frac{3}{4}$ (Pint)	$2\frac{1}{2}$	4
GTW9A/2 GT9A/2 GTZ9A/2	9	51	58	1 (Pint)	$3\frac{1}{2}$	5

(b) Specific Gravity of Electrolyte

The specific gravity of the electrolyte varies with the temperature, therefore, for convenience in comparing specific gravities, this is always corrected to 60°F., which is adopted as a reference temperature. The method of correction is as follows :

For every 5°F. *beolw* 60°F., *deduct* .002 from the observed reading to obtain the true specific gravity at 60°F. *For* every 5°F. *above* 60°F., *add* .002 to the observed reading to obtain the true specific gravity at 60°F

The temperature must be that indicated by a thermometer actually immersed in the electrolyte, and not the air temperature.

Home Trade and Climates ordinarily below 90°F. (32°C.). Specific Gravity of Acid (corrected to 60°F.)		Climates frequently over 90°F. (32°C.). Specific Gravity of Acid (corrected to 60°F.)	
Filling	Fully Charged	Filling	Fully Charged
1.270	1.270—1.290	1.210	1.210—1.230

1

(c) Maximum Permissible Electrolyte Temperature During Charge

Climates normally below 80°F. (27°C.)	Climates between 80°—100°F. (27°—38°C.)	Climates frequently above 100°F. (38°C.)
100°F. (38°C.)	110°F. (43°C.)	120°F. (49°C.)

Fig. 1   **Topping up Battery.**

N.B.—Never use a naked light when examining a battery, as the mixture of oxygen and hydrogen given off by the battery when on charge, and to a lesser extent when standing idle, can be dangerously explosive.

Examine the terminals and, if necessary, clean them and coat them with petroleum jelly. Wipe away any foreign matter or moisture from the top of the battery, and ensure that the connections and the fixings are clean and tight.

3. SERVICING
(a) Battery Persists in Low State of Charge

First consider the conditions under which the battery is used If the battery is subjected to long periods of discharge without suitable opportunities for re-charging, a low state of charge can be expected. A fault in the dynamo or regulator, or neglect of the battery during a period of low or zero mileage may also be responsible for the trouble.

Vent Plugs
See that the ventilating holes in each vent plug are clear.

Level of Electrolyte
The surface of the electrolyte should be level with the tops of the separators. If necessary, top up with distilled water. Any loss of acid from spilling or spraying (as opposed to the normal loss of *water* by evaporation) should be made good by dilute acid of the same specific gravity as that already in the cell.

Cleanliness
See that the top of the battery is free from dirt or moisture which might provide a discharge path. Ensure that the battery connections are clean and tight.

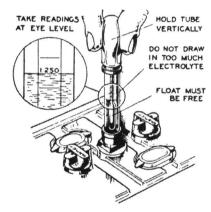

Fig. 2   **Taking Hydrometer Readings.**

Hydrometer Tests
Measure the specific gravity of the acid in each cell in turn, with a hydrometer. The reading given by each cell should be approximately the same; if one cell differs appreciably from the other, an internal fault in that cell is indicated. This will probably be confirmed by the heavy discharge test described below.

The appearance of the electrolyte drawn into the hydrometer when taking a reading gives a useful indication of the state of the plates; if it is very dirty, or contains small particles in suspension, it is possible that the plates are in a bad condition.

Discharge Test
A heavy discharge tester consists of a voltmeter, 2 or 3 volts full scale, across which is connected a shunt resistance capable of carrying a current of several hundred amperes. Pointed prongs are provided for making contact with the inter-cell connectors.

2

Press the contact prongs against the exposed positive and negative terminals of each cell. A good cell will maintain a reading of 1.2—1.5 volts, depending on the state of charge, for at least 6 seconds. If, however, the reading rapidly falls off, the cell is probably faulty, and a new plate assembly may have to be fitted.

(b) Recharging from an External Supply

If the above tests indicate that the battery is merely discharged, and is otherwise in a good condition, it should be recharged, either on the vehicle by a period of daytime running or on the bench from an external supply.

If the latter, the battery should be charged at the rate given in Para. 2 (a) until the specific gravity and voltage show no increase over three successive hourly readings. During the charge the electrolyte must be kept level with the tops of the separators by the addition of distilled water.

A battery that shows a general falling-off in efficiency, common to all cells, will often respond to the process known as "cycling." This process consists of fully charging the battery as described above, and then discharging it by connecting to a lamp board, or other load, taking a current equal to its 10-hour rate. The battery should be capable of providing this current for at least 7 hours before it is fully discharged, as indicated by the voltage of each cell falling to 1.8. If the battery discharges in a shorter time, repeat the "cycle" of charge and discharge.

4. PREPARING NEW UNFILLED, UNCHARGED BATTERIES FOR SERVICE

(a) Preparation of Electrolyte

Batteries should not be filled with acid until required for initial charging. Electrolyte of the specific gravity given in Para. 2 (b) is prepared by mixing distilled water and concentrated sulphuric acid, usually of 1.835 S.G.

The mixing must be carried out either in a lead-lined tank or in suitable glass or earthenware vessels. Slowly add the acid to the water, stirring with a glass rod. *Never add the water to the acid*, as the resulting chemical reaction causes violent and dangerous spurting of the concentrated acid.

Heat is produced by the mixture of acid and water, and the electrolyte should be allowed to cool before taking hydrometer readings—unless a thermometer is used to measure the actual temperature, and a correction applied to the reading as described in Para. 2 (b)—and before pouring the electrolyte into the battery.

The total volume of electrolyte required can be estimated from the figures quoted in Para. 2 (a).

(b) Filling the Battery

The temperature of the acid, battery and filling-in room must not be below 32°F.

Carefully break the seals in the filling holes and *half-fill* each cell with electrolyte of the appropriate specific gravity. Allow the battery to stand for at least six hours, in order to dissipate the heat generated by the chemical action of the acid on the plates and separators, and then add sufficient electrolyte to fill each cell to the top of the separators. Allow to stand for a further two hours and then proceed with the initial charge.

(c) Initial Charge

The initial charging rate is given in Para. 2 (a). Charge at this rate until the voltage and specific gravity readings show no increase over five successive hourly readings. This will take from 40 to 80 hours, depending on the length of time the battery has been stored before charging.

3

Keep the current constant by varying the series resistance of the circuit or the generator output. *This charge should not be broken by long rest periods.* If, however, the temperature of any cell rises above the permissible maximum quoted in Para. 2 (d), the charge must be interrupted until the temperature has fallen at least 10°F. below that figure. Throughout the charge, the electrolyte must be kept level with the top of the separators by the addition of acid solution of the same specific gravity as the original filling-in acid, until specific gravity and voltage readings have remained constant for five successive hourly readings. If the charge is continued beyond that point, top up with distilled water.

At the end of the charge carefully check the specific gravity in each cell to ensure that, when corrected to 60°F., it lies within the specified limits. If any cell requires adjustment, some of the electrolyte must be siphoned off and replaced either by distilled water or by acid of the strength originally used for filling-in, depending on whether the specific gravity is too high or too low Continue the charge for an hour or so to ensure adequate mixing of the electrolyte and again check the specific gravity readings. If necessary, repeat the adjustment process until the desired reading is obtained in each cell.

Finally, allow the battery to cool, and siphon off any electrolyte above the tops of the separators.

5. **PREPARING GTZ "DRY-CHARGED" BATTERIES FOR SERVICE**

"Dry-charged" batteries are supplied without electrolyte but with the plates in a charged condition. No initial charging is required.

When they are required for service it is only necessary to fill each cell with sulphuric acid of the correct specific gravity. No initial charging is required. This procedure ensures that there is no deterioration of the efficiency of the battery during the storage period before the battery is required for use.

(a) **Preparation of Electrolyte**

The electrolyte is prepared by mixing together distilled water and concentrated sulphuric acid, usually of specific gravity 1.835. This mixing must be carried out in a lead-lined tank or a glass or earthenware vessel. The acid must be added slowly to the water while the mixture is stirred with a glass rod. NEVER ADD THE WATER TO THE ACID, as the resulting chemical reaction may cause violent and dangerous spurting of the concentrated acid.

The total quantity of electrolyte needed to fill the battery can be calculated by reference to para. 2 (a).

The specific gravity of the filling electrolyte depends on the climate in which the battery is to be used. If the temperature of the battery and its surroundings will not normally rise above 90°F. (32°C.), electrolyte of specific gravity 1.270 is required. Electrolyte of this specific gravity is prepared by adding 1 part (by volume) of 1.835 specific gravity sulphuric acid to 2.8 parts of distilled water.

On the other hand, in tropical climates where the temperature may frequently rise above 90°F., the electrolyte should be of specific gravity 1.210, and is prepared by adding 1 part of 1.835 acid to 4 parts of distilled water.

N.B.—All specific gravity figures are given for an electrolyte temperature of 60°F., which is adopted as a reference temperature. Hydrometer readings taken at other temperatures can be corrected to this reference temperature as follows :—

For every 5°F. BELOW 60°F., DEDUCT .002 from the observed reading to obtain true reading at 60°F. For every 5°F. ABOVE 60°F., ADD .002 to the observed reading to obtain true reading at 60°F.

Heat is produced by the mixture of acid and water, and the electrolyte should be allowed to cool before pouring it into the battery.

(b) Filling the Cells

Remove the seals from the cell filling holes and fill each cell with correct specific gravity electrolyte to the top of the separators IN ONE OPERATION. The temperature of the filling room, battery and electrolyte should be maintained at between 60°F. and 100°F. If the battery has been stored in a cool place it should be allowed to warm up to room temperature before filling.

(c) Batteries filled in this way are up to 90 per cent charged,

and capable of giving a starting discharge ONE HOUR AFTER FILLING. When time permits, however, a short freshening charge will ensure that the battery is fully charged. Such a freshening charge should last for no more than 4 hours at the normal re-charge rate of the battery.

During the charge the electrolyte must be kept level with the top edge of the separators by the addition of distilled water. Check the specific gravity of the acid at the end of the charge; if 1.270 acid was used to fill the battery the specific gravity should now be between 1.270 and 1.290; if 1.210, between 1.210 and 1.230.

(d) Maintenance in Service

After filling, a dry-charged battery needs only the attention normally given to lead-acid type batteries.

6. BATTERY CABLE CONNECTORS

When fitting the diecast cable connectors, smear the inside of the tapered hole with petroleum jelly and push on the connector by hand. Insert the self-tapping screw and tighten with medium pressure only; fill in the recess around the screw with more petroleum jelly.

If the connectors are fitted dry and driven home on the tapered battery posts too tightly, difficulty may be experienced when it is required to remove them.

GENERATOR—MODEL C.39 PV/2

1. GENERAL

The generator is a shunt-wound two-pole two-brush machine, arranged to work in conjunction with a compensated voltage control regulator unit. A fan, integral with the driving pulley, draws cooling air through the generator, inlet and outlet holes being provided in the end brackets of the unit.

The output of the generator is controlled by the regulator and is dependent on the state of charge of the battery and the loading of the electrical equipment in use. When the battery is in a low state of charge, the generator gives a high output, whereas if the battery is fully charged, the generator gives only sufficient output to keep the battery in good condition without any possibility of overcharging. An increase in output is given to balance the current taken by lamps and other accessories when in use. Further, a high boosting charge is given for a few minutes immediately after starting.

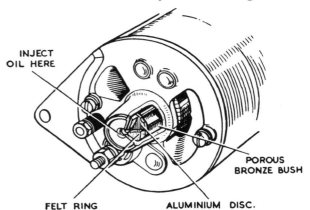

Fig. 3   Commutator End Bearing Lubrication.

2. ROUTINE MAINTENANCE

(a) Lubrication

Every 12,000 miles, inject a few drops of Oiline BBB, or any high quality medium viscosity (S.A.E.30) engine oil into the hole marked " oil " in the end of the bearing housing.

On earlier models, unscrew the cap of the lubricator on the side of the bearing housing, lift out the felt pad and spring and about half-fill the lubricator cap with high melting point grease (H.M.P. Grease). Replace the spring and felt pad and screw the lubricator cap back into position.

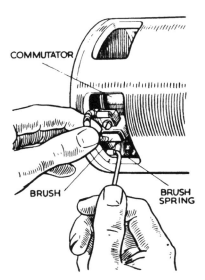

Fig. 4   Checking Brush Gear.

(b) Inspection of Brush Gear and Commutator

At the same time, remove the metal band cover to inspect the brushgear and commutator. Check that the brushes move freely in their holders by holding back the brush springs and pulling gently on the flexible connectors. If a brush is inclined to stick, remove it from its holder and clean its sides with a petrol-moistened cloth. Be careful to replace brushes in their original positions in order to retain the " bedding."

In service, brush wear takes place and the brushes become shorter. If the brushes are permitted to wear down until the embedded ends of the flexible connectors are exposed at the running surface, serious damage can occur to the commutator. It is therefore important to measure from time to time the length of each brush. If this length, measured from the running surface to the top edge of the brush, has decreased to $\frac{11}{32}''$ the brush (or brushes) should be replaced.

The commutator should be clean, free from oil or dirt and should have a polished appearence. If it is dirty, clean it by pressing a fine dry cloth against it while the engine is slowly turned over by hand. If the commutator is very dirty, moisten the cloth with petrol.

(c) Belt Adjustment

Occasionally inspect the generator driving belt. If necessary, adjust to take up any undue slackness by turning the generator on its mounting.

Care should be taken to avoid over-tightening the belt, which should have just sufficient tension to drive without slipping.

See that the machine is properly aligned, otherwise undue strain will be thrown on the generator bearings.

3. PERFORMANCE DATA

Cutting-in speed 1.050-1,200 r.p.m. at 13 generator volts. Maximum output:— 19 amps at 1,900-2,150 r.p.m. at 13.5 generator volts (on resistance load of 0.7 ohm). Field resistance 6.1 ohms.

4. SERVICING

(a) Testing in position to locate fault in charging circuit

In the event of a fault in the charging circuit, adopt the following procedure to locate the cause of the trouble.

(i) Inspect the driving belt and adjust if necessary (see Para. **2** (c)).

(ii) Check that the generator and control box are connected correctly. The larger generator terminal must be connected to control box terminal " D " and the smaller generator terminal to control box terminal " F ". Check the earth connection to control box terminal " E ".

(iii) Switch off all lights and accessories, disconnect the cables from terminals of generator and connect the two terminals with a short length of wire.

(iv) Start the engine and set to run at normal idling speed.

(v) Clip the negative lead of a moving coil type voltmeter, calibrated 0— 20 volts, to one generator terminal and the other lead to a good earthing point on the yoke.

(vi) Gradually increase the engine speed, when the voltmeter reading should rise rapidly and without fluctuation. Do not allow the voltmeter reading to reach 20 volts and do not race the engine in an attempt to increase the voltage. It is sufficient to run the generator up to a speed of 1,000 r.p.m.

If there is no reading, check the brush gear as described in (vii) below. If there is a low reading of approximately $\frac{1}{2}$—1 volt, the field winding may be at fault (see Para. **4** (**e**)). If there is a reading of approximately half the nominal voltage the armature winding may be at fault (see Para. **4** (**d**)).

(vii) Remove the cover band and examine the brushes and commutator. Hold back each brush spring and move the brush by pulling gently on its flexible connector. If the movement is sluggish, remove the brush from its holder and ease the sides by lightly polishing on a smooth file. Always replace the brush in its original position. If a brush has worn to $\frac{11}{32}''$ in length a new brush must be fitted and bedded to the commutator.

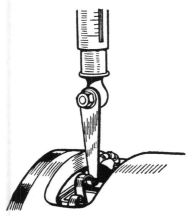

Fig. 5   Testing Brush Spring Tension.

Test the brush spring tension with a spring scale. The tension of the springs when new is 22—25 oz. In service it is permissible for this value to fall to 15 oz. before performance may be affected. Fit new springs if the tension is low. If the commutator is blackened or dirty, clean it by holding a petrol-moistened cloth against it while the engine is turned slowly by hand cranking. Re-test the generator as in (**vi**); if there is still no reading on the voltmeter, there is an internal fault and the complete unit, if a spare is available, should be replaced. Otherwise the unit must be dismantled (see Para. **4** (**b**)) for internal examination.

(viii) If the generator is in good order, remove the link from between the terminals and restore the original connections, taking care to connect the larger generator terminal to control box terminal "D" and the smaller generator terminal to control box terminal "F".

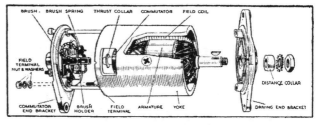

Fig. 6   Dismantled View of Generator (Yoke cut away to show Interior).

(b) To Dismantle

(i) Take off the driving pulley.

(ii) Remove the cover band, hold back the brush springs and remove the brushes from their holders.

(iii) Unscrew and withdraw the two through bolts.

(iv) The commutator end bracket can now be withdrawn from the generator yoke.

(v) The driving end bracket together with the armature can now be lifted out of the yoke.

7

(vi) The driving end bracket, which on removal from the yoke has withdrawn with it the armature and armature shaft ball-bearing, need not be separated from the shaft unless the bearing is suspected and requires examination, or the armature is to be replaced; in this event the armature should be removed from the end bracket by means of a hand press.

(c) Commutator

A commutator in good condition will be smooth and free from pits or burned spots. Clean the commutator with a petrol-moistened cloth. If this is ineffective, carefully polish with a strip of fine glass paper while rotating the armature. To remedy a

INSULATOR   SEGMENTS   INSULATOR

Fig. 7   Showing the Correct and Incorrect Ways of undercutting Commutator Insulation.

badly worn commutator, mount the armature, with or without the drive end bracket, in a lathe, rotate at high speed and take a light cut with a very sharp tool. Do not remove more metal than is necessary. Polish the commutator with very fine glass paper. Undercut the insulators between the segments to a depth of $\frac{1}{32}''$ with a hack saw blade ground down to the thickness of the insulator

Fig. 8   Method to be used when undercutting Commutator Insulation.

(d) Armature

The testing of the armature winding requires the use of a volt-drop test and growler. If these are not available the armature should be checked by substitution. No attempt should be made to machine the armature core or to true up a distorted armature shaft. To remove the armature shaft from the drive end bracket and bearing, support the bearing retaining plate firmly and press the shaft out of the drive end bracket. When fitting the new armature, support the inner journal of the ball bearing whilst pressing the armature shaft firmly home.

(e) Field Coils

Measure the resistance of the field coils, without removing them from the generator yoke, by means of an ohm meter connected between the field terminal and yoke. The ohm meter should read 6.1 ohms approximately. If an ohm meter is not available, connect a 12 volt D.C. supply with an ammeter in series between the field terminal and generator yoke. The ammeter reading should be approximately 2 amperes. No reading on the ammeter, or an infinite ohm meter reading, indicates an open circuit in the field winding. If the current reading is much more than 2 amperes, or the ohm meter reading much below 6.1 ohms, it is an indication that the insulation of one of the field coils has broken down.

In either case, unless a substitute generator is available, the field coils must be replaced. To do this, carry out the procedure outlined below, using a wheel-operated screwdriver.

(i) Drill out the rivet securing the field coil terminal block to the yoke and unsolder the field coil connections.

(ii) Remove the insulation piece which is provided to prevent the junction of the field coils contacting with the yoke.

(iii) Mark the yoke and pole shoes so that the latter can be fitted in their original positions.

(iv) Unscrew the pole shoe retaining screws by means of the wheel-operated screwdriver.

(v) Draw the pole shoes and coils out of the yoke and lift off the coils.

(vi) Fit the new field coils over the pole shoes and place them in position inside the yoke. Take care to ensure that the taping of the field coils is not trapped between the pole shoes and the yoke.

(vii) Locate the pole shoes and field coils by lightly tightening the fixing screws.

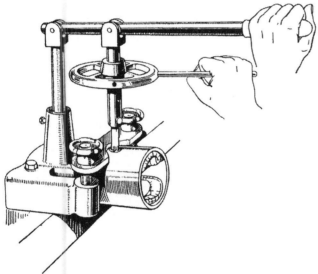

Fig. 9   Tightening Pole Shoe Retaining Screws.

(viii) Fully tighten the screws by means of the wheel-operated screwdriver and lock them by caulking.

(ix) Replace the insulation piece between the field coil connections and the yoke

(x) Resolder the field coil connections to the field coil terminal block and re-rivet to the yoke.

(f) **Bearings**

Bearings which have worn to such an extent that they will allow side movement of the armature shaft must be replaced.

To replace the bearing bush in a commutator end bracket, proceed as follows:—

(i) Remove the old bearing bush form the end bracket, the bearing should be removed by screwing a $\frac{5}{8}$ inch tap into the bush for a few turns and pulling out the bush with the tap. Screw the tap squarely into the bush to avoid damage to the bracket. Insert the felt ring and aluminium disc in the bearing housing, then press the new bearing bush into the end bracket (using a shouldered, highly polished mandrel of

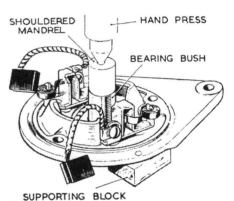

Fig. 10   Method of fitting Porous Bronze Bearing Bush.

the same diameter as the shaft which is to fit in the bearing) until the bearing is flush with the inner face of the bracket. Earlier models, fitted with screw-cap type lubricators, do not have a felt ring or aluminium disc in the bearing housing.

(ii) Porous bronze bushes must not be opened out after fitting, or the porosity of the bush may be impaired. Before fitting the new bearing bush it should be allowed to stand for 24 hours completely immersed in thin engine oil; this will allow the pores of the bush to be filled with lubricant. In cases of extreme urgency, this period may be shortened by heating the oil to 100°C. for 2 hours, then allowing to cool before removing the bearing bush.

The ball bearing at the driving end is replaced as follows:—

(i) Drill out the rivets which secure the bearing retaining plate to the end bracket and remove the plate.

(ii) Press the bearing out of the end bracket and remove the corrugated washer, felt washer and oil retaining washer.

(iii) Before fitting the replacement bearing see that it is clean and pack it with high melting point grease.

(iv) Place the oil retaining washer, felt washer and corrugated washer in the bearing housing in the end bracket.

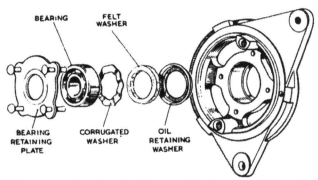

Fig. 11   Exploded View of Drive End Bearing.

(v) Locate the bearing in the housing and press it home. On earlier models the outer journal should be pressed home by means of a hand press.

(vi) Fit the bearing retaining plate. Insert the new rivets from the inside of the end bracket and open the rivets by means of a punch to secure the plate rigidly in position.

(g) **Re-assembly**
In the main, the re-assembly of the generator is a reversal of the operations described in Para. 4 (b). After re-assembly, lubricate the commutator end bearing, referring to Para. 2 (a) for the correct procedure.

STARTING MOTOR—MODEL M418G
(Outboard Drive)

1. **GENERAL**
The electric starting motor is a series-parallel connected four-pole, four-brush machine having an extended shaft which carries the engine engagement gear, or starter drive as it is more usually named. The diameter of the yoke is $4\frac{1}{8}''$.

The starting motor is of similar construction to the generator, except that heavier copper wire is used in the construction of the armature and field coils.

2. **ROUTINE MAINTENANCE**
About every 12,000 miles take the cover band off the starting motor and carry out the following procedure :

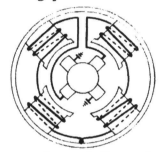

Fig. 12   Internal Connections of the Starting Motor.

(a) Check that the brushes move freely in their holders by holding back the brush springs and pulling gently on the flexible connectors. If movement is sluggish, remove the brush from its holder and clean its sides with fluffless

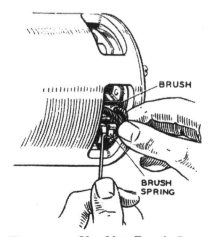

Fig. 13   Checking Brush Gear.

petrol-moistened cloth. Replace the brush in its original position. Brushes which are worn must be replaced, see Para. **4** (**d**) (**i**).

(**b**) Check the tension of the brush springs, using a spring scale. The correct tension is 30-40 oz. Fit new springs if the tension is low, see Para. **4** (**d**) (**i**).

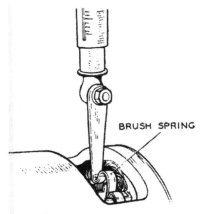

BRUSH SPRING

Fig. 14   Testing Brush Spring Tension.

(**c**) The commutator must be clean and have a polished appearance. If necessary clean it by pressing a fine dry cloth against it while the starter is turned by applying a spanner to the squared extension of the shaft. Access to the squared shaft is gained by removing the thimble-shaped metal cover. If the commutator is very dirty, moisten the cloth with petrol.

(**d**) Keep all electrical connections clean and tight, any which may have become dirty must be cleaned and the contacting surfaces lightly smeared with petroleum jelly.

3.   PERFORMANCE DATA

Lock torque 17 lb./ft. with 440—460 amps at 7.4—7.0 volts.

Torque at 1,000 r.p.m., 8 lb./ft. with 250—270 amps at 9.4—9.0 volts.

Light running current 4.5 amps at 7,400—8,500 r.p.m.

4.   SERVICING

(**a**) **Testing in Position**

If the motor does not operate or fails to crank the engine when the starting button is used, switch on the lamps and again use the starting button.

(**i**) **The lamps dim and the motor does not crank the engine :**

Before examining the starter check by hand-cranking that the engine is not abnormally stiff.

Sluggish action of the starting motor may be due to a discharged battery. Check by disconnecting the existing cables and re-connecting the motor to a battery known to be fully charged.

If the starting motor now gives normal cranking of the engine the vehicle battery must be examined and the motor circuit cables checked for damaged insulation.

If the motor does not operate satisfactorily it must be removed from the engine for examination, see Para. **4** (**b**).

(**ii**) **The lamps do not dim and the motor does not crank the engine :**

Check by means of a voltmeter or low voltage test lamp that the circuit up to the supply terminal on the motor is in order.

If no voltage is indicated, check the circuit from battery to motor via the starter switch. Ensure that all connections are clean and tight.

A voltage at the supply terminal indicates that the motor has an internal fault and must be removed from the engine for examination, see Para. **4** (**b**).

If the motor operates but does not crank the engine, the drive mechanism is probably faulty.

(b) Bench-testing

(i) Removing the starting motor from the engine :

Disconnect the earth terminal on the battery to avoid any danger of short circuits. Remove the heavy cable from the starting motor.

Remove the mounting bolts and withdraw the starting motor from the engine.

(ii) Measuring the light running current :

Secure the starting motor in a vice. Connect the motor in series with a starter switch, an ammeter capable of measuring 600 amperes and 12-volt voltage supply. Use cables of a similar size to those in the vehicle motor circuit. One of the fixing lugs on the drive end bracket is a suitable earthing point on the starting motor. Connect a voltmeter between the motor terminal and the yoke.

Operate the switch and note the speed of armature rotation, using a tachometer, and the readings given by the ammeter and voltmeter.

While the motor is running at speed, examine the brushgear and check if there is any undue sparking at the commutator or excessive brush movement.

(iii) Measuring lock torque and lock current

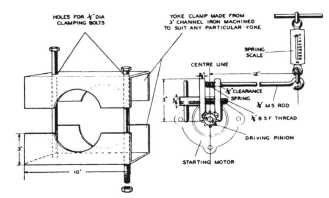

Fig. 15 Method of measuring stall torque and current.

With the motor firmly clamped in a vice, attach a brake arm to the driving pinion. Connect the free end of this arm to a spring scale. Operate the switch and note the current consumption, voltage and the reading on the spring scale.

The measure of torque can be calculated by multiplying the reading on the spring scale in pounds by the length of the brake arm in feet.

If a constant-voltage bus-bar supply is used when carrying out the lock torque test, a higher lock voltage may be shown on the voltmeter than the appropriate value given in Para. 3. In this event a variable resistor of suitable current-carrying capacity should be connected in the battery circuit and adjusted until the lock voltage is the same as that given in Para. 3. Take readings of current and torque at this value.

(iv) Fault Diagnosis :

An indication of the nature of the fault or faults may be deduced from the results of the no load and lock torque tests.

SYMPTOM	PROBABLE FAULT
Speed, torque and current consumption correct.	Assume motor to be in normal operating condition.

(iv) Fault Diagnosis—*(cont'd)*

SYMPTOM	PROBABLE FAULT
Speed, torque and current consumption low.	High resistance in brushgear, *e.g.,* faulty connections, dirty or burned commutator causing bad brush contact.
Speed and torque low, current consumption high.	Tight or worn bearings, bent shaft, insufficient end play, armature fouling a pole shoe, or cracked spigot on drive end bracket.
	Short-circuited armature, earthed armature or short-circuited field coils.
Speed and current consumption high, torque low.	Short-circuited field coils.
Armature does not rotate, no current consumption.	Open-circuited armature or field coils. If the commutator is badly burned there may be poor contact between brushes and commutator.
Armature does not rotate, high current consumption.	Earthed field winding. Armature prevented mechanically from rotating.
Excessive brush movement causing arcing at commutator.	Low brush spring tension, worn or out-of-round commutator. " Thrown " or high segment on commutator.
Excessive arcing at the commutator.	Defective armature windings.

If any fault is indicated, the motor must be dismantled, see Para. **4** **(c)** and a further check made.

(c) Dismantling

Remove the cover band, hold back the brush springs and lift the brushes from their holders.

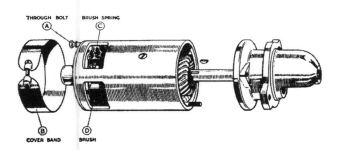

THROUGH BOLT   BRUSH SPRING

(A)   (C)

(B)   (D)

COVER BAND   BRUSH

Fig. 16   Showing Starter Motor dismantled.

Unscrew the terminal nuts from the field coil terminal post protruding from the commutator end bracket.

Unscrew the two through bolts from the commutator end bracket and remove the commutator end bracket from the yoke.

Remove the driving end bracket complete with armature and drive from the starting motor yoke.

(d) Bench Inspection

After the motor has been dismantled individual items must be examined, as follows :—

13

(i) Brushgear

Where necessary the brushes and brush-holders must be cleaned, using a fluffless petrol-moistened cloth.

To prevent damage to the commutator, brushes must be replaced when worn to $\frac{5}{16}''$ in length. The flexible connectors can be removed by unsoldering, and the connectors of the new brushes secured in place by re-soldering. The brushes are pre-formed so that bedding to the commutator is unnecessary.

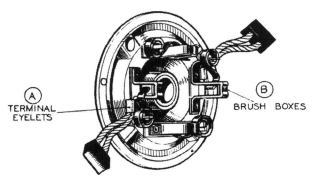

Fig. 17   Commutator End Bracket Brush Connections.

Check the brush springs, as in Para. 2 (**b**). To fit a new spring, prise open the spring anchor slot in the brush spring support post and lift the old spring away. Place the new spring in the slot in the same position as occupied by the old spring. Re-close the slot. Check the tension of the new spring and ensure that it makes contact with the centre of the brush.

(ii) Commutator

The commutator must be clean and have a polished appearance. If it is dirty it must be cleaned, using a fluffless petrol-moistened cloth or, if necessary, by polishing it with a strip of very fine emery cloth.

To remedy a badly worn commutator, dismantle the starter drive and remove the armature from the end bracket. Mount the armature in a lathe, rotate at a high speed and take a light cut with a very sharp tool.

Do not remove any more metal than is necessary. Finally polish with very fine glass paper. The insulators between the commutator segments MUST NOT BE UNDERCUT.

(iii) Armature

Check for lifted commutator segments and loose turns in the armature winding. These may be due to the starting motor having remained engaged while the engine is running, thus causing the armature to be rotated at excessive speed.

A damaged armature must always be replaced—no attempts should be made to machine the armature core or to true a distorted armature shaft. An indication of a bent shaft or a loose pole shoe may be given by scored armature laminations.

To check armature insulation, use an ohm meter or a 110-volt a.c. test lamp. A high reading should be shown on the meter when connected between the armature shaft and the commutator segments. If a test lamp is used, it must not light when connected as above. Faulty insulation will be indicated by a low ohmic reading or by lighting of the test lamp.

If a short circuit is suspected, check the armature on a " growler." The motor overheating may cause blobs of solder to short circuit the commutator segments.

If an armature fault cannot be located and remedied, a replacement armature must be fitted.

(iv) Field Coils

Continuity Test :

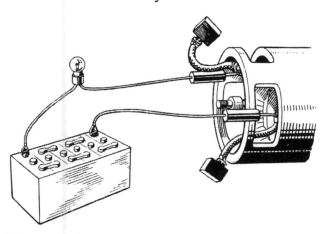

Fig. 18   Testing for Open Circuit in the field coils.

Connect a battery and suitable bulb in series with two pointed probes.

If the lamp fails to light in the following test an open circuit in the field coils is indicated and the defective coils must be replaced.

Place the probes on the brush tappings. The bulb should light. Lighting of the lamps does not necessarily indicate that the field coils are in order. It is possible that a field coil may be earthed to a pole shoe or to the yoke.

Insulation Test :

Connect an ohm meter or a 110-volt a.c. test lamp between the terminal post and a clean part of the yoke.

Lighting of the test lamp or a low ohmic reading indicates that the field coils are earthed to the yoke and must be replaced.

Replacing the field coils :

Unscrew the four pole-shoe retaining screws, using a wheel-operated screwdriver.

Remove the insulation piece which is fitted to prevent the inter-coil connectors from contacting with the yoke. Mark the yoke and pole-

shoes in order that they may be refitted in their original positions.

Draw the pole-shoes and coils out of the yoke and lift off the coils.

Fit the new field coils over the pole-shoes and place them in position inside the yoke. Ensure that the taping of the field coils is not trapped between the pole-shoes and the yoke.

Locate the pole-shoes and field coils by lightly tightening the fixing screws.

Replace the insulation piece between the field coil connections and the yoke.

Finally, tighten the screws by means of the wheel-operated screwdriver.

(v) Bearings

Bearings which are worn to such an extent that they will allow excessive side play of the armature shaft, must be replaced. To replace the bearing bushes proceed as follows :

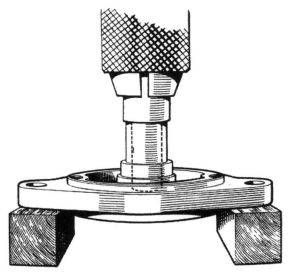

Fig. 19   Method of fitting Bearing Bushes.

Press the bearing bush out of the end bracket.

Press the new bearing bush into the end bracket, using a shouldered, highly polished mandrel of the same diameter as the shaft which is to fit in the bearing.

15

Porous bronze bushes must not be opened out after fitting or the porosity of the bush may be impaired.

NOTE: Before fitting a new porous bronze bearing bush it should be completely immersed for 24 hours in clean, thin engine oil. In cases of extreme urgency this period may be shortened by heating the oil to 100°C. for 2 hours, then allowing to cool before removing the bearing bush.

(e) Re-assembly

This is, in the main, a reversal of the procedure given in Para. **4** (**c**) for dismantling.

Commutator end bracket replacement: The starting motor is designed for clockwise rotation, indicated by the arrow on the yoke. Press out the through bolt indentations marked " C " on the replacement bracket.

Press the locating dowel into the appropriate hole marked " C."

Insert the through bolts into the holes made in the bracket and tighten the bracket to the yoke.

STARTING MOTOR DRIVE

1. GENERAL

The drive embodies a combination of rubber torsion member and friction clutch in order to control the torque transmitted from the starter to the engine flywheel and to dissipate the energy in the rotating armature of the starter at the moment when the pinion engages with the flywheel.

It also embodies an overload release mechanism which functions in the event of extreme stress, such as may occur in the event of a very heavy backfire, or if the starter is inadvertently meshed into a flywheel, rotating in the reverse direction.

When the starter is energised, the torque is transmitted by two paths, one via the outer sleeve of the rubber coupling and through the friction washer to the screwed sleeve, while the other path is from the outer to the inner sleeve through the rubber coupling and then directly to the screwed sleeve.

The torque through the rubber limits the total torque which the drive transmits and since the rubber is bonded to the inner sleeve, under overload conditions slipping will occur between the rubber bush and the outer sleeve of the coupling. Slipping does not take place under normal engagement conditions, when the rubber acts merely as a spring with a limiting relative twist on the two members of approximately 30°.

Under conditions of unduly severe overload 'which might cause damage to the drive or its mounting, the rubber slips in its housing so that a definite upper limit is set to the torque transmitted and to the stresses which may occur.

2. ROUTINE MAINTENANCE

If any difficulty is experienced with the starting motor not meshing correctly with the flywheel, it may be that the drive requires cleaning. The pinion should move freely on the screwed sleeve; if there is any dirt or other foreign matter on the sleeve it must be washed off with paraffin.

In the event of the pinion becoming jammed in mesh with the flywheel, it can usually be freed by turning the starter motor armature by means of a spanner applied to the shaft extension at the commutator end. This is accessible by removing the cap which is a push fit.

3. CONSTRUCTION

The construction of the drive will be clear from the illustration. The pinion is carried on a barrel type assembly which is mounted on a screwed sleeve.

The screwed sleeve is secured to the armature shaft by means of a location nut and is also keyed to the inner sleeve of the rubber coupling by a centre coupling plate. A friction washer is fitted between the coupling plate and rubber assembly and the outer sleeve of the rubber coupling is keyed at the armature end of the starter by means of a transmission plate.

A pinion restraining spring is fitted in the barrel assembly to prevent the pinion vibrating into mesh when the engine is running.

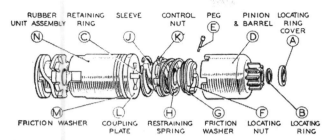

RUBBER RETAINING SLEEVE CONTROL PEG PINION LOCATING
UNIT ASSEMBLY RING NUT & BARREL RING COVER
(N) (C) (J) (K) (E) (D) (A)

FRICTION WASHER COUPLING RESTRAINING FRICTION LOCATING LOCATING
(M) PLATE SPRING WASHER NUT RING
(L) (H) (G) (F) (B)

Fig. 20   Exploded view of Starter Motor Drive Assembly.

4. DISMANTLING

Having removed the armature as described in the section dealing with starting motors, the drive can be dismantled as follows :—
Remove the locating cover (A) and then withdraw the locating ring (B) from the starter shaft at the end of the starter drive.

Remove the retaining ring (C) from inside the end of the pinion and barrel assembly (D) and then withdraw the pinion and barrel assembly.

Take out the peg (E) securing the locating nut (F) to the shaft, hold the squared starter shaft extension at the commutator end by means of a spanner and unscrew the locating nut.

Withdraw the friction washer (G), restraining spring (H). Slide the sleeve (J) and control nut (K) off the splined shaft.
Finally remove coupling plate (L), friction washer (M) and rubber unit assembly (N).

NOTE : On some models the locating nut is secured by caulking the nut into the keyway provided in the shaft and therefore no peg (E) is fitted. When re-assembling it will be necessary to fit a new locating nut.

5. RE-ASSEMBLY

The re-assembly of the drive is a reversal of the dismantling procedure.

DISTRIBUTOR—Model DM2

1. GENERAL

Mounted on the distributor driving shaft, immediately beneath the contact breaker, is a centrifugally operated timing control mechanism. It consists of a pair of spring-loaded governor weights, linked by lever action to the contact breaker cam. Under the centrifugal force imparted by increasing engine speed, the governor weights swing out against the spring pressure to advance the contact breaker cam and thereby the spark, to suit engine conditions at the greater speed.

A built-in vacuum-operated timing control is also included, designed to give additional advance under part-throttle conditions. The inlet manifold of the engine is in direct communication with one side of a spring-loaded diaphragm. This diaphragm acts through a lever mechanism to rotate the heel of the contact breaker about the cam, thus advancing the spark for part-throttle

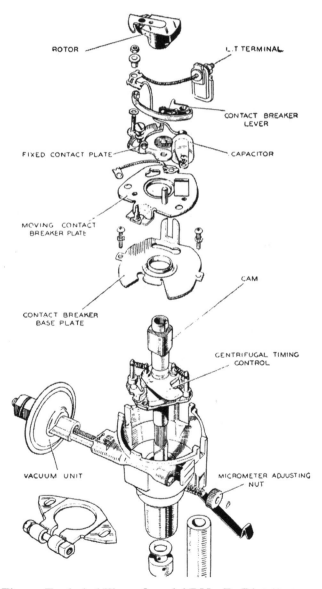

ROTOR

L.T TERMINAL

CONTACT BREAKER LEVER

FIXED CONTACT PLATE

CAPACITOR

MOVING CONTACT BREAKER PLATE

CONTACT BREAKER BASE PLATE

CAM

CENTRIFUGAL TIMING CONTROL

VACUUM UNIT

MICROMETER ADJUSTING NUT

Fig. 21 Exploded View of model DM2. P4 Distributor.

17

operating conditions. There is also a micrometer adjustment by means of which fine alterations in timing can be made to allow for changes in running conditions, *e.g.*, state of carbonisation, change of fuel, etc.

A completely sealed metallised paper capacitor is utilised. This has the property of being self-healing ; should the capacitor break down, the metallic film around the point of rupture is vaporised away by the heat of the spark, so preventing a permanent short circuit. Capacitor failure will be found to be most infrequent.

The H.T. pick-up brush is of a composite construction, the top portion consisting of a resistive compound and the lower of softer carbon to prevent wear taking place on the rotor electrode. The resistive portion of this carbon brush which is in circuit between the coil and the distributor gives a measure of radio interference suppression. Under no circumstances must a short non-resistive brush be used as a replacement for one of these longer resistive brushes.

The Pre-tilted Contact Breaker Unit
During 1955 an improved contact breaker unit was introduced on the DM2P4 distributor. Important features of this pre-tilted contact breaker unit are : improved sensitivity of vacuum control and elimination of any tendency for the moving contact breaker plate to rock at high cam speeds. Contact adjustment has also been simplified.

2. ROUTINE MAINTENANCE
In general, lubrication and cleaning constitute normal maintenance procedure.

(a) Lubrication—every 3,000 miles
Take great care to prevent oil or grease from getting on or near the contacts.

Add a few drops of thin machine oil through the aperture at the edge of the contact breaker to lubricate the centrifugal timing control.

Smear the cam with Mobilgrease No. 2.

Lift off the rotor arm and apply to the spindle a few drops of Ragosine Molybdenised non-creep oil or thin machine oil to lubricate the cam

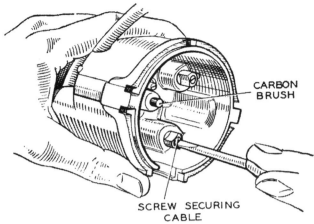

Fig. 22 **Fitting H.T. Cables.**

bearing. It is not necessary to remove the exposed screw, since it affords a clearance to permit the passage of oil. Replace the rotor arm carefully, locating its moulded projection in the keyway in the spindle and pushing it on as far as it will go.

(b) Cleaning—every 6,000 miles
Thoroughly clean the moulded distributor cover, inside and out, with a soft dry cloth, paying particular attention to the spaces between the metal electrodes. Ensure that the carbon brush moves freely in its holder.

Examine the contact breaker. The contacts must be quite free from grease or oil. If they are burned or blackened, clean them with very fine carborundum stone or emery cloth, then wipe with a petrol-moistened cloth. Cleaning is facilitated by removing the contact breaker lever. To do this, remove the nut, washer, insulating piece and connections from the post to which the end of the contact breaker spring is anchored. The contact breaker lever may now be removed from its pivot. Before refitting the contact breaker, smear the pivot post with Ragosine Molybdenised non-creep oil or Mobilgrease No. 2. After cleaning, check the contact breaker setting. Turn the engine by hand until the contacts show the maximum opening. This should measure 0.014″ to 0.016″. If the measurement is incorrect, keep the engine in the position giving maximum opening, slacken the screw(s) securing the fixed contact plate and adjust its position to give the required gap. Tighten the screw(s). Recheck the setting for other positions of the engine giving maximum opening.

3. DESIGN DATA

(a) Firing angles : 0°, 90°, 180°, 270°, ±1°.
Closed period : 60° ± 3°.
Open period : 30° ± 3°.

(b) Contact breaker gap : 0.014" to 0.016".

(c) Contact breaker spring tension, measured at contacts : 20—24 ozs.

(d) Capacitor : 0.2 microfarad.

(e) Rotation : Anti-clockwise.

(f) Checking Automatic timing control :

(i) Advance due to centrifugal control :

Set to spark at zero degrees at minimum r.p.m.

Run distributor at 2,700 r.p.m. Advance should lie between 13° and 15°.

Check advance at following decelerating speeds :—

Speed r.p.m.	Advance (degrees)
2,000	$12\frac{1}{2}$—$14\frac{1}{2}$
750	$8\frac{1}{2}$—$10\frac{1}{2}$
600	$6\frac{1}{2}$— 9
200	0 — 2

Part No(s). of auto advance springs : 421218, 421219.

(ii) Advance due to vacuum control :
Apply a vacuum of 18" of mercury. Advance to lie between 6° and 8°. Check advance at the following points, as the vacuum is reduced :

Vacuum (in hg.)	Advance (degrees)
$9\frac{1}{2}$	5 — 7
$4\frac{3}{4}$	$\frac{1}{2}$—$2\frac{1}{2}$

No advance below 2" of mercury.

4. SERVICING

Before starting to test, make sure that the battery is not fully discharged, as this will often produce the same symptoms as a fault in the ignition circuit.

(a) **Testing in Position to Locate Cause of Uneven Firing**

Run the engine at a fairly fast idling speed.

If possible, short circuit each plug in turn with the blade of an insulated screwdriver or a hammer head placed across the terminal to contact the cylinder head. Short circuiting the plug in the defective cylinder will cause no noticeable change in the running note. On the others, however, there will be a pronounced increase in roughness. If this is not possible, due to the sparking plug being fitted with a shrouded cable connector, remove each plug connector in turn. Again, removal of the connection to the defective cylinder will cause no noticeable change in the running note, but there will be a definite increase in roughness when the other plugs are disconnected. Having thus located the defective cylinder, stop the engine and remove the cable from the sparking plug terminal.

Restart the engine and hold the cable end about $\frac{3}{16}$" from the cylinder head. If sparking is strong and regular, the fault lies with the sparking plug, and it should be removed, cleaned and adjusted, or a replacement fitted. If, however, there is no spark, or only weak irregular sparking, examine the cable from the plug to the distributor cover for deterioration of the insulation, renewing the cable if the rubber is cracked or perished. Clean and examine the distributor moulded cover for free movement of the carbon brush. If a replacement brush is necessary, it is important that the correct type is used. If tracking has occurred, indicated by a thin black line between two or more electrodes or between one of the electrodes and the body, a replacement distributor cover must be fitted.

19

(b) Testing in Position to Locate Cause of Ignition Failure

Spring back the clips on the distributor head and remove the moulded cover. Lift off the rotor, carefully levering with a screwdriver if necessary.

Switch on the ignition and whilst the engine is slowly cranked, observe the reading on the car ammeter, or on an ammeter connected in series with the battery supply cable.

The reading should rise and fall with the closing and opening of the contacts if the low tension wiring is in order. When a reading is given which does not fluctuate, a short circuit, or contacts remaining closed, is indicated. No reading indicates an open circuit in the low tension circuit, or badly adjusted or dirty contacts.

Check the contacts for cleanliness and correct gap setting as described in Para. 2 (b). Ensure that the moving arm moves freely on the pivot. If sluggish, remove the arm and polish the pivot post with a strip of fine emery cloth. Smear the post with Ragosine Molybdenised non-creep oil or Mobilgrease No. 2, replace the arm. If the fault persists, proceed as follows :

(c) Low Tension Circuit — Fault Location

(i) **No reading in ammeter test.** Refer to wiring diagram and check circuit for broken or loose connections, including ignition switch. Check the ignition coil by substitution.

(ii) **Steady reading in ammeter test**

Refer to wiring diagram and check wiring for indications of a short circuit.

Check capacitor (either by substitution or on a suitable tester). Check ignition coil by substitution. Examine insulation of contact breaker.

(d) High Tension Circuit

If the low tension circuit is in order, remove the high tension lead from the centre terminal of the distributor cover. Switch on the ignition and turn the engine until the contacts close. Flick open the contact breaker lever whilst the high tension lead from the coil is held about $\frac{3}{16}''$ from the cylinder block. If the ignition equipment is in good order, a strong spark will be obtained. If no spark occurs, a fault in the circuit of the secondary winding of the coil is indicated and the coil must be replaced.

The high tension cables must be carefully examined and replaced if the rubber insulation is cracked or perished, using 7 mm. rubber covered ignition cable.

The cables from the distributor to the sparking plugs must be connected in the correct firing order, i.e. 1.3.4.2.

(e) Dismantling

When dismantling, carefully note the positions in which the various components are fitted, in order to ensure their correct replacement on re-assembly. If the driving dog or gear is offset, or marked in some way for convenience in timing, note the relation between it and the rotor electrode and maintain this relation when re-assembling the distributor. The amount of dismantling necessary will obviously depend on the repair required.

Spring back the securing clips and remove the moulded cover. Lift the rotor arm off the spindle, carefully levering with a screwdriver if it is tight.

Disconnect the vacuum unit link to the moving contact breaker plate and remove the two screws at the edge of the contact breaker base. The contact breaker assembly, complete with external terminal, can now be lifted off (see (i) below). Remove the circlip on the end of the micrometer timing screw and turn the micrometer nut until the screw and the vacuum unit

20

assembly are freed. Take care not to lose the ratchet and coil type springs located under the micrometer nut.

The complete shaft assembly, with automatic timing control and cam foot can now be removed from the distributor body (see (ii) below).

(i) Contact Breaker

To dismantle the assembly further, remove the nut, insulating piece and connections from the pillar on which the contact breaker spring is anchored. Slide out the terminal moulding. Lift off the contact breaker lever and the insulating washers beneath it. Remove the screw(s) securing the fixed contact plate, together with the spring and plain steel washers and take off the plate. Withdraw the single screw securing the capacitor and, on earlier models, the contact breaker earthing lead.

Dismantle the contact breaker base assembly by turning the base plate clockwise and pulling to release it from the moving contact breaker plate. On earlier models remove the circlip and star washer located under the base plate.

(ii) Shaft and Action Plate

To dismantle the assembly further, take out the screw inside the cam and remove the cam and cam foot. The weights, springs and toggles (when fitted) of the automatic timing control can now be lifted off the action plate. Note that a distance collar is fitted on the shaft underneath the action plate.

(f) Bearing Replacement

The single long bearing bush used in this distributor can be pressed out of the shank by means of a shouldered mandrel.

If the bearing has been removed the distributor must be assembled with a new bush fitted. The bush should be prepared for fitting by allowing it to stand completely immersed in medium viscosity (S.A.E.30—40) engine oil for at least 24 hours. In cases of extreme urgency, this period of soaking may be shortened by heating the oil to 100°C. for two hours, then allowing to cool before removing the bush.

Press the bearing into the shank, using a shouldered, polished mandrel of the same diameter as the shaft.

Under no circumstances should the bush be overbored by reamering or any other means, since this will impair the porosity and thereby the effective lubricating quality of the bush.

(g) Re-assembly

The following instructions assume that **complete** dismantling has been undertaken.

(i) Place the distance collar over the shaft, smear the shaft with Ragosine Molybdenised non-creep oil or clean engine oil, and fit it into its bearing.

(ii) Refit the vacuum unit into its housing and replace the springs, milled adjusting nut and securing circlip.

(iii) Re-assemble the centrifugal timing control. See that the springs are not stretched or damaged. Place the cam and cam foot assembly over the shaft, engaging the projections on the cam foot with the toggles, and fit the securing screw.

(iv) Before re-assembling the contact breaker base assembly, lightly smear the base plate with Ragosine Molybdenised non-creep oil or Mobilgrease No. 2. On earlier distributors, the felt pad under the rotating contact breaker plate should be moistened with a few drops of thin machine oil.

Fit the rotating plate to the contact breaker base plate and secure with the star washer and circlip. Refit the contact breaker base into the distributor body. Engage the link from the vacuum unit with the bearing bush in the rotating plate and secure with the split pin. Insert the two base plates securing screws, one of which also secures one end of the earthing lead.

(v) Fit the capacitor into position, on earlier models the eyelet on the other end of the contact breaker earthing lead is held under the capacitor fixing screw. Place the fixed contact plate in position and secure lightly with securing screw(s). One plain and one spring washer must be fitted under each of these screws.

(vi) Place the insulating washers on the contact breaker pivot post and on the pillar on which the end of the contact breaker spring locates. Refit the contact breaker lever and spring.

(vii) Slide the rubber terminal block into its slot.

(viii) Thread the low tension connector and capacitor eyelets on to the insulating piece, and place these on to the pillar which secures the end of the contact breaker spring. Refit the washer and securing nut.

(ix) Set the contact gap to 0.014″ to 0.016″ and tighten the securing screw(s) of the fixed contact plate.

(x) Refit the rotor arm, locating the moulded projection in the rotor arm with the keyway in the shaft and pushing fully home. Refit the moulded cover.

(h) Replacement Contacts

If the contacts are so badly worn that replacement is necessary, they must be renewed as a pair and not individually. The contact gap must be set to 0.014″ to 0.016″; after the first 500 miles running with new contacts fitted, the setting should be checked and the gap reset to 0.014″ to 0.016″. This procedure allows for the initial " bedding-in " of the heel.

HEADLAMPS—MODEL F700 MK/VI

1. General Description

The lamps incorporate a combined reflector and front lens assembly known as the Lucas Light Unit. They are fitted with a " prefocus " bulb which ensures that the filament is always positioned correctly with respect to the focal point of the reflector.

(a) Light Unit

The construction of the Light Unit ensures that the reflector surface is

effectively protected. The outer surface of the " Block-pattern " lens is smooth, to facilitate cleaning, but the inner surface has formed in it a series of small lenses which determine the spread and pattern of the light.

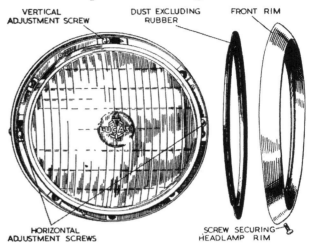

Fig. 23   **Headlamp with Front Rim and Dust-excluding Rubber removed.**

(b) Bulbs

The " prefocus " bulb eliminates the need for any focusing device in the lamp. The bulb cap is carried on a flange accurately positioned in relation to the filament during manufacture. A slot in the flange engages with a projection on the inside of the bulb holder at the back of the reflector,

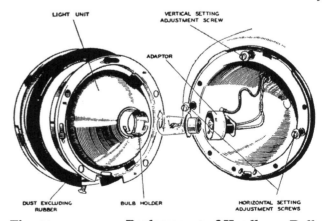

Fig. 24   **Replacement of Headlamp Bulb.**

thus ensuring the correct positioning of the filament. A bayonet-fitting cap with spring-loaded contacts secures the bulb firmly in position and also carries the supply to the bulb contacts.

2. BULB REPLACEMENT

Slacken the captive securing screw at the bottom of the front rim and remove the front rim and dust-excluding rubber. To remove the Light Unit assembly from the three spring-loaded screws, press the Unit inwards, turning it anti-clockwise to disengage the slotted holes in the seating rim from the setting adjustment screws. Disengage the bayonet fitting cap and withdraw the defective bulb from the Light Unit.

Re-assembly of the Light Unit to the lamp is a reversal of the above procedure.

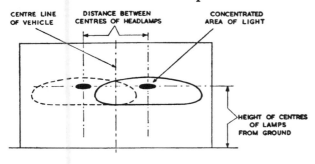

(A) FRONT OF VEHICLE TO BE SQUARE WITH SCREEN

(B) VEHICLE TO BE LOADED AND STANDING ON LEVEL GROUND

(C) RECOMMENDED DISTANCE FOR SETTING IS AT LEAST 25 FT

(D) FOR EASE OF SETTING ONE HEADLAMP SHOULD BE COVERED

Fig. 25 Diagram showing Headlamp Beam Setting.

3. SETTING

In overseas markets, lamps must be set to comply with local lighting regulations.

(a) Ministry of Transport Lighting Regulations (United Kingdom)

The Lighting Regulations state that a lighting system must be arranged so that it can give a light which is "incapable of dazzling any person standing on the same horizontal plane as the vehicle at a greater distance than twenty-five feet from the lamp, whose eye-level is not less than three feet six inches above that plane". The headlamp must therefore be set so that the main beams of light are parallel with the road and with each other.

(b) Adjustment of Setting

Slacken the captive securing screw at the bottom of the front rim and remove the rim and dust-excluding rubber. The spring-loaded adjustment screws are now accessible.

To adjust the vertical setting, turn the screw at the top of the lamp clockwise to raise the beam and anti-clockwise to lower the beam. Adjustment in the horizontal plane is effected by turning the two spring-loaded screws at the sides of the Light Unit.

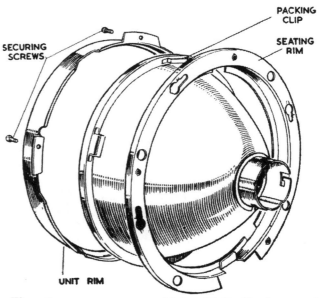

Fig. 26   Light Unit Replacement.

4. RENEWAL OF LIGHT UNIT

Remove the Light Unit and bulb. Withdraw the three small screws from the unit rim to separate the unit rim and seating rim from the Light Unit.

Position the replacement Light Unit on the seating rim, taking care to see that the locating clips at the edge of the Light Unit fit into the slots in the rim. Ensure that the unit rim is correctly positioned before securing in position by means of the three small screws. Refit the bulb, adapter, etc.

CONTROL BOX—Model RB106-1

1. GENERAL

The control box shown in Fig. 27, contains two units—a voltage regulator and a cut-out. Although combined structurally, the regulator and cut-out are electrically separate. Both are accurately adjusted during manufacture, and the cover protecting them should not be removed unnecessarily. Cable connections are secured by grub screw terminals.

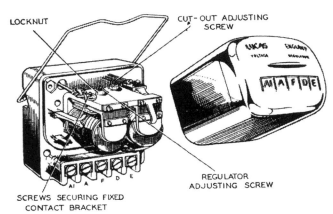

Fig. 27   **Control Box with Cover removed.**

The Regulator

The regulator is set to maintain the generator terminal voltage between close limits at all speeds above the regulating point, the field strength being controlled by the automatic insertion and withdrawal of a resistance in the generator field circuit. When the generator voltage reaches a pre-determined value, the magnetic flux in the regulator core due to the shunt or voltage winding becomes sufficiently strong to attract the armature to the core. This causes the contacts to open, thereby inserting the resistance in the generator field circuit.

The consequent reduction in the generator field current lowers the generator terminal voltage and this, in turn, weakens the magnetic flux in the regulator core. The armature therefore returns to its original position, and the contacts closing allow the generator voltage to rise again to its maximum value. This cycle is then repeated and an oscillation of the armature is maintained.

As the speed of the generator rises above that at which the regulator comes into operation, the periods of contact separation increase in length and, as a result, the mean value of the generator voltage undergoes practically no increase once this regulating speed has been attained.

The series or current winding provides a compensation on this system of control, for if the control were arranged entirely on the basis of voltage there would be a risk of seriously overloading the generator when the battery was in a low state of charge, particularly if the lamps were simultaneously in use.

Under these conditions of reduced battery voltage, the output to the battery rises and, but for the series winding, would exceed the normal rating of the generator. The magnetism due to the series winding assists the shunt winding, so that when the generator is delivering a heavy current into a discharged battery the regulator comes into operation at a somewhat reduced voltage, thus limiting the output accordingly. As

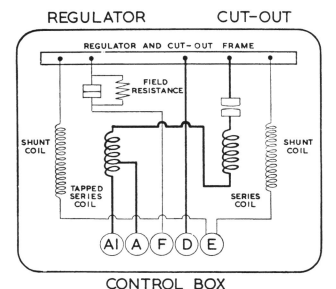

Fig. 28   **Internal Connections.**

shown in Fig. 28, a split series winding is used, terminal A being connected to the battery and terminal A1 to the lighting and ignition switch.

By means of a temperature compensation device the voltage characteristic of the generator is caused to conform more closely to that of the battery under all climatic conditions. In cold weather the voltage required to charge the battery increases, whilst in warm weather the voltage of the battery is lower. The method of compensation takes the form of a bi-metallic spring located behind the tensioning spring of the regulator armature. This bi-metallic spring, by causing the operating voltage of the regulator to be increased in cold weather and reduced in hot weather, compensates for the changing temperature-characteristics of the battery and prevents undue variation of the charging current which would otherwise occur.

24

The bi-metallic spring also compensates for effects due to increases in resistance of the copper windings from cold to working values.

The Cut-out

The cut-out is an electro-magnetically operated switch connected in the charging circuit between the generator and the battery. Its function is automatically to connect the generator with the battery when the voltage of the generator is sufficient to charge the battery, and to disconnect it when the generator is not running, or when its voltage falls below that of the battery, and so prevent the battery from discharging through and possibly damaging the generator windings.

The cut-out consists of an electro-magnet fitted with an armature which operates a pair of contacts. The electro-magnet employs two windings, a shunt winding of many turns of fine wire, and a series winding of a few turns of heavier gauge wire. The contacts are normally held open and are closed only when the magnetic pull of the magnet on the armature is sufficient to overcome the tension of the adjusting spring.

The operation of the cut-out is as follows : The shunt coil is connected across the generator. When the vehicle is starting, the speed of the engine, and thus the voltage of the generator, rises until the electro-magnet is sufficiently magnetised to overcome the spring tension and close the cut-out contacts. This completes the circuit between the generator and the battery through the series winding of the cut-out and the contacts. The effect of the charging current flowing through the cut-out windings creates a magnetic field in the same direction as that produced by the shunt winding. This increases the magnetic pull on the armature so that the contacts are firmly closed and cannot be separated by vibration. When the vehicle is stopping the speed of the generator is decreased until the generator voltage is lower than that of the battery. Current then flows from the battery through the cut-out series winding and generator in a reverse direction to the charging current. This reverse current through the cut-out will produce a differential action between the two windings and partly demagnetise the electro-magnet. The spring, which is under constant tension, then pulls the armature away from the magnet and opens the circuit. The contacts opening prevent further discharging of the battery through the generator.

Like the regulator, operation of the cut-out is temperature-controlled by means of a bi-metallic tensioning spring.

2. SETTING DATA

(a) Regulator

Open-circuit setting at 20°C. and 1,500 dynamo r.p.m. : 15.6—16.2 volts.

Note : For ambient temperatures other than 20°C., the following allowances should be made to the above setting :

For every 10°C. (18°F.) above 20°C., subtract 0.3 volt.

For every 10°C. below 20°C., add 0.3 volt.

(b) Cut-out

Cut-in voltage : 12.7—13.3
Drop-off voltage : 8.5—10.0
Reverse current : 3.5—5.0 amp.

3. SERVICING

(a) Testing in Position to Locate Fault in Charging Circuit

If the generator and battery are in order, check as follows :—

(i) Ensure that the wiring between battery and regulator is in order. To do this, disconnect the wire from the A terminal of the control box and connect the end of the wire removed to the negative terminal of a voltmeter. Connect the positive voltmeter terminal to an earthing point on the chassis. If a voltmeter reading is given, the wiring is in order and the regulator must be examined.

(ii) If there is no reading, examine the wiring between battery and control box for defective cables or loose connections.

(iii) Re-connect the wire to terminal A.

(b) Regulator Adjustment

The regulator is carefully set during manufacture and, in general, it should not be necessary to make further adjustment. If, however, the battery does not keep in a charged condition, or if the generator output does not fall when the battery is fully charged, the setting should be checked and, if necessary, corrected.

It is important before altering the regulator setting to check that the low state of charge of the battery is not due to a battery defect or to slipping of the generator belt.

(i) Electrical Setting

It is important that only a good quality MOVING COIL VOLT-METER (0—20 volts) is used when checking the regulator. The electrical setting can be checked without removing the cover from the control box.

Withdraw the cables from terminals A and A1 at the control box and connect these cables together.

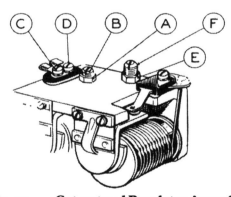

Fig. 29   Cut-out and Regulator Assembly.

Connect the negative lead of the voltmeter to control box terminal D and connect the other lead to terminal E.

Slowly increase the speed of the engine until the voltmeter needle " flicks " and then steadies. This should occur at a voltmeter reading between the appropriate limits given in Para. 2 (a) according to the ambient temperature.

If the voltage at which the reading becomes steady occurs outside these limits, the regulator must be adjusted.

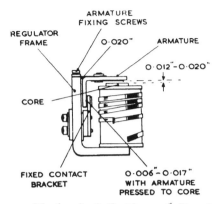

Fig. 30   Mechanical Setting of Regulator.

Shut off the engine and remove the control box cover.

Release locknut A (see Fig. 29) of adjusting screw B and turn the screw in a clockwise direction to raise the setting or in an anti-clockwise direction to lower the setting. Turn the screw only a fraction of a turn at a time and then tighten the locknut. Repeat as above until the correct setting is obtained.

Adjustment of regulator open-circuit voltage should be completed within 30 seconds, otherwise heating of the shunt winding will cause false settings to be made.

Remake the original connections. A generator run at high speed on open circuit will build up a high voltage. Therefore, when adjusting the regulator, do not run the engine up to more than half throttle or a false setting will be made.

(ii) Mechanical Setting

The mechanical or air-gap settings of the regulator, shown in Fig. 30, are accurately adjusted before leaving the works and, provided that the armature carrying the moving contact is not removed, these settings should not be tampered with. If, however, the armature has been removed, the regulator will have to be reset.

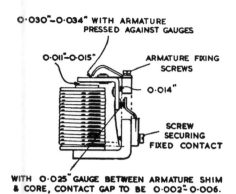

O·030"–O·034" WITH ARMATURE PRESSED AGAINST GAUGES

O·011"–O·015"

ARMATURE FIXING SCREWS

O·014"

SCREW SECURING FIXED CONTACT

WITH O·025" GAUGE BETWEEN ARMATURE SHIM & CORE, CONTACT GAP TO BE O·002–O·006.

Fig. 31 Mechanical Setting of Cut-out.

To do this proceed as follows :— Slacken the two armature fixing screws and also adjusting screw B. Insert a 0.020" feeler gauge between the back of the armature and the regulator frame. It is permissible for this gap to taper, either upwards or downwards, between the limits of 0.018" to 0.022".

With gauge in position, press back the armature against the regulator frame and tighten the two armature fixing screws. Remove the gauge and check the gap between the **shim** on the underside of the armature and the top of the core. This gap should be 0.012"—0.020". If the gap is outside these limits, correct by carefully bending the fixed contact bracket. Remove the gauge and press the armature down, when the gap between the contacts should be 0.006"—0.017".

(iii) Cleaning Contacts

After long periods of service it may be found necessary to clean the regulator contacts. The contacts are made accessible by slackening the screws securing the fixed contact bracket. It will be necessary to slacken screw C a little more than screw D (see Fig. 29) so that

the contact bracket can be swung outwards. Clean the contacts by means of fine carborundum stone or fine emery cloth.

Carefully wipe away all traces of dust or other foreign matter with methylated spirits (de-natured alcohol). Re-position the fixed contact bracket and tighten the securing screws.

(c) Cut-out Adjustment

(i) Electrical Setting

If the regulator is correctly set but the battery is still not being charged, the cut-out may be out of adjustment. To check the voltage at which the cut-out operates, remove the control box cover and connect the voltmeter between terminals D and E. Start the engine and slowly increase its speed until the cut-out contacts are seen to close, noting the voltage at which this occurs. This should be 12.7—13.3 volts.

If operation of the cut-out takes place outside these limits, it will be necessary to adjust. To do this, slacken locknut E (Fig. 29) and turn screw F in a clockwise direction to raise the voltage setting or in an anti-clockwise direction to reduce the setting. Turn the screw only a fraction of a turn at a time and then tighten the locknut. Test after each adjustment by increasing the engine speed and noting the voltmeter readings at the instant of contact closure. Electrical settings of the cut-out, like the regulator, must be made as quickly as possible because of the temperature-rise effects. Tighten the locknut after making the adjustment. If the cut-out does not operate, there may be an open circuit in the wiring of the cut-out and regulator unit, in which case the unit should be removed for examination or replacement.

(ii) Mechanical Setting

If for any reason the cut-out armature has to be removed from the frame, care must be taken to obtain the correct air-gap settings on re-assembly (see Fig. 31). These can be obtained as follows :—

Slacken the two armature fixing screws, adjusting screw F and the screw securing the fixed contact.

Insert a 0.014″ gauge between the back of the armature and the cut-out frame. (The air gap between the core face and the armature shim should now measure 0.011″—0.015″. If it does not, fit a new armature assembly.) Press the armature back against the gauge and tighten the armature fixing screws. With the gauge still in position, set the gap between the armature and the stop plate arm to 0.030″—0.034″ by carefully bending the stop plate arm. Remove the gauge and tighten the screw securing the fixed contact.

Insert a 0.025″ gauge between the core face and the armature. Press the armature down on to the gauge. The gap between the contacts should now measure 0.002″ to 0.006″ and the drop-off voltage should be between the limits given in Para. **2** (**b**). If necessary, adjust the gap by carefully bending the fixed contact bracket.

(iii) Cleaning Contacts

If the cut-out contacts appear rough or burnt, place a strip of fine glass paper between the contacts—then, with the contacts closed by hand, draw the paper through. This should be done two or three times with the rough side towards each contact. Wipe away all dust or other foreign matter, using a clean fluffless cloth moistened with methylated spirits (de-natured alcohol).

Do not use emery cloth or a carborundum stone for cleaning cut-out contacts.

WINDSCREEN WIPER CRT15

1. GENERAL

Normally the windscreen wiper will not require any servicing apart from the occasional renewal of the rubber blades. In the event of irregular working, first check for loose connections, chafed insulation, discharged battery, etc., before removing the gearbox or commutator covers.

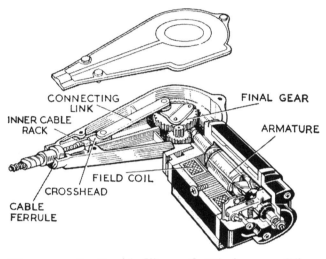

Fig. 32 Sectioned View of Windscreen Wiper Motor with Gearbox Cover removed.

(a) To Detach the Cable Rack from the Motor and Gearbox

Remove the gearbox cover.

Lift off the connecting link.

Disengage the outer casing, cable rack and crosshead from the gearbox.

Replace the gearbox cover to prevent the ingress of foreign matter.

(b) To Detach the Cable Rack from the Wheelboxes

Remove the wiper arms from the wheelbox spindles by slackening the collet nuts and continuing to rotate them until the arms are freed from the spindles. The cable rack can then be withdrawn from the outer casing for inspection. Before refitting the cable into the outer casing, see that the wheelbox gears are undamaged and thoroughly lubricate the cable rack with Duckham's HBB or an equivalent grease.

(c) Inspection of Commutator

Disconnect the wiper at its terminals and withdraw the three screws securing the cover at the commutator end. Lift off the cover. Clean the commutator, using a petrol-moistened cloth, taking care to remove any carbon dust from between the commutator segments.

(d) Inspection of Brush Gear

Check that the brushes bear freely on the commutator. If they are loose or do not make contact, a replacement tension spring is necessary. The brush levers must be free on their pivots. If they are stiff, they should be freed by working them backwards and forwards. Brushes which are considerably worn must be replaced.

(e) Motor Operates but does not Transmit Motion to Spindles

Remove the gearbox cover. A push-pull motion should be transmitted to the inner cable of the flexible rack. If the crosshead moves sluggishly between the guides, lightly smear a small amount of medium grade engine oil in the groove formed in the die-cast housing.

When overhauling, the gearbox must be lubricated by packing it with a grease of the zinc oxide base type.

2. FLASHING LIGHT DIRECTION INDICATORS

In the event of irregular operation of the flasher system, the following procedure should be followed :—

(a) Check the bulbs for broken filaments.

(b) Refer to the wiring diagram and check all flasher circuit connections.

(c) Switch on the ignition and :—

(i) Check with a voltmeter that flasher unit terminal B is at twelve volts with respect to the chassis.

(ii) Connect together flasher unit terminals B and L and operate the direction indicator switch.

If the lamps now light, the flasher unit is defective and must be replaced.

If the lamps do not light, the indicator switch is defective and must be replaced.

ELECTRIC WINDTONE HORNS — Models WT614 and WT618

1. GENERAL

Windtone horns depend for their operation on the vibration of an air column, excited at its resonant frequency, or a harmonic of it, by an electrically energised diaphragm. The horns are fitted in pairs, one horn having a higher note than the other. The horns differ in note by an interval of a major third. Earlier fitment WT614 and later WT618 horns are recognisable from each other by the different shape of their trumpet flares. High and low note horns can be distinguished by the letters " H " or " L " marked inside the trumpet flares.

(a) Note of Horn Unsatisfactory or Operation Intermittent

(i) Check that the bolts securing the horn bracket are tight and that the body or flare of the horn does not foul any other fixture. See that any units fitted near the horn are rigidly mounted, and do not vibrate when the horn is blown. Examine the cables of the horn circuit, renewing any that are badly worn or chafed. Ensure that all connections are tight, and that the connecting eyelets or nipples are firmly soldered to the cables.

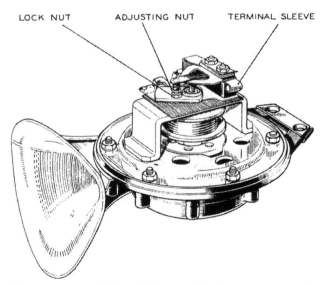

LOCK NUT   ADJUSTING NUT   TERMINAL SLEEVE

Fig. 33   WT.618 Horn with Cover removed.

(ii) Adjustment

Adjustment of the horn does not alter the pitch of the note, but takes up wear of the moving parts which if uncorrected, would result in loss of power and roughness of tone.

The horn must not be used repeatedly when out of adjustment, as the resulting excessive current may damage it. The maximum current consumption of a horn in correct adjustment is 6½ amps. for WT614 horns and 8 amps. for WT618 horns (the total current, taken by both horns together, will naturally be twice the figure quoted).

If it is desired to check the current consumption of the horns, break the circuit at some convenient point and connect an ammeter, 0—30 or 0—50 amps., in series with the horns.

If the consumption is in excess of 13 amps. for WT614 horns or 16 amps. for WT618 horns, it will be necessary to adjust the horns, even if they are apparently operating correctly. Horns will normally be tested with the car stationary and the battery at roughly its nominal voltage, but under running conditions with the battery charging the voltage may be appreciably higher, and may overload the horns if the latter are not in correct adjustment.

If the horns are badly out of adjustment, it will be necessary to short circuit the horn fuse, A1-A2, as otherwise the excessive current taken by the horns during the process of adjustment might result in its repeated blowing.

Withdraw the cover securing screws and remove the covers. Disconnect the supply lead from one horn, taking care that it cannot touch any part of the car and so cause a short circuit.

Horns must always be securely bolted down when carrying out an adjustment, and if it is necessary to remove a horn from the car for testing, it must always be firmly clamped by its securing bracket for the test or adjustment to be effective.

Slacken the locking nut on the fixed contact and rotate the adjusting nut in a clockwise direction until the contacts are just separated, as indicated by the horn failing to sound. Turn the adjusting nut **half a turn** in the opposite direction, and hold it while tightening down the locking nut. Check the current consumption of the horn, if the current is incorrect, make further very fine adjustments to the contact breaker, turning the adjusting screw in a clockwise direction in order to decrease the current, and *vice versa*.

Adjust the other horn in a similar manner.

(b) Internal Faults

If the note cannot be improved by adjustment of the contact breaker, examine the movement for the following faults :

(i) **Contacts badly worn,** so that correct adjustment is impossible. A new set of contacts, *i.e.*, moving contact and spring, and fixed contact and adjusting screw, must be fitted, and the horn adjusted as described above.

(ii) **Faulty resistance.** To prevent excessive sparking as the horn contacts separate, a carbon resistance is connected across the horn coil. The correct resistance valve is 8 ohms. On model WT618 horns the contact breaker terminal block is manufactured from a resistance material and this serves as the spark suppressing resistance. If the resistance becomes open circuited the horn note will become rough and fierce sparking will occur as the horn contacts separate.

(iii) **Steel push rod stiff** or jammed in its bush. Remove the contact breaker spring and work the push rod up and down to ease it. If necessary, clean the rod and bush with petrol to remove any accumulations of dirt or grease. The exposed portion of the rod should be smeared with a fairly thin grease (Duckham's H.B.B., or its equivalent), which will work down into the bush when the horn is blown.

(iv) **Push rod too slack,** causing rattle when the horn is blown. This will be due to the push rod having run dry of grease, with consequent excessive wear. A new push rod must be fitted. If, due to wear of the bush, the new push rod is also slack, no repair is possible and the horn must be replaced.

(v) **Armature fouling base plate.** There should be a clearance of approximately .020″ between the armature and the base plate. If the armature touches the base plate at any point, slacken the six screws securing the base plate and move the armature until it is centrally placed in the aperture. It is advisable to fit shims round the armature to hold it central while the securing screws are tightened.

(c) **Both Horns Fail to Operate**

Examine the fuse protecting the horn circuit. If it has blown, examine the wiring and horns for evidence of a short circuit. Renew any damaged leads, covering them with extra protective sleeving if necessary, and fit a new fuse into position.

If the fuse still blows, it is possible that the adjustment of one or both horns is badly out, and that as a result the current consumption is very greatly increased.

(d) **One Horn Fails to Operate**

Disconnect one lead from the terminal block of the second horn, taking care that it is not allowed to touch any part of the car.

Remove the cover of the faulty horn and examine the movement for the faults enumerated in Para. (**b**).

Pay particular attention to the internal wiring of the horn, which may have broken or become unsoldered as a result of vibration, and see that chafed insulation does not cause a partial or complete short circuit.

Note—All joints in the internal wiring of the horn must be firmly soldered using a non-corrosive flux.

(i) If the horns are removed for bench testing or adjustment, it is advisable to carry out an insulation test before replacement, testing between each terminal and the body with a 500-volt test set or similar equipment.

(ii) Under no circumstances must the movement be dismantled. If, after carrying out the above testing procedure, the fault has not been located, a new horn must be fitted.

ELECTRICAL EQUIPMENT

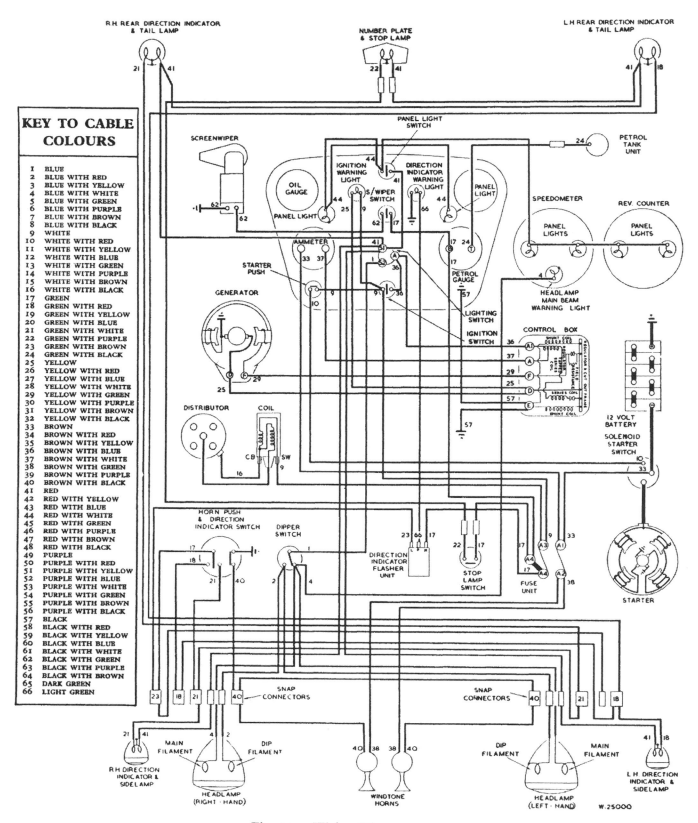

KEY TO CABLE COLOURS

1 BLUE
2 BLUE WITH RED
3 BLUE WITH YELLOW
4 BLUE WITH WHITE
5 BLUE WITH GREEN
6 BLUE WITH PURPLE
7 BLUE WITH BROWN
8 BLUE WITH BLACK
9 WHITE
10 WHITE WITH RED
11 WHITE WITH YELLOW
12 WHITE WITH BLUE
13 WHITE WITH GREEN
14 WHITE WITH PURPLE
15 WHITE WITH BROWN
16 WHITE WITH BLACK
17 GREEN
18 GREEN WITH RED
19 GREEN WITH YELLOW
20 GREEN WITH BLUE
21 GREEN WITH WHITE
22 GREEN WITH PURPLE
23 GREEN WITH BROWN
24 GREEN WITH BLACK
25 YELLOW
26 YELLOW WITH RED
27 YELLOW WITH BLUE
28 YELLOW WITH WHITE
29 YELLOW WITH GREEN
30 YELLOW WITH PURPLE
31 YELLOW WITH BROWN
32 YELLOW WITH BLACK
33 BROWN
34 BROWN WITH RED
35 BROWN WITH YELLOW
36 BROWN WITH BLUE
37 BROWN WITH WHITE
38 BROWN WITH GREEN
39 BROWN WITH PURPLE
40 BROWN WITH BLACK
41 RED
42 RED WITH YELLOW
43 RED WITH BLUE
44 RED WITH WHITE
45 RED WITH GREEN
46 RED WITH PURPLE
47 RED WITH BROWN
48 RED WITH BLACK
49 PURPLE
50 PURPLE WITH RED
51 PURPLE WITH YELLOW
52 PURPLE WITH BLUE
53 PURPLE WITH WHITE
54 PURPLE WITH GREEN
55 PURPLE WITH BROWN
56 PURPLE WITH BLACK
57 BLACK
58 BLACK WITH RED
59 BLACK WITH YELLOW
60 BLACK WITH BLUE
61 BLACK WITH WHITE
62 BLACK WITH GREEN
63 BLACK WITH PURPLE
64 BLACK WITH BROWN
65 DARK GREEN
66 LIGHT GREEN

Fig. 34   Wiring Diagram.

AUTOMATIC ADVANCE CURVE

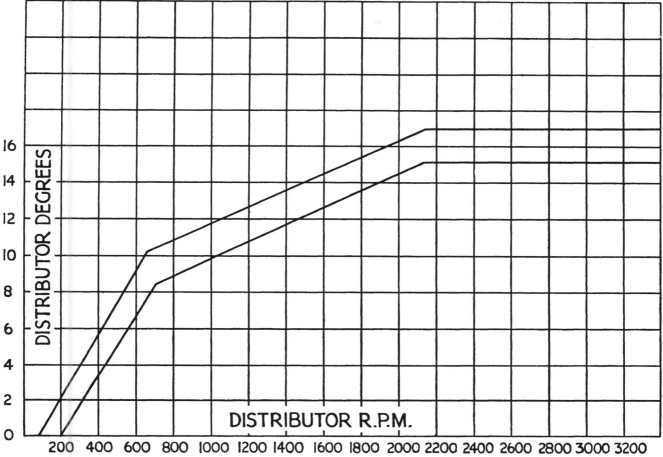

Fig. 35   Automatic Advance Curve.

CONTROL BOX. MODEL RB106-2

Later production cars were fitted with this control box, the function of which is identical to its predecessor, RB106/1.

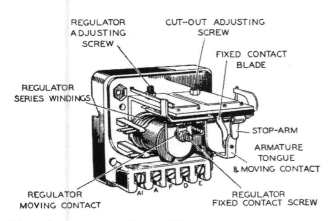

REGULATOR ADJUSTING SCREW

CUT-OUT ADJUSTING SCREW

FIXED CONTACT BLADE

REGULATOR SERIES WINDINGS

STOP-ARM

ARMATURE TONGUE & MOVING CONTACT

REGULATOR MOVING CONTACT

REGULATOR FIXED CONTACT SCREW

Fig. 36   **Control Box with cover removed.**

I. GENERAL

The control box, shown in Fig. 36, contains two units—a voltage regulator and a cut-out. Although combined structurally, the regulator and cut-out are electrically separate. Both are accurately adjusted during manufacture, and the cover protecting them should not be removed unnecessarily. Cable connections are secured by grub screw type terminals.

The Regulator

The regulator is set to maintain the generator terminal voltage between close limits at all speeds above the regulating point, the field strength being controlled by the automatic insertion and withdrawal of a resistance in the generator field circuit. When the generator voltage reaches a predetermined value, the magnetic flux in

33

the regulator core, due to the shunt or voltage winding, becomes sufficiently strong to attract the armature to the core. This causes the contacts to open, thereby inserting the resistance in the generator field circuit.

The consequent reduction in the generator field current lowers the generator terminal voltage, and this, in turn, weakens the magnetic flux in the regulator core. The armature therefore returns to its original position, and the contacts closing allow the generator voltage to rise again to its maximum value. This cycle is then repeated and an oscillation of the armature is maintained.

As the speed of the generator rises above that at which the regulator comes into operation, the periods of contact separation increase in length and, as a result, the mean value of the generator voltage undergoes practically no increase once this regulating speed has been attained.

The series or current winding provides a compensation on this system of control, for if the control were arranged entirely on the basis of voltage there would be a risk of seriously overloading the generator when the battery was in a low state of charge, particularly if the lamps were simultaneously in use.

Under these conditions of reduced battery voltage, the output to the battery rises and, but for the series winding, would exceed the normal rating of the generator. The magnetism due to the series winding assists the shunt winding, so that when the generator is delivering a heavy current into a discharged battery the regulator comes into operation at a somewhat reduced voltage, thus limiting the output accordingly. As shown in Fig. 37, a split series winding is used, terminal A being connected to the battery and terminal A1 to the lighting and ignition switch.

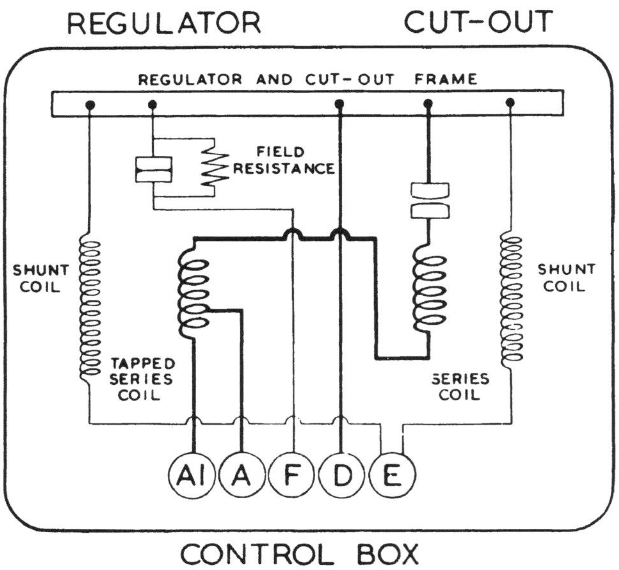

By means of a temperature compensation device, the voltage characteristic of the generator is caused to conform more closely to that of the battery under all climatic conditions. In cold weather the voltage required to charge the battery increases, whilst in warm weather the voltage required is lower. The method of compensation takes the form of a bi-metallic spring located behind the tensioning spring of the regulator armature. This bi-metallic spring, by causing the operating voltage of the regulator to be increased in cold weather and reduced in hot weather, compensates for the changing temperature-characteristics of the battery and prevents undue variation of the charging current which would otherwise occur.

The bi-metallic spring also compensates for effects due to increases in resistance of the copper windings from cold to working values.

The Cut-out

The cut-out is an electro-magnetically operated switch connected in the charging circuit between the generator and the battery. Its function is automatically to connect the generator with the battery when the voltage of the generator is sufficient to charge the battery, and to disconnect it when the generator is not running, or when its voltage falls below that of the battery, and so prevent the battery from discharging through and possibly damaging the generator windings.

The cut-out consists of an electro-magnet fitted with an armature which operates a pair of contacts. The electro-magnet employs two windings, a shunt winding of many turns of fine wire, and a series winding of a few turns of heavier gauge wire. The contacts are normally held open and are closed only when the magnetic pull

REGULATOR   CUT-OUT

CONTROL BOX

Fig. 37   **Internal connections of Control Box.**

of the magnet on the armature is sufficient to overcome the tension of the adjusting spring.

The operation of the cut-out is as follows :

The shunt coil is connected across the generator. When the vehicle is starting, the speed of the engine and thus the voltage of the generator, rises until the electro-magnet is sufficiently magnetised to overcome the spring tension and close the cut-out contacts. This completes the circuit between the generator and the battery through the series winding of the cut-out and the contacts. The effect of the charging current flowing through the cut-out windings creates a magnetic field in the same direction as that produced by the shunt winding. This increases the magnetic pull on the armature so that the contacts are firmly closed and cannot be separated by vibration. When the vehicle is stopping the speed of the generator falls until the generator voltage is lower than that of the battery. Current then flows from the battery through the cut-out series winding and generator in a reverse direction to the charging current. This reverse current through the cut-out will produce a differential action between the two windings and partly de-magnetise the electro-magnet. The spring, which is under constant tension, then pulls the armature away from the magnet and so separates the contacts and opens the circuit.

Like the regulator, operation of the cut-out is temperature-controlled by means of a bi-metallic tensioning spring.

2. SETTING DATA

(a) Regulator

Open-circuit setting at 20°C. and 1500 dynamo r.p.m. :   15.6—16.2 volts.

NOTE : For ambient temperatures other than 20°C. the following allowances should be made to the above setting :—

For every 10°C. (18°F.) above 20°C. subtract 0.3 volt.

For every 10°C. below 20°C. add 0.3 volt.

(b) Cut-out

Cut-in voltage :   12.7—13.3
Drop-off voltage :   8.5—11.0
Reverse current :   3.5— 5.0 amp.

3. SERVICING

(a) Testing in position to locate fault in charging circuit

If the generator and battery are in order, check as follows :—

(i) Ensure that the wiring between battery and regulator is in order. To do this, disconnect the wire from control box terminal " A " and connect the end of the wire removed to the negative terminal of a voltmeter.

Connect the positive voltmeter terminal to an earthing point on the chassis. If a voltmeter reading is given, the wiring is in order and the regulator must be examined.

(ii) If there is no reading, examine the wiring between battery and control box for defective cables or loose connections.

(iii) Re-connect the wire to control box terminal " A."

(b) Regulator Adjustment

The regulator is carefully set during manufacture and, in general, it should not be necessary to make further adjustment. If, however, the battery does not keep in a charged condition, or if the generator output does not fall when the battery is fully charged, the setting should be checked and, if necessary, corrected.

It is important before altering the regulator setting to check that the low state of charge of the battery is not due to a battery defect or to slipping of the generator belt.

(i) Electrical Setting

It is important that only a good quality MOVING COIL VOLT-METER (0-20 volts) is used when checking the regulator. The electrical setting can be checked without removing the cover from the control box.

Withdraw the cables from control box terminals A and A1 and connect these cables together.

Connect the negative lead of the voltmeter to control box terminal D, and connect the other lead to terminal E.

Slowly increase the speed of the engine until the voltmeter needle " flicks " and then steadies. This should occur at a voltmeter reading between the appropriate limits given in Para. 2 (**a**) according to the ambient temperature.

If the voltage at which the reading becomes steady occurs outside these limits, the regulator must be adjusted.

Shut off the engine and remove the control box cover.

Slacken the locknut of the voltage adjusting screw (see Fig. 38) and turn the screw in a clockwise direction to raise the setting or in an anti-clockwise direction to

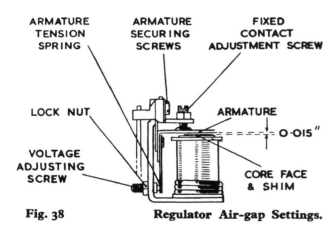

ARMATURE TENSION SPRING

ARMATURE SECURING SCREWS

FIXED CONTACT ADJUSTMENT SCREW

LOCK NUT

ARMATURE

0·015″

VOLTAGE ADJUSTING SCREW

CORE FACE & SHIM

Fig. 38　　　**Regulator Air-gap Settings.**

lower the setting. Turn the screw only a fraction of a turn at a time and then tighten the locknut. Repeat as above until the correct setting is obtained.

Adjustment of regulator open-circuit voltage should be completed within 30 seconds, otherwise heating of the shunt winding will cause false settings to be made.

Re-make the original connections.

A generator run at high speed on open circuit will build up a high voltage. Therefore, when adjusting the regulator, do not run the engine up to more than half throttle or a false setting will be made.

(ii) Mechanical Setting

The mechanical or air-gap settings of the regulator, shown in Fig. 38, are accurately adjusted before leaving the works and, provided that the armature carrying the moving contact is not removed, these settings should not be tampered with. If, however, the armature has been removed, the regulator will have to be reset. To do this proceed as follows :

Slacken the fixed contact locking nut and unscrew the contact screw until it is well clear of the armature moving contact.

Slacken the voltage adjusting screw locking nut and unscrew the adjuster until it is well clear of the armature tension spring.

Slacken the two armature assembly securing screws.

Using a 0.015″ thick feeler gauge, wide enough to cover completely the core face, insert the gauge between the armature and core shim, taking care not to turn up or damage the edge of the shim.

Press the armature **squarely** down against the gauge and re-tighten the two armature assembly securing screws.

With the gauge still in position, screw the adjustable contact down until it just touches the armature contact. Re-tighten the locking nut.

Reset the voltage adjusting screw as described under Para. 3 (**b**) (**i**).

(iii) Cleaning Contacts

After long periods of service it may be found necessary to clean the regulator contacts. Clean the

ELECTRICAL EQUIPMENT

contacts by means of fine carborundum stone or fine emery cloth.

Carefully wipe away all traces of dust or other foreign matter with methylated spirits (de - natured alcohol).

(c) Cut-out Adjustment

(i) Electrical Setting

If the regulator is correctly set but the battery is still not being charged, the cut-out may be out of adjustment. To check the voltage at which the cut-out operates, remove the control box cover and connect the voltmeter between terminals D and E. Start the engine and slowly increase its speed until the cut-out contacts are seen to close, noting the voltage at which this occurs. This should be 12.7—13.3 volts.

If operation of the cut-out takes place outside these limits, it will be necessary to adjust. To do this, slacken the locknut securing the cut-out adjusting screw (see Fig. 39) and turn this screw in a

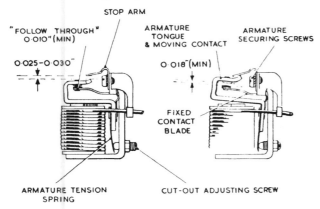

Fig. 39   **Cut-out Air-gap Settings.**

clockwise direction to raise the voltage setting or in an anticlockwise direction to reduce the setting. Turn the screw only a fraction of a turn at a time and then tighten the locknut. Test after each adjustment by increasing the engine speed and noting the voltmeter readings at the

instant of contact closure. Electrical settings of the cut-out, like the regulator, must be made as quickly as possible because of temperature-rise effects. Tighten the locknut after making the adjustment. If the cut-out does not operate, there may be an open circuit in the wiring of the cut-out and regulator unit, in which case the unit should be removed for examination or replacement.

(ii) Mechanical Setting

If for any reason the cut-out armature has to be removed from the frame, care must be taken to obtain the correct air-gap settings on re-assembly. These can be obtained as follows :

Slacken the adjusting screw locking nut and unscrew the cut-out adjusting screw until it is well clear of the armature tension spring.

Slacken the two armature securing screws.

Press the armature **squarely** down against the copper-sprayed core face and re-tighten the armature securing screws.

Using a pair of suitable pliers, adjust the gap between the armature stop arm and the armature tongue by bending the stop-arm. The gap must be 0.025"—0.030" when the armature is pressed squarely down against the core face.

Similarly, the fixed contact blade must be bent so that when the armature is pressed **squarely** down against the core face there is a **minimum** " follow - through," or blade deflection, of 0.010".

The contact gap, when the armature is in the free position, must be 0.018" minimum.

Reset the cut-out adjusting screw as described under Para. **3** (**c**) (**i**).

(iii) Cleaning Contacts

If the cut-out contacts appear rough or burnt, place a strip of fine glass paper between the contacts—then, with the contacts closed by hand, draw the paper through. This should be done two or three times with the rough side towards each contact. Wipe away all dust or other foreign matter, using a clean fluffless cloth moistened with methylated spirits (de-natured alcohol).

Do not use emery cloth or a carborundum stone for cleaning cut-out contacts.

Service Instruction Manual

BODY

SECTION N

BODY

INDEX

LIST OF ILLUSTRATIONS

BODY

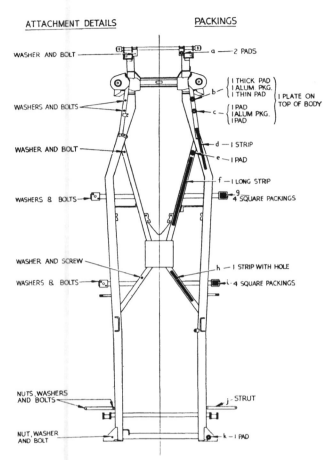

ATTACHMENT DETAILS   PACKINGS

WASHER AND BOLT

WASHERS AND BOLTS

WASHER AND BOLT

WASHERS & BOLTS

WASHER AND SCREW

WASHERS & BOLTS

NUTS, WASHERS
AND BOLTS

NUT, WASHER
AND BOLT

a — 2 PADS

b {
1 THICK PAD
1 ALUM. PKG.
1 THIN PAD
} 1 PLATE ON
TOP OF BODY

c {
1 PAD
1 ALUM PKG.
1 PAD
}

d — 1 STRIP

e — 1 PAD

f — 1 LONG STRIP

g
4 SQUARE PACKINGS

h — 1 STRIP WITH HOLE

i · 4 SQUARE PACKINGS

j - STRUT

k — 1 PAD

Fig. 1 Body Mounting Points. For clarification the attachment details and packing are shown on one side only.

1. BODY MOUNTING POINTS (Fig. 1)

(a) Point at front of chassis.
Two pads at each side.

(b) Upper points in side brace. An aluminium block sandwiched between a thick and thin pad, each side.

(c) Lower points in side brace.
Aluminium block with pads either side, each side.

(d) Along chassis side member and along side brace for approximately 2 inches.
One strip laid each side.

(e) On cruciform adjacent to clutch bell housing.
Two pads at each side.

(f) Along front cruciform member.
One strip each side.

(g) Front outrigger brackets.
Four square pads each side.

(h) Along rear cruciform member.
One strip with hole each side.

(i) Rear outrigger brackets.
Four square pads each side.

(j) Rear of rear wheel. Wing valance to chassis frame.
A metal stay secured to wing and chassis frame bracket by bolts, nuts and lock washers at each side.

(k) Rear end of chassis frame.
One pad at each side.

2. TO REMOVE BODY

(a) Working under the car.

(i) Remove centre tie rod assembly from drop arm.

(ii) Drain both hydraulic systems.

(iii) Drain petrol tank.

(iv) Disconnect petrol pipe at tank union.

(v) Free petrol vent pipe from clip at R.H. side chassis member.

(b) Working under the bonnet.

(i) Disconnect and remove battery.

(ii) Disconnect oil pressure pipe.

(iii) Disconnect clutch hydraulic union.

(iv) Remove L.T. cable from ignition coil.

(v) Withdraw rev. counter drive.

(vi) Disconnect the brake stop light cable.

(vii) Remove dip stick from engine sump.

(viii) Disconnect electrical connections at L.H. wing valance and wires from steering column centre if the car is L.H.S.

(ix) Remove water temperature gauge and free capilliary tube from petrol pipe.

(x) Remove radiator stays from corners of radiator.

(xi) Disconnect electrical connections at R.H. wing valance and wires from steering column centre if the car is R.H.S.

1

(**xii**) Remove carburettors after disconnecting control linkage.

(**xiii**) Remove cables from dynamo and starter motor.

(**xiv**) Disconnect brake hydraulic union.

(**xv**) Loosen steering column draught excluder clip.

(**c**) Working inside the car.
(**i**) Remove the seat cushions followed by the seat frames.

(**ii**) Remove the carpets.

(**iii**) Disconnect the electrical control wires for the overdrive (if one is fitted).

(**iv**) Free the gear lever grommet and push the rim through the tunnel aperture.

(**v**) Remove the gearbox tunnel after withdrawing battery box drain pipe.

(**vi**) Remove speedometer drive.

(**vii**) Remove control head and steering wheel.

(**viii**) Loosen steering column bracing.

(**ix**) Remove brake handle grip and protect thread with tape to prevent damage when body is lifted.

(**d**) Working at the front of the car.
(**i**) Remove front cowling. (See page 5.)

(**ii**) Remove front bumpers.

(**iii**) Remove steering column. (See " Steering " Section G.)

(**e**) Working at the rear of the car.
Remove o v e r - r i d e r s complete with brackets.

(**f**) Ensuring that the hand brake is on, the body can be lifted when the securing bolts or screws as shown in Fig. 1 have been withdrawn.

3. **TO FIT BODY**

The fitting of the body is the reversal of the removal but the following points should be noted.

(**a**) New packing pieces as detailed in " Body Mounting Points " (page 1) should be used and positioned on the chassis frame as shown in Fig. 1, a smear of "Bostick" C or similar compound to adhere packings to chassis will assist this operation.

(**b**) The thread of the handbrake lever should be protected with tape and the lever placed in the " On " position.

(**c**) It may be considered desirable to feed guide pins through the extreme front and rear mounting points of the body before lowering it to the chassis. Attachment bolts and screws are shown in Fig. 1.

(**d**) It is essential that sufficient sealing compound is used to effect a 100 % seal at the gearbox tunnel and floor inside the car.

(**e**) Care must be taken to connect the overdrive electric cable correctly as damage will result if this instruction is not followed.

(**f**) Both clutch and brake hydraulic systems must be bled at the completion of body replacement.

(**g**) The twin carburettors will need tuning before the car can be used.

4. **BATTERY BOX DRAIN (Fig. 2)**

A battery box drain tube has now been incorporated in normal manufacture and was introduced at Commission No. TS 3288. Retrospective action can be taken on earlier cars if so desired as shown in the illustration.

5. **TO REMOVE AND DISMANTLE FRONT BUMPER.**

It is possible to remove the front bumper from its four support brackets without first removing the latter from the chassis.

(**a**) Remove the over-riders by loosening the two nuts behind the inner support brackets. The over-riders can now be lifted free of the bolt head and the four mouldings collected.

(**b**) Remove the loosened nut followed by the lock and plain washer. It is suggested that the bolt remains loose at this juncture.

(**c**) The two outer support bracket nuts are now removed together with the lock and plain washers and the bolts withdrawn.

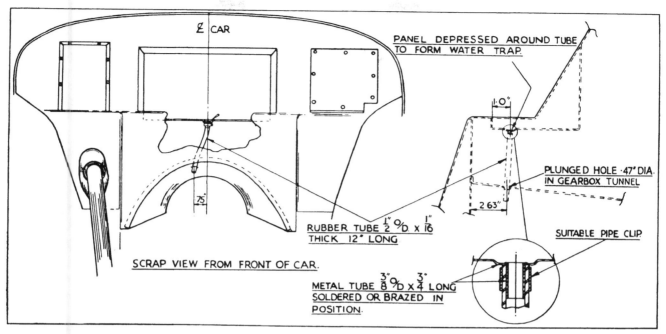

Fig. 2 Illustration giving details of Battery Box Drain for modifying cars prior to Commission No. T.S. 3288.

(**d**) The bumper can now be lifted away from its support brackets and the four metal packings and the two centre bolts collected.

(**e**) Withdraw two bolts from each pair of support brackets and chassis frame to release the four brackets. The two brackets on the steering column side have a secondary support from the lower steering column trunnion bracket bolt, and it may be necessary to loosen this bolt before the bumper support brackets on that side can be removed

6. TO FIT FRONT BUMPERS

Whilst it is possible to build the bumper assembly on the bench and then fit it to the car as a unit, it may be considered desirable to fit the support brackets to the chassis frame and then fit the bumper to the brackets.

The fitting procedure is the reverse of that for dismantling, but the following points should be noted.

(**a**) That an additional support is fitted to the brackets on the steering column side. This is a short plate with holes at each end. One end is fitted under the head of the lower steering column trunnion bracket bolt and the other end under the head of the front bumper support bracket bolt.

(**b**) The four strips of moulding should be placed between the contact edges of the over-riders and the bumper bar.

7. TO REMOVE REAR OVER-RIDERS AND BRACKETS

(**a**) Release the over-riders by loosening the nuts and then slide the over-riders off.

(**b**) Hold the head of the lower attachment bolt under the car and remove the nut, lock and plain washer and bolt.

(**c**) Hold the nut of the upper attachment bolt and withdraw bolt through the distance piece and support bracket. Collect the nut and plain washer and remove distance piece from body of car.

8. TO FIT REAR OVER-RIDERS

(**a**) Attach the support bracket to the chassis frame first at its bottom point by feeding the attachment bolt through the chassis frame into the bracket and attaching a plain and lock washer, but leave the nut loose at this juncture.

3

(b) Position the distance piece in the car body. Feed the bolt through the support bracket and a plain washer and thence into the distance piece, following with a second plain washer and then secure with a nut.

(c) The lower attachment can now be tightened.

(d) The over-rider attachment bolts are positioned in the brackets together with the plain and lock washers and nuts. The over-rider has a " key-hole " shaped aperture to accommodate the head of the attachment bolt, the nut of which is tightened when the over-rider is in position.

9. TO REMOVE FRONT WING

(a) Jack up the car and remove the appropriate road wheel.

(b) Withdraw the six bolts securing front wing to apron and the five bolts, the heads of which face the tyre tread.

(c) Remove the six bolts from on top of the wing, these are situated just beneath the side of the bonnet lid.

(d) Remove the door by withdrawing the seven bolts attaching the hinges to the door post and withdraw the nut and bolt from the door check strut. This gives access to six bolts at the extreme rear of the wing, these can now be removed.

(e) Remove the rubber grommet from inside the car and withdraw the bolt from inside the aperture.

(f) Remove the bulkhead sealer plate after withdrawing the five bolts from under the wing at rear of arch. Withdraw the three bolts situated underneath the sill and behind the arch opening.

(g) Free the lower rear end of the wing by pulling outward, then lift to disengage the flange of the wing abutting the dash panel.

10. TO FIT FRONT WING

This is the reversal of the removal but care should be taken to ensure all joints are watertight and that the door closes correctly. The sealing bead strip between the wing and apron is fitted with its hole uppermost.

11. TO REMOVE REAR WING

(a) Disconnect battery.

(b) Remove rear light unit by withdrawing two fixing screws and disconnecting the wires at the snap connectors. These will need identification marks if the code colours of the harness are not distinguishable.

(c) Jack up the car and remove the appropriate road wheel.

(d) Withdraw nine bolts from inside the wing running from the top of the wing to the lower front edge.

(e) Remove five bolts from inside the rear luggage compartment.

(f) Release wing/chassis stay by removing nut, bolt, lock and plain washer.

(g) Loosen three bolts on fixing flange of wing at extreme rear end.

(h) The wing can now be removed in a backward direction and the sealing strip collected.

12. TO FIT REAR WING

This is the reversal of the removal but care should be taken when replacing the sealing strip and the electrical wires, the latter should be carried out with regard to the diagram in the " Electrical Equipment ", Section M, or to the special identification markings.

13. TO REMOVE THE BONNET LID

(a) Release the bonnet locks either side by cable or by turning the Dzuz fastener and leave the bonnet resting in this lower position.

(b) Remove the four nuts and washers (two to each hinge) from under the dash inside the car.

(c) With an operator each side of the car lift the lid squarely upwards.

14. TO FIT THE BONNET LID

The fitting is the reversal of the removal. If the locks are cable operated the instructions on " Adjustment of Bonnet Locks ", page 5, should be followed.

15. TO REMOVE FRONT APRON

(a) Open the bonnet by releasing the locks from inside the car, or cars after Commission No. TS.4229 fitted with Dzuz fasteners at the forward corners of the bonnet lid by use of the carriage key. Prop the bonnet open and disconnect battery.

(b) Remove four bolts (two each side), which secure the top apron re-enforcement bar to the "U" brackets, situated on top of the front wings.

(c) Disconnect the electrical wires at their snap connectors after suitably identifying them if the colours are not distinguishable.

(d) If the car is earlier than Commission No. TS.4229, release the cable which connects the two locks from its clip. This clip is fitted at the centre of and forward of the apron re-inforcement bar. On cars later than TS.4229 this instruction can be disregarded.

(e) Remove the twelve bolts (six each side) which secure the outer edges of the apron to the wings. These bolts are those which are fitted horizontally from inside the wheel arches. The other series of bolts, fitted vertically into the wheel arch, are **NOT** to be touched.

(f) Remove the chassis frame to apron steady stay, at the apron end, by removal of the nut and bolt with lock washer.

(g) Withdraw the bolt from the starting handle guide bracket. There is no necessity to remove the bracket itself.

(h) The apron can now be removed by lifting the lower portion upward and forward to break the water seal and then lifting it bodily out of its brackets on top of the wing. The sealing beadings can now be removed.

16. TO FIT FRONT APRON

The fitting is the reversal of its removal but care should be taken over the following points.

(a) The sealing beading is adhered to the apron in such a manner that the hole is adjacent to the uppermost hole of the apron and the remaining slotted holes are adjacent to the lower holes.

(b) The electrical wires are connected with regard to their colour identifications and the wiring diagram as found in the "Electrical Equipment", Section M, or the special identifications if the colours are not distinguishable.

(c) On completion of the fitting the bonnet lid must be lowered gently to ascertain that the lock plungers and locks align correctly. (See notes below).

17. ADJUSTMENT TO BONNET LOCKS

On cars prior to Commission No. TS.4229 the bonnet locks were cable operated. It is essential when the bonnet lid or front apron have been removed that the bonnet locks are checked for alignment and the operating cables are correctly set.

(a) It must be positively determined that when the bonnet release knob is operated the release levers of the locks are pulled clear of the plunger apertures. This can be ascertained by an operator in the car and an observer at the locks. If the release lever is not fully clear the cable must be adjusted.

(b) Plunger centres and apertures must be identical. Longtitudal positioning of the plungers can be approximated by positioning on the lock centres. First attempt at closing the bonnet lid should be done with gentle pressure and the locking mechanism released. Any fouling of the plungers can be easily felt and adjustments made.

18. TO REMOVE WINDSCREEN

(a) Release the hood from the top of the windscreen.

(b) Remove windscreen wiper blades and arms.

(c) Turn the windscreen stanchion securing screws 90° anti-clockwise. Although these screws are spring loaded it may be necessary to ease the head outwards to ensure that the bolts are quite free.

5

(d) With operators each side of the car gently ease the windscreen assembly forward allowing the draught excluder to slide over the wiper blade spindles. The windscreen can be withdrawn and lifted from the car.

19. TO FIT WINDSCREEN

This is the reversal of the removal but the following points should be noted.

(a) The stanchion guides should be greased to prevent corrosion.

(b) After fitting the screen ensure that the draught excluder are in good condition and position correctly.

(c) Fit the windscreen wiper arms and blades and test for correct arcuate movement.

20. TO FIT AERO-WINDSCREEN

(a) Remove winsdcreen as described on page 5. The steady bracket can also be removed if desired.

(b) Withdraw the two chrome headed bolts on each side of the scuttle panel. Using these bolts attach the aero windscreen. The toe of the mounting bracket should point forward.

(c) If it is so desired the normal windscreen can be replaced with the aero-screens still in position.

21. TO REMOVE DOOR

(a) Withdraw the nut and bolt securing door check strap to the front door-post

(b) Withdraw the screws securing the two hinges to the front door post, four in upper hinge, three in lower hinge.

(c) The door can be lifted away.

22. TO FIT DOOR

The fitting of the door is the reversal of its removal but care should be taken to ensure that it hangs correctly and the lock engages with the dovetail on the rear post. It is suggested that the two hinges are not fully tightened and the door is closed slowly and gently. Any fouling will be immediately ascertained and the appropriate corrective action taken.

23. FRONT DOOR WATER SEALING

Additional water sealing at the top forward end of the doors was introduced in manufacture at Commission No. TS.5251. This sealing can be fitted to cars prior to this number. (Fig. 3)

This additional seal has been effected by the introduction of a rubber seal (Pt. No. 603257). This seal is fitted to the underside face of the front door post by six clips (Pt. No. 552901) in $\frac{1}{4}''$ diameter holes drilled in this face .19″ from the edge. A seventh and similar clip is fitted in the outward face of the pillar above the top of the hinge.

24. TO REMOVE DOOR LOCK

(a) Withdraw four screws securing front side screen retainer bracket, identify the component and its position.

(b) Remove upper end of trimmed lock pull strap by withdrawing screw.

(c) Remove rear side screen retainer bracket and identify.

(d) Remove dome nut from door lever and withdraw two screws to remove lock plate.

(e) Withdraw the screws and cup washers from edge of door trim and remove trim.

(f) The lock can be detached by removing the four screws holding the plate to the door frame.

(g) The door check can be removed by first removing the nut and bolt attaching the strap to the door post. Then remove from the door by withdrawing the two attachment screws.

25. TO FIT DOOR LOCK

The fitting of the door lock is the reversal of the removal. The following points should be noted.

(a) To ensure satisfactory operation of the lock it should be greased before fitting.

(b) After fitting the lock to the door frame it should be set in conjunction with the striker dovetail.

6

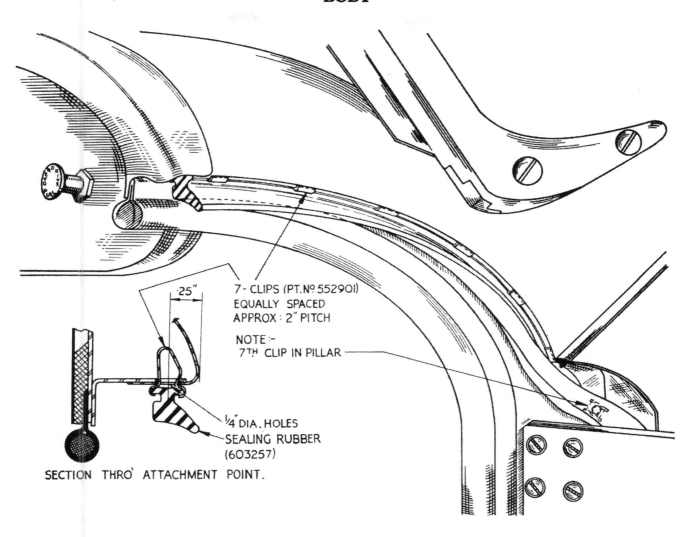

Fig. 3   Front Door Water Sealing.   For illustration purposes, only the right-hand door is shown.

(c) When fitting the side screen retainer brackets the correct position is only obtained by fitting them so that the heads of the locking screws face inwards. Having replaced the brackets it is a wise precaution to check the fitting of the side curtain.

26. REMOVAL OF GEARBOX TUNNEL

(a) Lift out seat cushions and remove eight nuts from each seat. Lift out seats.

(b) Remove front carpets and underfelts.

(c) Release hand brake and speedo drive draught excluder and slide this up the brake lever.

(d) Withdraw the sixteen fixing bolts around the flange of the tunnel. On early R.H.S. production cars it is necessary to remove the dipper switch and bracket (3 bolts).

(e) If the car is fitted with overdrive disconnect the electric control wires at their snap connectors and feed them through the aperture in the tunnel.

(f) Withdraw the drain pipe from fron portion of tunnel.

(g) Remove screws from gear leve grommet and push the rim of the grommet through the aperture.

7

(**h**) The tunnel can now be removed by levering up the rear end to break the water seal.

27. TO FIT THE GEARBOX TUNNEL

The fitting is the reversal of the removal, but the following points should be noted.

(**a**) It is essential that sufficient compound is used around the periphery of the tunnel to effect a good water seal.

(**b**) If the car is fitted with overdrive it will be necessary to feed the control wires through the aperture in the tunnel before finally bolting the latter in position. These wires must be correctly matched.

(**c**) On replacing the carpets an adhesive will be necessary.

(**d**) The dipper switch will need replacing on early production cars.

28. TO REMOVE HOOD AND FITTINGS

(**a**) Remove the hood by lifting the fasteners around the edge starting at the screen rail.

(**b**) The metal frame can be removed by withdrawing the screws and fastener pegs and aluminium plate securing the webbing strap to the rear elbow rail.

(**c**) Withdraw the two dome headed screws (one each side) securing the frame to the pivot bracket. The bracket can then be detached from the body by the withdrawal of four countersunk screws (2 each side).

(**d**) The webbing strap can be removed by withdrawing the two screws and aluminium plate at each attachment point.

(**e**) The frame is a riveted construction and unless any servicing is required the frame rivets should not be disturbed.

(**f**) The fastener pegs may be withdrawn from the body by turning the hexagon head. The canopy fasteners can be withdrawn by removal of the nut on the inside of the canopy, utilising a forked tool.

29. TO FIT HOOD AND FITTINGS

The fitting is the reversal of removal, but care should be taken with the following points.

(**a**) That the front draught excluder is in good condition.

(**b**) All canopy fasteners are securely fitted and operate correctly.

(**c**) All seams are fully watertight and if any new panels fitted or stitching carried out the stitching should be coated with " Everflex " Stitch Sealing Lacquer. See below.

30. WATER SEALING OF HOOD SEAMS

When panels have been replaced in the hood or tonneau cover it is essential that the stitching should be sealed. Failure to observe this instruction may cause water leaks not only at the seam itself but by the inner backing material acting as a wick and spreading the water to other parts of the component.

The sealing compound recommended is " Everflex " Stitch Sealing Lacquer. This is obtainable from our Spares Department in 4 oz. tins. It should be noted that the lacquer is highly inflammable and as such must comply with the limitations imposed upon transport and storing of such materials. The seams or stitching to be treated should be first carefully cleaned with a small nail brush using soap and water and then left to dry.

The " Everflex " Stitch Sealing Lacquer must be applied in a warm work shop, to dry material and to both sides of the seams. In no circumstances must it be allowed to come into contact with the transparent plastic windows owing to the solvent effect of this lacquer upon such material.

The lacquer should be applied by a brush with light even strokes and as it dries quickly excessive brushing must be avoided. Two coats are usually sufficient, allowing ten minutes drying time at room temperature between each coat.

Immediately upon completion of the lacquering the component should be heat treated to improve the bonding of the coating. Thirty minutes heat treatment at 220° F is recommended and should not be exceeded. The use of an infra red lamp should be avoided.

A lower temperature than that recommended in the previous paragraph may be used, or a hot air blast can be directed to the lacquer. If neither oven nor hot air blast is available the component can be left undisturbed in a warm atmosphere for 24 hours. Although reasonable sealing will be obtained by the instructions contained in this paragraph, the proper heat treatment at the higher temperature will provide the best possible water proofing.

31. ADJUSTMENT OF SIDE CURTAINS

An aluminium wedge with two tapped holes is attached to each side screen support stay by a single screw which fits in slotted apertures providing the adjustment.

It is by moving these wedges up or down the support stays that adjustment is obtained. When adjustments have been completed ensure that the press studs of the curtain align with those on the door panel and the support stays are secured in their sockets by knurled screws.

32. TO PREPARE CAR FOR FIBRE-GLASS HARD TOP CANOPY

(a) Remove hood and fittings as described on page 8.

(b) Withdraw the screws securing the three cappings to the rear elbow rail and the fixing screw of the front petrol tank trim. Protect the exterior of the car adjacent to the elbow rail with masking tape.

(c) Remove the millboard from the rear of the petrol tank by withdrawing the screws.

(d) Assemble the windscreen bracket and bridge pieces to the canopy.

(e) Position canopy on the windscreen and elbow rail of the car—windscreen *first*. Mark the position of the windscreen bracket holes on the flange.

(f) Remove the canopy from the car and drill the windscreen beading. Transfer windscreen brackets from the canopy to the beading and secure with the fixing screws. (Fig. 4).

Fig. 4   **Hard top Attachment Brackets fitted to Windscreen.**

(g) Reposition the canopy on the car and secure it to the windscreen. Check the position of the bridge pieces relative to the fixing holes in the elbow rail. If the holes do not align correctly it may be necessary to elongate the holes in the body. On cars previous to Commission No. TS.6820 these holes will need to be drilled. Mark the position of the brackets on the elbow rail and identify them to these positions. Release the canopy at the windscreen and remove from car.

(h) Remove the bridge pieces from the canopy and secure them to the elbow rail with screws (in accordance with their position and identification markings) to a tapping plate fed in from the rear luggage compartment. (Fig. 5).

In order to simplify this operation it is suggested that the shank of a 2BA bolt is brazed to one end of a carburettor choke control cable or similar piece of wire.

To this assembly, when fed through a bridge piece toward the rear of the car, can be attached a tapping plate. The wire is now drawn back into the car until the plate is positioned under the elbow rail. The plate can now be secured to the bridge piece by one screw and the second screw fitted when the wire has been removed.

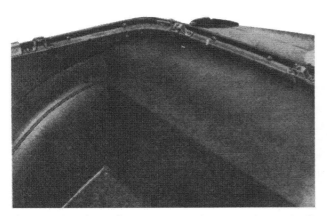

Fig. 5 **Bridge Pieces in position on Elbow Rail Channel.**

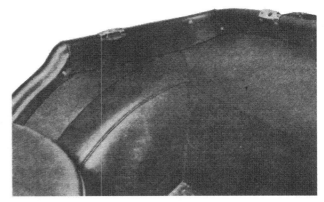

Fig. 6 **Protection Plates in position on R.H. and Centre Cappings.**

(i) Reposition the canopy on the car and secure to the windscreen brackets. Secure at the rear, setting the bridge pieces so that the bolts enter them correctly and obviating any possibility of cross threading

(j) Remove canopy, rear end *first*. Pencil on the body protection tape lines which correspond to the threaded centres of the bridge pieces.

(k) Position the cappings and transfer the markings on the body. On removing the capping drill a $\frac{3}{8}''$ dia. hole on each line to align with the tapping of the bridge piece.

(l) Fit the petrol tank trim in the rear luggage compartment. Remove protecting tape from the body of the car.

(m) Fit the screw securing the front petrol tank trim and secure the three cappings to the elbow rail. Fit four counter sunk screws and chromium washers (two each side) in the holes previously accommodating the hood bracket screws.

(n) Select the three norrow protection caps and position these on the rear cappings, aligning the apertures with the threaded centres of the bridge pieces : the $\frac{3}{8}''$ dia. holes may need elongating to permit this adjustment. Drill the cappings through the protection caps. Secure with two screws each. The two larger caps are fitted similarly to the side elbow rails. (Fig. 6).

33 TO FIT FIBREGLASS HARD TOP CANOPY

(a) The canopy is positioned on the car and secured to the windscreen first.

(b) The rear of the canopy is then secured to the elbow rail with five bolts. (Fig 7).

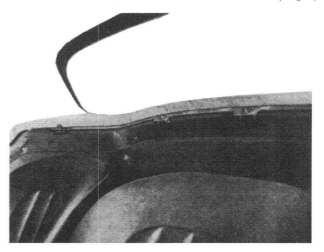

Fig. 7 **Hard Top positioned on Elbow Rail at side and rear.**

To position the canopy correctly it may be necessary to spring it over the rear elbow rail. This is permissible owing to the flexible nature of the fibreglass material.

(c) The sidescreens are adjusted (see page 9), so that their front edges fit inside the windscreen side beading and the top and rear edges fit as close to the canopy as possible.

On initial fitting of the fibreglass canopy it may be necessary to remove and reposition the sidescreen retainer brackets.

34. TO REMOVE FIBREGLASS HARD TOP CANOPY

It is essential that the following instructions are carried out in the sequence mentioned, difficulty may be experienced if operations (**b**) and (**c**) are reversed.

(**a**) Remove the side screens from the doors by loosening knurled nuts and lifting side screens.

(**b**) Withdraw the five bolts securing the rear of the canopy to the elbow rail. These bolts are " waisted " to retain them in the mounting flange of the canopy and care must be exercised during their removal to ensure that the shank below the " waist " does not become locked in the mounting flange.

(**c**) Similarly, withdraw the three bolts securing the front of the canopy to the windscreen flange.

(**d**) With an operator either side of the car, lift the canopy and carry it rearwards to effect its final removal.

35. TO REMOVE LUGGAGE BOOT LID

Before dismantling, the hinges and carriage locks should be marked as they are handed.

(**a**) The lid is opened and the two nuts and shakeproof washers removed from each hinge. The right-hand hinge also accommodates the boot lid stay rod. The lid is now moved clear.

(**b**) The hinges can be removed by first removing the front trim of the luggage boot to gain access to their attachment nuts. Two nuts and shakeproof washers are removed to withdraw each hinge.

(**c**) The two carriage locks are removed by withdrawing the two fixing screws each. These locks should be marked as they are handed.

(**d**) The escutcheons are removed by withdrawing two screws from each.

(**e**) The centre lock is removed from the lid by first withdrawing the bolt securing the lock latch to the lock shaft and collecting shake proof washer, then removing the nut securing the lock barrel to the boot lid.

36. TO FIT LUGGAGE BOOT LID

The fitting is the reversal of the removal but care should be taken over the following points.

(**a**) The hinges and carriage locks are handed and should be fitted to their appropriate sides.

(**b**) The aperture rubber seal should be in good condition. The drain pipes at the rearmost corners should also be inspected for condition. It is a wise precaution to feed a thin wire through these pipes to ensure that the passage way is clear.

(**c**) On replacing the lid to the hinges the attachment nuts should be loose at this juncture. The lid should then be lowered into position to ascertain that it is central in its aperture. The nuts are then fully tightened.

37. TO DISMANTLE SPARE WHEEL LID

(**a**) The lid is removed by turning the carriage locks.

(**b**) The locks are removed by withdrawing the four attachment screws (two to each lock). These locks should be marked as they are handed.

(**c**) The escutcheon plates are removed by withdrawing four screws (two to each plate).

(**d**) The wheel and tool securing straps are removed from inside the wheel compartment by withdrawing the two screws for each strap staple.

38. TO ASSEMBLE SPARE WHEEL LID

The assembly and fitting of the spare wheel lid is the reversal of the removal and dismantling. The following points should be noted.

(**a**) The buckle end of the strap should always be fitted to the floor.

(**b**) The locks are handed and should be fitted to the correct side.

TO FIT SMITHS CIRCULAR HEATER C.H.S. 920/4

The following procedure for carrying out this installation is recommended :—

1. Disconnect the battery lead.

2. Drain the cooling system and remove the two square headed plugs, one from the rear of the cylinder head and the other from the water pump housing.

3. Fit the taper threaded tap (28)(Fig. 16) into the tapped hole at the rear of the cylinder head and screw into the tap the special extension (27), so that this protrudes from the engine on the R.H. side of the unit (Fig. 8).

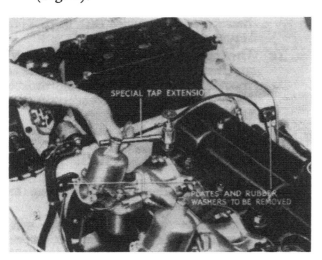

Fig. 8   **Fitting Tap Extension**

4. Install the taper threaded end of the female adapter (32) into the back of the pump housing. Attach the metal return pipe (29) to this adapter with the olive and union nut. Secure the pipe steady bracket to the rear of the two ignition coil fixing bolts.

5. Remove the two plates and rubber washers —one from each side of the bulkhead— after withdrawing the chamfer headed screws. Assemble the metal water pipe connecters (22) with their rubber washers (21) into these two apertures, securing each with two chamfer headed screws (Fig. 9).

6. Attach the two short lengths of rubber water hose (26) to the forward ends of these metal connecters, fitting the other ends of

Fig. 9   **Assembling Water Pipe Connecters**

these hoses to the previously installed tap adapter tube and the metal return pipe on the right and left sides of the car respectively.

7. Remove the trimmed glove casing after the withdrawal of the four P.K. screws.

8. Working underneath the dashboard, remove the four nuts, spring and plain washers—two from each side—those on the steering side of the car secure the "U" shaped steering bracket support rod.

Fig. 10 Releasing Steering Support Rod when fitting Demister Nozzle to steering side of car

It will be necessary to drop this support rod clear of the studs and this will be facilitated by slackening off the upper nuts on each arm of the "U" (Fig. 10). Locate the demister nozzles (1) on the two pairs of studs ensuring that they are above and clear of the screen wiper drive cable. Reposition "U" shaped rod and refit nuts and washers on the four studs and retighten with a suitable spanner.

9. At this point it is advisable to install the electrical control switch (Fig. 11), a hole for which is already provided in the dashboard. For the sake of appearance the hole in the dashboard is covered by trim material until it is required. This covering of trim can easily be cut away with a small sharp blade, after location of the hole with the tip of a finger, its position is approximately 4″ from the steering end of the dashboard at a point $2\frac{5}{8}$″ from the lower edge of the dash panel.

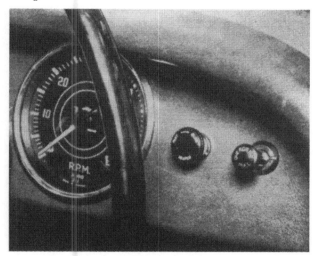

Fig. 11 Showing location of Heater Control Switch.

10. One side of the control switch (13) should be connected to the live side (L.H.) of the windscreen wiper switch. Attach the length of wire (12) supplied at one end with a snap connecter nipple to the other side of the switch leaving the completion of the circuit until operation 16.

11. Attach the mounting bracket (15) to the heater unit securing it with three spring washers and nuts. Assemble the two longer lengths of water hose (20) on to the adapters on the heater and secure with clips (19). Fit and secure alloy elbow piece (5) to heater unit (Fig. 12).

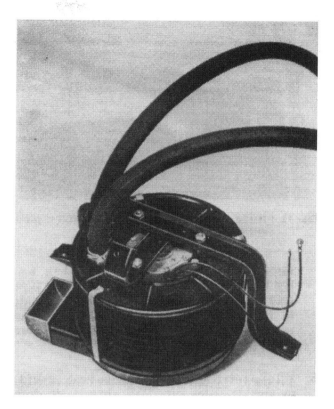

Fig. 12 The Heater Unit ready for assembling into position

12. Working under the bonnet, remove the centrally positioned rubber grommet from above the battery.

13. Install the Heater Unit, after fitting the two P.K. spire nuts (16) on either side of the Heater Unit mounting bracket, and position the unit so that the stud on the forward stay of the bracket protrudes through the hole from which the grommet was removed (Operation 12), securing with nut and a spring washer (18). Next attach the transverse portion of the heater attachment bracket with the two bolts to the forward of two central slots in each of two panel stays.

NOTE—When fitting this equipment to an early car which is equipped with an electrically operated overdrive ensure that the heater unit does not foul the overdrive relay and cause a short circuit. If such a condition arises suitably reposition the relay.

14. Assemble the free ends of the two longer hoses, already fitted to the heater unit, on

13

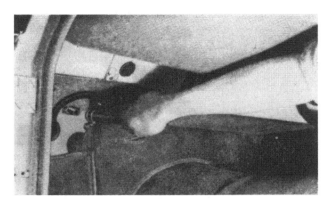

Fig. 13 Fitting Heater Hoses on water pipe connecters

their respective connecters (Fig. 13), *i.e.*, the hose on the L.H. side to the water pipe return connecter and that on the other side to the connecters for the feed hose, and secure with clips. These connecters were fitted in operation 5.

15. Fit the two lengths of demister hose (2)&(3) to the demister pipe "Y" shaped air duct (4) and install into the alloy elbow piece (5) (fitted in No. 11) on the heater unit (Fig. 14). The longer length of hose should be attached to the L.H. side demister nozzle and the shorter to the R.H. side demister nozzle.

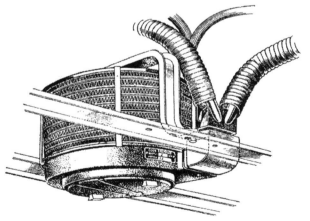

Fig. 14   Heater Unit in assembled position

16. To complete the electrical circuit connect the nipple on the free end of the cable attached to the control switch into a snap connecter (11) on the feed wire (10) already attached to the heater unit. The earth wire (9) from the heater unit should then be secured to the L.H. dash bracket by one of its forward screws.

17. Replace the trimmed glove box casing.

18. Replenish cooling system, ensuring that the heater tap (28) is turned on and the cooling system drain taps are turned off.

19. Reconnect detached battery lead.

20. If, when the engine is warm, the heater and demister nozzles still blow cold air it is probably due to air in the water system. To overcome this it will be necessary to slacken off the water pipes one at a time from their connecters, working in the direction of circulation, increasing the revolutions of the engine occasionally to help circulate the water. This operation should be carried out with the radiator filler cap removed.

Fig. 15 Showing position of Delivery and Return water pipes

NOTE—The Heater Kit for this Model is supplied under Part No. 551877, and a copy of these instructions will be packed in each carton.

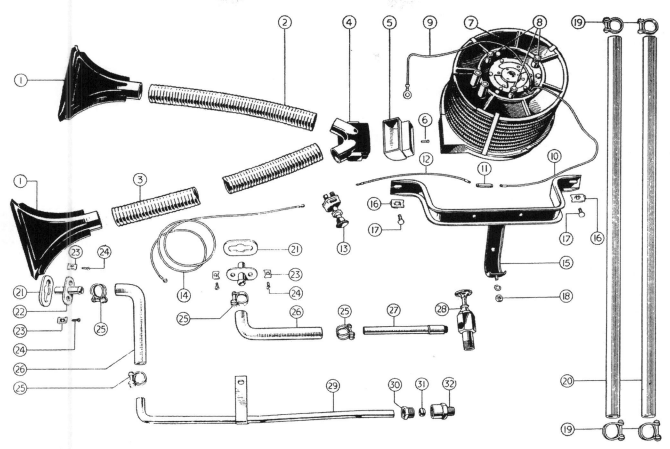

Fig. 16 Exploded view of Heater kit

NOTATIONS

Ref. No.	Description	Ref. No.	Description
1.	Demister Nozzle (2 off).	18.	Nut with Spring Washer for securing Forward Stay of Attachment Bracket.
2.	Demister Hose, R.H.		
3.	Demister Hose, L.H.		
4.	Demister Pipe "Y" shaped Air Duct.	19.	Large Diameter Pipe Clip (4 off).
5.	Alloy Elbow Piece.	20.	Long lengths of Heater Hose (2 off).
6.	Elbow Piece Securing Screw.	21.	Rubber Washer (2 off).
7.	Heater Unit.	22.	Metal Water Pipe Connecter (2 off).
8.	Securing Nuts for Attachment Bracket (3 off), Spring Washers (3 off)	23.	P.K. Spire Nuts, Small (4 off).
		24.	P.K. Spire Screws (4 off).
9.	Earth Wire.	25.	Heater Pipe Clip, Small Size (4 off).
10.	Feed Wire to Heater Unit.	26.	Short length of Rubber Water Hose (2 off).
11.	Snap Connecter.		
12.	Feed Wire from Control Switch.	27.	Special Tap Extension.
13.	Control Switch.	28.	Taper threaded Tap.
14.	Feed Wire from Live Side of Windscreen Wiper Switch.	29.	Metal Water Return Pipe.
		30.	Union Nut.
15.	Heater Unit Mounting Bracket.	31.	Olive.
16.	P.K. Spire Nuts, Large (2 off).	32.	Taper threaded Female Adapter.
17.	P.K. Spire Bolts (2 off).		

Service Instruction Manual

FUEL SYSTEM

SECTION P

FUEL SYSTEM

INDEX

LIST OF ILLUSTRATIONS

FUEL SYSTEM

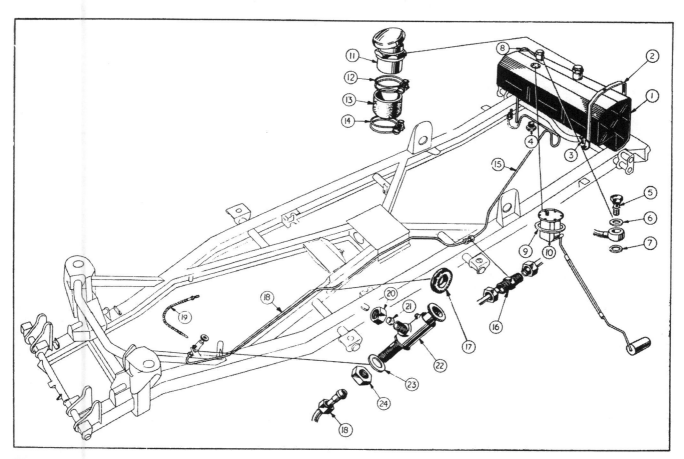

Fig. 1   Exploded view of Petrol Tank and Pipe Lines.

NOTATION FOR Fig. 1			
Ref. No.	Description	Ref. No.	Description
1	Petrol tank	13	Rubber hose connection
2	Petrol tank strap	14	Lower hose clip
3	Petrol tank strap fixing blot	15	Petrol pipe tank to connection
4	Drain plug	16	Pipe connection
5	Banjo bolt for vent pipe	17	Rubber grommet
6	Fibre washer above banjo connection	18	Petrol pipe (connection to stop tap)
7	Fibre washer below banjo connection	19	Flexible hose
8	Vent pipe	20	Stop tap outlet union nut
9	Cork washer	21	Brass olive
10	Petrol tank gauge unit	22	Petrol stop tap
11	Petrol filler cap and neck assembly	23	Plain washer
12	Upper hose clip	24	Jam nut for top attachment

1

1. DATA AND DESCRIPTION

(a) Tank capacity

12½ gallons (no reserve).

(b) Petrol Stop Tap

Situated on the left-hand side of the chassis frame and is connected to the petrol pump by a flexible hose.

(c) Petrol Pump

A.C. type " UE " camshaft driven situated on left-hand side of engine.

(d) Carburettors

Twin S.U. type H4 fitted to interconnected manifold on right-hand side of engine.

Standard needle FV.

For high speed and competition work GC needles.

(e) Air Cleaners

A.C. Shpinx type 7222575. Oil damped. One fitted to each carburettor.

The petrol tank is situated forward of the luggage boot and access is gained by removing the trim from the rear of the driver's cockpit. The filler cap is a press button release type centrally situated forward of the luggage boot. Looking forward from the rear, the vent pipe and capacity gauge tank unit are situated on the upper right-hand side of the tank and the pipe feed is taken from the lower right-hand side. Provision is made for draining, the plug being centrally situated on the underside of the tank.

The petrol feed pipe is brought forward and to the left-hand side of the chassis. As the level of the fuel is above that of the petrol pump union a petrol stop tap is incorporated in the pipe line. This will facilitate the disconnection of this union without first draining the petrol tank. The tap is fitted to a welded fork bracket on the left-hand chassis frame member. A flexible hose connects the tap to the fuel lift pump.

From the petrol pump a metal pipe passes round the front of the engine, to the twin S.U. carburettors.

Each carburettor is fitted with an individual oil damped A.C. air filter.

2. TO REMOVE PETROL TANK

(a) Drain the petrol from the tank by the centrally situated drain plug in the underside of the tank.

(b) Remove the centre capping of the rear elbow rail by withdrawing the securing screws. Slide this capping to one side until its other end is clear of the side capping. The centre can now be withdrawn.

(c) Remove the carpet fixing screws and ease up carpet to withdraw tank cover board fixing screws, by removing the latter the board can be eased away from the side capping and the upper retaining clips.

(d) Remove the rear cover board from inside the luggage boot. The lower fixing screws are under the front edge of the carpet.

(e) Loosen hose clips on filler pipe assembly and unscrew filler cap. Ease the short hose from the filler neck of the tank. Remove banjo bolt securing vent pipe to tank.

(f) Remove cable from petrol gauge tank unit.

(g) Remove petrol feed pipe from underside of tank. This may have already been disconnected to facilitate draining.

(h) Remove the four tank securing bolts and the lock washers followed by the tank straps and felts.

(i) The tank can be removed from the car in a forward direction. Tape the opening of the tank as a precaution against the entry of dirt.

2

3. TO FIT PETROL TANK

After ensuring that the tank is perfectly sound and clean, it can be replaced in the car.

The recommended method of testing the tank is to clean the exterior with a wire brush, blank off the filler pipe and all but one union then connect to a compressed air line. Submerge the tank in water and slowly fill the tank with air. Faults will clearly be seen by escaping air.

The replacement of the tank is the reversal of the removal.

It is a wise precaution to run the engine for a short time to observe the connections for leaks before replacing the trim.

4. PETROL GAUGE

Description

The petrol gauge comprises two components, the dashboard meter and the tank unit.

The dashboard meter consists of a metal case, containing the coils and shaped knob pieces which operate the gauge, also a bezel with a calibrated dial and indicator needle.

The coils are wound on bakelite bobbins with soft iron cores and the shaped knob pieces exert a magnetic force on a pivotted iron armature which is attached to the indicator. The magnetic force of the two coils cause the armature to be deflected in accordance with the amount of petrol in the tank. The connections of these coils and a resistance mounted below the armature are shown in the wiring diagram, Fig. 2.

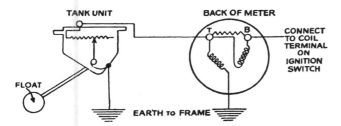

Fig. 2 Theoretical circuit of Fuel Gauge, Tank Unit and Motor.

The voltage across each coil is varied according to the position of the tank unit float arm.

The tank unit consists of a float and float arm mounted in a zinc based die casting. The float arm carries a contact arm which travels over a resistance wound on a bakelite former. The contact arm takes up a position according to the quantity of petrol in the tank and so varies the current through to the meter.

5. PRECAUTION WHEN CARRYING OUT TESTS

In no circumstances should the battery supply be connected directly to the terminal of the tank unit.

On no account should the float arm be bent or set to any other shape than that when it is supplied.

The float arm is provided with top and bottom stops which prevent the contact arm over-riding the resistance.

6. TO TEST DASH METER

The following tests will indicate whether the dash meter is functioning satisfactorily.

(a) Disconnect the wire from terminal "T" and switch on ignition. The dash meter should read full.

(b) With the wire to terminal "T" still disconnected, connect the wire to the car or connect to earth by a similar method. The meter should read empty when the ignition is switched on.

7. TO TEST TANK UNIT

(a) Remove unit from tank.

(b) Check the float arm for freedom of movement.

(c) Having checked the dash meter and found it to be satisfactory, connect terminal "T" of the tank to terminal "T" of the meter.

(d) Connect tank unit body casting to body of dash meter.

(e) Switch on ignition and the reading of the meter will vary according to the position of the float arm. If the dash meter indicates "full" irrespective to the position of the float arm, the tank unit is faulty and should be replaced.

3

FUEL SYSTEM

FUEL GAUGE FAULT LOCATION

SYMPTOM	CAUSE		REMEDY
No Reading	(1)	Meter supply interrupted.	Reconnect wires.
	(2)	Meter case not earthed.	Connect case or fix to earth
	(3)	Tank unit cable earthed.	Replace cable.
Meter reads full.	(4)	Tank unit cable broken or disconnected.	Reconnect.

8. TO REMOVE FLEXIBLE PETROL FEED PIPE

In no circumstances must an attempt be made to remove this hose from the lift pump without first diconnecting it from the petrol stop tap.

(**a**) Turn off petrol at the stop tap.

(**b**) Loosen the union nut securing the flexible hose to the tap and withdraw its rigid end together with olive and union nut.

(**c**) Remove hose from the pump by turning the entire length of the hose.

9. TO FIT FLEXIBLE PETROL FEED HOSE

Do not attempt to twist the hose without allowing its entire length to turn.

(**a**) Attach the hose to the petrol pump and secure to make a petrol tight joint.

(**b**) To the rigid end feed on the union nut and the olive.

(**c**) Position this rigid end in the petrol stop tap so that it reaches the bottom of its bore. Secure with union nut, the tightening of the union nut seat the olive and make a petrol tight joint.

(**d**) Open petrol tap and using hand primer on the petrol pump prime the system to ensure carburettor float chambers are full.

(**e**) Start engine and run for a little while observing the connections for leaks.

10. PETROL STOP TAP
Description
The tap, fitted at the end of the rigid petrol line, is secured to the chassis by a special welded fork bracket to the L.H. side chassis frame brace.

It is an Ewarts " pull and push " type which can be locked in the " on " position by turning the plunger head in an anti-clockwise direction approximately $\frac{1}{8}''$ of a turn.

The purpose of this tap is to facilitate the disconnection of the petrol pipe at the pump without first draining the petrol tank as the level of the petrol in the tank is above that of the pump.

11. TO REMOVE PETROL STOP TAP

(**a**) Drain the petrol tank.

(**b**) Remove the union of the flexible hose and withdraw from outlet connection of the tap body.

(**c**) Remove the union nut from the lower extremity of the tap and ease out the rigid petrol supply pipe.

(**d**) Loosen the jam nut situated on the underside of the welded fork bracket. The tap can now be lifted out of the fork.

12. TO FIT PETROL STOP TAP

(**a**) To the threaded stem of the tap attach the securing nut and plain washer. Screw the nut until it is approximately $\frac{1}{4}''$ from the abutment shoulder.

(b) Fit the tap into the fork bracket so that the feed to the pump is uppermost. The two flats on the tap body will assist in locating its position. Secure the tap to bracket by tightening the jam nut.

(c) Position the rigid petrol feed pipe from tank into lower portion of tap and ensure that the olive is seated before the union nut is attached and tightened.

(d) Attach the flexible hose from pump to outlet connection of the tap and secure to give a petrol tight joint.

(e) Fill petrol tank, open tap and prime pump by hand until the carburettor chambers are full.

(f) Start the engine and allow it to run for a short time while inspecting the connections for leaks.

13. SERVICING THE PETROL STOP TAP

In practice the tap will require little attention apart from a periodical inspection to ensure that it is leak proof.

The tap has a cork plunger which can be expanded to increase the interference and so improve the seal.

The cork is expanded by loosening the lock nut at the top of the plunger and the centre rod in an anti-clockwise direction, retighten the locknut. It will be noticed that increased resistance is felt when the tap is operated.

14. TO DISMANTLE PETROL STOP TAP

(a) Loosen the round headed screw at the side of the tap body sufficiently to allow the plunger to be withdrawn.

(b) Remove the lock nut in the head of the plunger. By turning the cork it can be removed together with the centre rod.

(c) The cork can now be pushed off the centre rod.

(d) Clean and inspect all parts and renew any that are believed to be defective.

15. TO ASSEMBLE PETROL STOP TAP

(a) Fit the cork seal on to the centre rod and screw the rod into the plunger head sufficiently to just nip the seal. Attach the lock nut to the centre rod protruding through the head of the plunger.

(b) Smear the cork and the inside of the tap body with a little oil or grease.

(c) Carefully feed the plunger into the tap body so that the groove in the plunger aligns with the round headed screw in the exterior of the body.

(d) Tighten the body screw so that the plunger is located in the tap body and has freedom of movement.

(e) Adjust the interference of the plunger to ensure that petrol will not seep past the cork seal. This is effected by turning the centre rod of the plunger anti-clockwise to increase or clockwise to decrease the interference.

(f) Lock the centre rod with the lock nut in the head of the plunger.

16. AC FUEL PUMP TYPE " UE "

Description (Fig. 3)

The AC fuel pump, type " UE ", is operated mechanically from an eccentric (H) on the engine camshaft (G). The illustration gives a sectional view of the pump, the method of operation is as follows :—

As the engine camshaft (G) revolves, the cam (H) lifts pump rocker arm (D) pivoted at (E) which pulls the pull rod (F) together with the diaphragm (A) downward against spring pressure (C) thus creating a vacuum in the pump chamber (M).

Petrol is drawn from the tank and enters at (J) into sediment chamber (K) through filter gauze (L), suction valve (N) into the pump chamber (M). On the return stroke the spring pressure (C) pushes the diaphragm (A) upwards, forcing petrol from the pump chamber (M) through the delvery valve (O) and outlet (P) to the carburettor feed pipe.

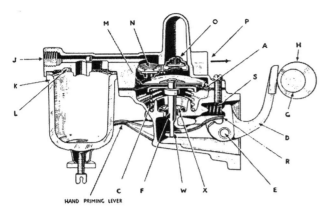

Fig. 3   **Section view of Petrol Pump.**

turns passes into the lower body of the petrol pump below the diaphragm assembly and by action of the latter is pumped out by way of the breather hole.

To obviate this condition an oil seal is fitted round the diaphragm assembly push rod and is prevented from rising with the action of the push rod by a metal retainer staked to the lower pump body.
Petrol pumps fitted with this oil seal were fitted to engines after No. TS.2074E.

During dismantling this oil seal should not be removed unless it is known to be defective.

When the carburettor float chambers are full the float will rise and shut the needle valve, thus preventing any flow of petrol from the pump chamber (M). This will hold diaphragm (A) downward against spring pressure (C), and it will remain in this position until the carburettors requires further petrol and the needle valve opens. The rocker arm (D) operates the connecting link by making contact at (R) and this

18. TO CLEAN THE PUMP FILTER

The pump filter should be examined every 1,000 miles and cleaned if necessary.

Access to the filter is gained by loosening the thumb nut situated below the glass sediment chamber at the side of the petrol pump body and swinging the wire frame to

	NOTATION FOR Fig. 3.			
Ref. No.	Description		Ref. No.	Description
A	Diaphragm assembly		N	Inlet or suction valve.
C	Diaphragm spring		O	Outlet or delivery valve.
D	Rocker arm		P	Outlet port
E	Rocker arm fulcrum pin.		R	Contact point between rocker arm and link lever.
F	Diaphragm pull rod.		S	Rocker arm spring.
G	Engine camshaft.		W	Link lever.
H	Fuel pump cam on camshaft.		X	Oil seal and retainer. Petrol pump with this oil seal were fitted to engines after No. TS.2074E.
J	Inlet port.			
K	Sediment chamber.			
L	Filter gauze.			
M	Pump chamber.			

construction allows idling movement of the rocker arm when there is no movement of the fuel pump diaphragm.

Spring (S) keeps the rocker arm (D) in constant contact with cam (H) and eliminates noise.

17. PETROL PUMP OIL SEAL

During very fast cornering oil rises up the cylinder block walls and during right-hand

one side. The sediment chamber can be removed followed by the cork gasket and gauze filter.

The gauze filter should be cleaned by a blast of air or washing it in clean petrol. The cork gasket should be inspected for condition and replaced if broken or hard. The glass sediment chamber should be cleaned and its upper rim inspected for chips.

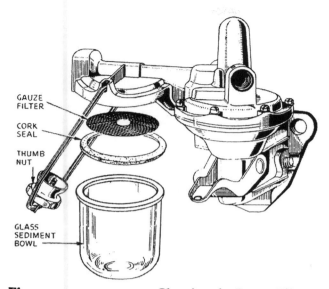

GAUZE
FILTER

CORK
SEAL

THUMB
NUT

GLASS
SEDIMENT
BOWL

Fig. 4   **Cleaning the Pump Filter.**

The replacement of the filter is the reversal of the removal. The thumb nut should be tightened sufficiently to make a good seal, overtightening tends to harden the seal which then looses its sealing properties. The hand priming pump should be used to fill the sediment chamber and carburettor float chambers. The engine should be started and run for a few minutes so that the pump may be observed for leaks.

19. TESTING WHILE ON ENGINE

With the engine stopped and switched off, the pipe to the carburettor pipe should be disconnected at the pump and replaced by a shorter tube, leaving a free outlet from the pump. The engine is then turned over by hand, when there should be a well defined spurt of petrol at every working stroke of the pump, namely, once every two revolutions of the engine.

20. TO REMOVE PETROL PUMP FROM ENGINE

(a) Turn off at petrol stop tap and remove the flexible hose from the tap first, then remove the hose from the pump.

(b) Remove fuel feed from pump to carburettor at its pump connection.

(c) Remove the two pump securing nuts and spring washers. Note the oil pressure pipe clip is attached to the rear stud.

(d) The pump can be removed from the cylinder block, together with the packing.

21. TO FIT PETROL PUMP TO ENGINE

(a) Place a new packing of correct thickness on the pump attachment studs followed by the pump. Secure with foremost nut and lock-washer finger tight.

(b) Position on the rear stud the oil pressure pipe clip, secure with nut and lock-washer. Tighten both nuts

(c) Attach the carburettor feed to pump and secure with union nut taking care to seat the pipe olive before attaching the union nut.

(d) Attach the flexible hose to the forward end of the pump. Attach and secure the rigid end to the petrol stop tap.

(e) Turn on petrol and prime pump with hand lever, until the glass sediment chamber and carburettor float chambers are full.

(f) Start and run the engine for a few moments and examine the connections for leaks.

22. TO DISMANTLE PETROL PUMP

For Notation see Fig 5.

(a) Clean the exterior of the pump and with a file mark the two flanges with a small cut.

(b) Loosen the thumb nut under the glass sediment chamber (6) and swing the frame (9) clear. The sediment chamber can now be lifted clear together with cork seal (4) and gauze filter (2).

(c) The wire frame can now be lifted out of the upper body (1) of the pump.

(d) Separate the two castings (1 and 16) by withdrawing the six securing screws (12) and lockwashers (13).

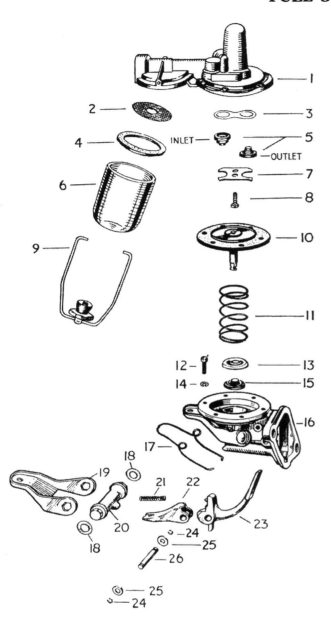

Ref. No.	Description
1	Upper body.
2	Gauze filter.
3	Valve gasket.
4	Cork seal
5★	Inlet and outlet valve assemblies.
6	Glass sediment bowl.
7	Valve retaining plate.
8	Screw for retaining plate.
9	Wire cage.
10	Diaphragm assembly.
11	Diaphragm spring.
12	Body securing screw.
13	Oil seal retainer.
14	Lock washer.
15	Oil seal.
16	Lower body.
17	Hand primer spring.
18	Cork washer.
19	Hand primer lever.
20	Hand primer lever shaft.
21	Rocker arm spring.
22	Link lever.
23	Rocker arm.
24	Retainer ring.
25	Washer.
26	Rocker arm pin.

★These valves are identical, but on fitting them to the upper body the spring of the inlet valve is pointing towards the diaphragm and the spring of the outlet valve away from the diaphragm, as shown in the illustration.

Fig. 5 Showing the "UE" type Fuel Pump in exploded form.

(**e**) To remove the diaphragm assembly (10) first turn it through 90° in an anti-clockwise direction and lift out of engagement with link lever (22). Collect the diaphragm spring (11). No attempt should be made to separate the four layers of the diaphragm as it is a riveted assembly. The oil seal (15) and retainer (13) can be prised out if known to be defective.

(**f**) Prise off hand primer lever (19) collecting cork washers (18) and hand lever spring (17) only if the hand primer is known to be defective. Drift out hand primer lever shaft (20).

(**g**) Remove circlips (24) from either end of rocker arm pin (26). Drift out rocker arm pin (26), collecting washers (25), rocker arm (23), link lever (22) and rocker arm spring (21).

(**h**) Invert the upper casting (1) and withdraw two valve retaining plate screws (8) followed by the retaining plate (7) valves (5) and valve gasket (3).

TO ASSEMBLE PETROL PUMP

(a) Place the figure of eight gasket (3) in position on the valve ports in the upper body (1). Position the inlet valve assembly (5) in the off centre and shallower port, with the spring of the valve pointing towards diaphragm. The outlet valve (5) is positioned in the centre port with the spring of the valve inside the port itself. The valve retainer (7) is secured, holding both valves in place, with two screws (8).

(b) Fit the diaphragm rod oil seal (15) and retainer (13) in the lower body (16) and stake over the wall of the seal recess. Position the hand primer shaft (20) with the offset uppermost and with its tongue pointing toward the pump mounting flange. Fit the cork washers (18) to the protruding ends of the shaft, on each side of the body (16).

(c) Fit the hand primer lever (19) and then peen over the ends of the shaft (20) to retain the lever (19).

(d) With the loops of the lever spring (17) upwards, feed the legs of the spring between the lever and the pump body so that it settles in its position on the upper side of the lever. The two legs are positioned above the lower body web adjacent to the outside of the pump mounting flange.

(e) Feed the rocker arm pin (26) partially into the pump body (16). Position one packing washer (25) on the pin following with one flange of the link lever (22).

(f) With the mounting flange uppermost position the rocker arm spring (21) on the cone-like protrusion in the pump body. The rocker arm (23) is fitted into the link lever and a protrusion allowed to engage the coil spring.

(g) The pin (26) is pressed through the link lever (22), the rocker arm (23) and a washer (25) situated between the second flange of the link lever (22) and the pump body (16). A retaining ring (24) is fitted when the pin (26) protrudes through the pump body (16).

(h) Position the diaphragm spring (11) on its base and fit the diaphragm (10) (with the tab toward the engine) by inserting the rod through the oil seal into the slot of the link lever (22) and turning it a quarter turn to the right (Fig. 6).

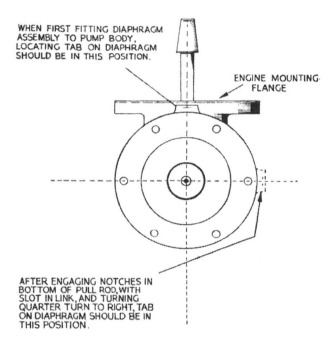

WHEN FIRST FITTING DIAPHRAGM ASSEMBLY TO PUMP BODY, LOCATING TAB ON DIAPHRAGM SHOULD BE IN THIS POSITION.

ENGINE MOUNTING FLANGE

AFTER ENGAGING NOTCHES IN BOTTOM OF PULL ROD, WITH SLOT IN LINK, AND TURNING QUARTER TURN TO RIGHT, TAB ON DIAPHRAGM SHOULD BE IN THIS POSITION.

Fig. 6   Fitting the Diaphragm to the Pump Body.

(i) The upper and lower bodies are secured with six bolts and lock washers, in such a manner that the sediment chamber (6) is on the opposite side to the diaphragm tab, or in accordance with the file marks.

(j) Position the gauze filter (2) in its housing, followed by the cork seal (4) and the glass sediment bowl (6). The wire cage (9) is attached and the thumb nut is tightened sufficiently to effect a petrol tight seal. Overtightening of this seal (4) will only harden the seal and destroy its properties.

24. INSPECTION OF PARTS

For Notation see Fig. 5.

Firstly, all parts must be thoroughly cleaned to ascertain their condition. Wash all parts in the locality of the valves in a clean paraffin bath separate from that employed for the other and dirtier components. Diaphragm and pull rod assemblies should normally be replaced unless in entirely sound condition without any signs of cracks or hardening.

Upper and lower castings should be examined for cracks or damage, and if diaphragm or engine mounting flanges are distorted these should be lapped to restore their flatness.

All badly worn parts should be replaced, and very little wear should be tolerated on rocker arm pins (26), the holes and engagement slot in links (22), holes in rocker arm (23). On the working surface of the rocker arm (23) which engages with the engine eccentric, slight wear is permissible but not exceeding .010″ in depth.

The valve assemblies (5) should not be replaced unless in perfect condition. Diaphragm springs (11) seldom call for replacement, but where necessary ensure that the replacement spring has the same identification colour and consequently the same strength as the original. Rocker arm springs (21) are occasionally found to be broken after service. All gaskets and joint washers should be replaced as a matter of routine. This also applies to oil seal (15) held in position by retainer (13).

25. AC AIR CLEANERS

Description

This cleaner is the wire gauze fitted metal canister type and is oil damped. The oil damping is carried out as a servicing operation.

Each carburettor has its own air cleaner functioning in such a manner that air drawn in by the engine first passes through the oiled gauze before entering the carburettor and so prolongs the life of the engine. Whenever the air cleaners are being replaced it is essential that the holes adjacent to the setscrew holes are uppermost so that they will align with those holes in the carburettor flange.

26. TO REMOVE AIR CLEANERS

This is required each time the carburettors are tuned or to service the cleaner itself.

(a) Loosen the cap nut on the top of the carburettor float chamber and turn the splash overflow pipe away from the air filter.

(b) Withdraw the two bolts securing the air cleaner to its mounting flange.

(c) The air cleaner and joint washer can now be removed.

27. TO FIT AIR CLEANERS TO CARBURETTORS

(a) Adhere the joint washer to the body of the air cleaner with a smear of grease.

(b) Ensuring the splash overflow pipe does not foul the air cleaner, offer the cleaner to the carburettor in such a manner that the holes adjacent to the setscrew holes are uppermost so that they will align with those holes in the carburettor flange.

(c) Secure air cleaners to carburettors with two setscrews and lockwashers each.

(d) Position the splash overflow pipe so that the open end is close to the filtering media and tighten the cap nut in centre of float chamber.

28. SERVICING AIR CLEANERS

Unless operating in a very dusty climate the AC air cleaners need only be serviced every 5,000 miles. It is suggested that in dusty climates, the cleaners are serviced at 2,500 miles and this period increased or diminished according to the dirt removed.

The cleaners should be washed in a bowl containing a mixture of paraffin and petrol until free from dirt.

After a thorough washing the units should be allowed to dry in clean air. When dry they should be filled with engine oil and the surplus oil allowed to drain in clean air. The cleaners must be dried and drained in clean air that is as free from dust as possible to ensure maximum cleanliness.

29. DISCONNECTION OF CARBURETTOR CONTROLS

There are nine throttle or carburettor control connections and it may be necessary to disconnect one or more to make adjustments, to effect removal of the carburettors or manifolds.

(a) The folding coupling on the throttle butterfly spindle. One pinch bolt.

(b) The outer Bowden cable at the front jet lever link. One pinch bolt.

(c) The inner Bowden cable at the cable swivel pin fitted to the front carburettor jet lever. One setscrew.

(d) Jet lever connection rod fitted between the two jet levers. The front fork end of the rod connects with the upper hole in the front jet lever. Clevis pin and split pin.

(e) The rear coupling of the long link rod assembly is attached to bulkhead lever assembly. Nut and washer.

(f) Front throttle and short rod assembly. Nut and Washer.

(g) On inlet manifold, pivot for bell crank. Setscrew and lock washer.

(h) Bell crank pivot. Washer and split pin.

30. TO REMOVE ACCELERATOR PEDAL, R.H.S. (Fig. 7)

(a) Remove the nut from the rear attachment of the long link rod assembly for the carburettor and withdraw end from lever assembly at the bulkhead.

(b) Release the spring from the lever assembly and drift out mills pin, utilising a thin shanked drift. The lever can now be withdrawn from the operating shaft.

(c) Withdraw the four self tapping screws securing the bearing housing to the bulkhead, collect housings and nylon bush bearing.

(d) From inside the car release the jam nut of the pedal limit stop and remove the screw stop from the fulcrum bracket on the toe board. Remove also the remaining three setscrews. The accelerator pedal assembly can now be withdrawn from inside the car.

(e) The L.H. Fulcrum bracket, double coil spring washer and plain washer can now be threaded off the operating shaft.

(f) By removal of the two split pins the R.H. fulcrum bracket can be withdrawn in a similar manner.

31. TO FIT ACCELERATOR PEDAL, R.H.S. (Fig. 7)

(a) Feed the right-hand fulcrum bracket on to the pedal shaft so that the mounting flange points towards the pedal pad, followed by two plain washers, a coil spring washer and the second mounting bracket, the mounting flange of which points away from the pedal. Fit the two split pins through the two holes in the shaft **between** the two plain washers.

(b) The pedal shaft is fed through the bulkhead bearing from inside the car. The assembly is secured to the toe board of the car by three bolts and lock washers, the lower right-hand fixing point is a pedal limit stop and jam nut.

(c) Feed a half bearing housing on to the pedal assembly shaft protuding into the engine compartment, followed by the nylon bearing and second half bearing housing. Secure bearing housings to bulkhead with four self tapping screws.

(d) The lever assembly is secured to the shaft by a mills pin from inside the engine compartment and the return spring is attached to the lever shank.

(e) The long link rod assembly is attached to the lever assembly by a nut and spring washer.

(f) Adjust pedal limit stop screw.

32. TO REMOVE ACCELERATOR PEDAL, L.H.S. (Fig. 7)

(a) Remove the nut from the rear attachment of the long link rod assembly and withdraw end from lever assembly at the bulkhead.

(b) Release the spring from the lever assembly.

(c) Drift out the two mills pins adjacent to the left-hand bearing.

(d) Remove the two bolts and lockwashers securing the support bracket to the bulkhead.

11

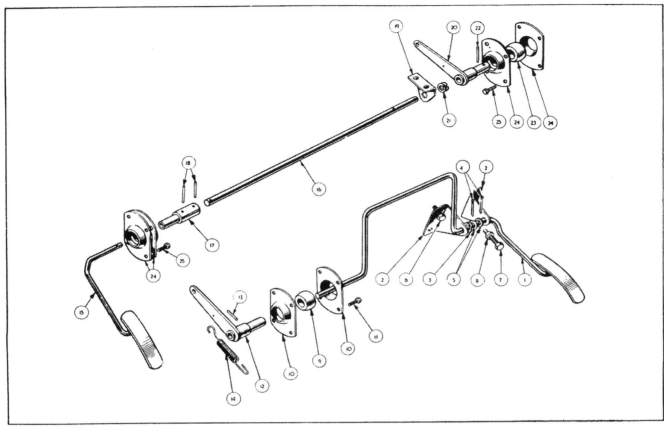

Fig. 7   Exploded view of R.H. and L.H.S. Accelerator Pedal Assemblies.

NOTATION FOR Fig. 7.			
Ref. No.	Description	Ref. No.	Description
1	R.H.S. pedal assembly.	14	Lever return spring.
2	Fulcrum bracket.	15	L.H.S. pedal assembly.
3	Double coil washer.	16	Pedal shaft.
4	Split pins.	17	Connecting bush.
5	Plain washers.	18	Mills pin.
6	Attachment bolts.	19	Support bracket.
7	Pedal limit stop bolt.	20	Lever assembly.
8	Jam nut.	21	Double coil spring.
9	Shaft bearing.	22	Mills pin.
10	Bearing housings.	23	Shaft bearing.
11	Self tapping screw.	24	Bearing housings.
12	Lever assembly.	25	Self tapping screws.
13	Mills pin.		

(e) Push the rod to the left of the car, this will eject the accelerator pedal in to the interior of the car and also free the shaft from its right-hand bearing. On drawing the shaft to the right it can be freed from the left-hand bearing.

(f) The bearings and housings can be removed by withdrawing the eight self tapping screws (four each bearing).

(g) The shaft can now be dismantled by drifting out the mills pin securing the lever assembly to the shaft and collecting a double coil washer and mounting bracket.
The split pin locating the bracket on the shaft can also be withdrawn.

33. TO FIT ACCELERATOR PEDAL L.H.S. (Fig. 7)

(a) Position the nylon bearing between the half housings and secure both to the bulkhead with eight self tapping screws (four each bearing).

(b) It will be observed that the shaft is drilled at each end; the single hole end is on the left-hand side and the end with two holes is the right-hand end.

(c) Fit the lever assembly to the right-hand end, with lever on left-hand side, and secure with a mills pin to the outer or extreme right-hand hole.

(d) Feed on the shaft the double coil spring washer followed by the support bracket, mounting holes to the left. Apply pressure to the support bracket to compress the spring and feed split pin through hole in shaft to position bracket.

(e) Feed metal bush on to left-hand end of shaft (larger end first). Feed shaft and bush into the left-hand bearing already fitted to car. Position fulcrum of lever assembly in the right-hand bearing, it may be necessary to withdraw the shaft from the left-hand bearing, and secure mounting bracket to bulkhead, utilising two bolts and lock washers. Secure the bush to the shaft by a mills pin, supporting bush and shaft with a small anvil.

(f) From inside the car feed the accelerator pedal into the bush and similarly secure with a mills pin.

(g) Couple up long carburettor link rod and secure with nut and lock washer.

34. TO REMOVE CARBURETTOR FROM MANIFOLD

(a) Remove air cleaners as described on page 9.

(b) Disconnect petrol supply pipe, taking care not to damage the conical filter and spring situated in the top of each float chamber body.

(c) Withdraw the split pin from the clevis pin at the rear end of the mixture control link and remove clevis pin.

(d) Disconnect the throttle spindle at the rear folded coupling by loosening the clamping bolt.

(e) By removing the two nuts at the mounting flange of the rear carburettor it can be removed from the manifold together with an asbestos insulating washer and two packings.

(f) Disconnect the Bowden inner cable from the swivel pin of the jet lever and the outer cable from the front jet lever link by loosening a clamp bolt.

(g) Remove the nut and lock washer of the short link rod assembly and disconnect the control linkage from the carburettor throttle lever.

(h) Remove the two nuts securing the carburettor to the manifold and remove carburettor together with the asbestos insulating washer and two packings.

35. TO FIT CARBURETTORS TO MANIFOLD

(a) Ensure that the joint washers and asbestos insulating washers are in good order. Fit two joint washers, one to each manifold flange, followed by an asbestos insulating washer and a second joint washer.

(b) Offer up and secure the rear carburettor to its mounting and secure with plain washers, lock washers and nuts.

(c) Ensure that the folding connection of the throttle spindle connecting rod will not foul the front carburettor when the latter is offered up to its position.

(d) Attach and secure front carburettor to its mounting, utilising plain and lock washers and nuts.

(e) Connect the outer Bowden cable to the front jet lever link.

(f) Connect the short link rod assembly to the throttle lever of the front carburettor.

(g) The inner cable, the throttle rods and jet levers are left disconnected until after the carburettors have been tuned. See page 23.

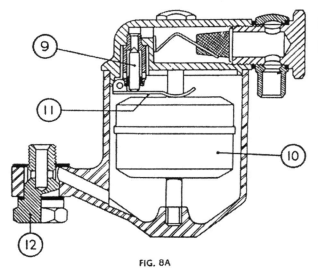

FIG. 8A

Fig. 8A Sectional view of Float Chamber.

Fig. 8B Showing the shoulder datum of the Jet Needles.

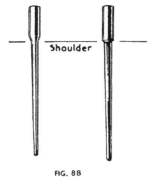

Shoulder

FIG. 8B

The shoulder of the needle should be flush with the under face of the piston. Two types of shoulder are in use, and the correct datum point for each is shown.

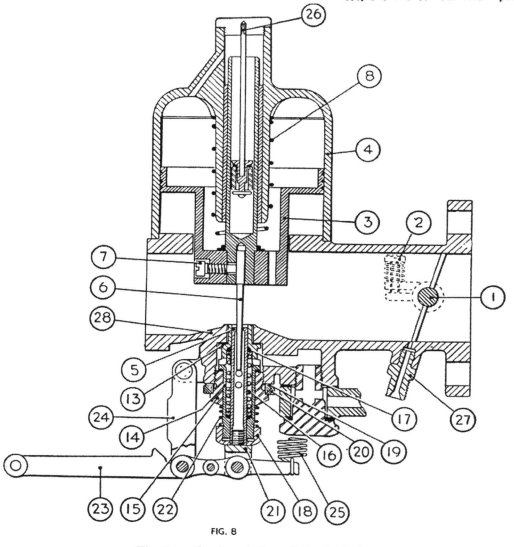

FIG. 8

Fig. 8 Sectional view of the S.U. Carburettor.

FUEL SYSTEM

THE S.U. CARBURETTOR

<table>
<tr><td colspan="4" align="center">NOTATION FOR Fig. 8.</td></tr>
<tr><td>Ref. No.</td><td>Description</td><td>Ref. No.</td><td>Description</td></tr>
<tr><td>1</td><td>Throttle butterfly and spindle.</td><td>15</td><td>Jet locking nut.</td></tr>
<tr><td>2</td><td>Throttle butterfly stop and adjusting screw</td><td>16</td><td>Compression spring.</td></tr>
<tr><td>3</td><td>Piston.</td><td>17</td><td>Sealing gland.</td></tr>
<tr><td>4</td><td>Suction chamber.</td><td>18</td><td>Jet adjusting nut.</td></tr>
<tr><td>5</td><td>Jet bore.</td><td>19</td><td>Sealing gland.</td></tr>
<tr><td>6</td><td>Needle.</td><td>20</td><td>Conical washer.</td></tr>
<tr><td>7</td><td>Needle locking screw.</td><td>21</td><td>Jet head.</td></tr>
<tr><td>8</td><td>Spring.</td><td>22</td><td>Loading spring.</td></tr>
<tr><td>9</td><td>Float chamber needle valve.</td><td>23</td><td>Jet lever.</td></tr>
<tr><td>10</td><td>Float.</td><td>24</td><td>Jet lever link.</td></tr>
<tr><td>11</td><td>Float lever.</td><td>25</td><td>Jet lever return spring.</td></tr>
<tr><td>12</td><td>Float chamber attachment bolt.</td><td>26</td><td>Damper piston.</td></tr>
<tr><td>13</td><td>Jet bush. Top half.</td><td>27</td><td>Ignition connection union.</td></tr>
<tr><td>14</td><td>Jet bush. Bottom half.</td><td>28</td><td>Bridge piece.</td></tr>
</table>

36. THE S.U. CARBURETTOR

(a) Description

The S.U. carburettor is of the automatically expanding choke type, in which the cross sectional area of the main air passage adjacent to the fuel jet, and the effective orifice of the jet, is variable. The variation takes place in accordance with the demand of the engine as determined by the degree of the throttle opening, the engine speed, and the load against which the engine is operating.

The distinguishing feature of the type of carburettor is that an approximately constant air velocity, and hence an approximately constant degree of depression, is at all times maintained in the region of the fuel jet. This velocity is such that the air flow demanded by the engine in order to develop its maximum power is not appreciably impeded, although good atomisation of the fuel is assured under all conditions of speed and load

The maintenance of a constant high air velocity across the jet, even under idling conditions, obviates the necessity for an idling jet. A single jet only is employed in the S.U. carburettor.

(b) Construction

For Notation see Fig. 8 and 8A.

The main constructional features of the carburettor in its simplest form are shown in Figs. 8 and 9, which illustrate the horizontal-type carburettor. The diagrams illustrate the main body, butterfly throttle, automatically expanding choke and variable fuel-jet arrangement. They also indicate the means whereby the jet is lowered by a manual control to effect enrichment of the mixture for starting and warming up. A float chamber of the type employed is illustrated in Fig. 8a.

Turning to Fig. 8 it will be seen that a butterfly throttle mounted on the spindle (1) is located close to the engine attachment flange, at one end of the main air passage, and that an adjustable idling stop screw (2) is arranged to prevent complete closure of the throttle, thus regulating the flow of mixture from the carburettor under idling conditions with the accelerator released. At the outer end of the main passage is mounted the piston (3), its lower part constituting a shutter, restricting the cross-sectional area of the main air passage in the vicinity of the fuel jet (5) as the piston falls. This component is enlarged at its upper end to form a piston of considerably greater diameter which moves axially within the bore of the suction chamber (4) and at the bottom of the piston is mounted the tapered needle (6) which is retained by means of the setscrew (7).

The piston component (3) is carried upon a central spindle which reciprocates and is mounted in a bush fitted in the central boss, forming the upper part of the suction chamber casting

An extremely accurate fit is provided between the spindle and the bush in the suction chamber so that the enlarged portion of the piston is held out of contact with the bore of the suction chamber, within which, nevertheless, it operates with an extremely fine clearance. Similarly, the needle (6) is restrained from contacting the bore of the jet (5) which it is seen to penetrate, moving axially therein to correspond with the rise and fall of the piston.

It will be appreciated that, as the piston rises, the air passage in the neighbourhood of the jet becomes enlarged, and passes an additional quantity of air. Provided that the needle (6) is of a suitably tapered form, its simultaneous withdrawal from the jet (5) ensures the delivery to the engine of the required quantity of fuel corresponding to any given position of the piston and hence to a given air flow.

The piston, under the influence of its own weight and assisted by the light compression spring (8) will tend to occupy its lowest position, two slight protuberences on its lower face contacting the bottom surface of the main air passage adjacent to the jet. The surface in this region is raised somewhat above the general level of the main bore of the carburettor, and is referred to as the " bridge " (28).

Levitation of the piston is achieved by means of the induction depression, which takes effect within the suction chamber, and thus upon the upper surface of the enlarged portion of the piston through drillings in the lower part of the piston which make communication between this region and that lying between the piston and the throttle. The annular space beneath the enlarged portion of the piston is completely vented to atmosphere by ducts not indicated in the diagram.

It will be appreciated that, since the weight of the piston assembly is constant, and the augmenting load of the spring (8) approximately so, a substantially constant degree of depression will prevail within the suction chamber, and consequently in the region between the piston and the throttle, for any given degree of lift of the piston between the extremities of its travel.

It will be clear that this floating condition of the piston will be stable for any given air-flow demand as imposed by the degree of throttle opening, the engine speed and the load ; thus, any tendency in the piston to fall momentarily will be accompanied by an increased restriction to air flow in the space bounded by the lower side of the piston and the bridge, and this will be accompanied by a corresponding increase in the depression between the piston and throttle, which is immediately communicated to the interior of the suction chamber, instantly counteracting the initial disturbance by raising the piston to an appropriate extent.

The float chamber, which is shown in Fig. 8A, is of orthodox construction, comprising a needle valve (9) located within a separate seating which, in turn, is screwed in the float chamber lid, and a float (10), the upward movement of which, in response to the rising fuel level, causes final closure of the needle upon its seating through the medium of the hinged fork (11).

The float-chamber is a unit separate from the main body of the carburettor to which it is attached by means of the bolt (12), suitable drillings being provided therein to lead the fuel from the lower part of the float chamber to the region surrounding the jet. It is steadied at its upper extremity by a suction chamber attachment screw.

The buoyancy of the float, in conjunction with the form of the lever (11) is such that a fuel level is maintained approximately $\frac{1}{8}''$ below the jet bridge (see page 23). This can easily be observed after first detaching the suction chamber and suction piston, and then lowering the jet to its full rich position. The level can vary a further $\frac{1}{4}''$ downwards without any ill effects on the functioning of the carburettor. The only parts of importance in Figs. 8 and 8A not so far described are those associated with the jet.

Under idling conditions the piston is completely dropped, being then supported by the two small protuberances provided on its lower surface, which are in contact with the bridge (28) ; the small gap thus formed between piston and bridge permits the flow of sufficient air to meet the idling demand of the engine without, however, creating enough depression on the induction side to raise the piston.

The fuel discharge required from the jet is very small under these conditions, hence the diameter of the portion of the needle now obstructing the mouth of the jet is very nearly equal to the jet bore. Initial manufacture of the complete carburettor assembly to the required degree of accuracy to ensure perfect concentricity between the needle and the jet bore under these conditions is impracticable, and an individual adjustment for this essential centralisation is therefore provided.

It will be seen that the jet is not mounted directly in the main body, but is housed in the parts (13) and (14) referred to as the jet bushes, or jet bearings.

The upper jet bush is provided with a flange which forms a face seal against a recess in the body, while the lower one carries a similar flange contacting the upper surface of the hollow hexagon locking nut (15).

The arrangement is such that tightening of the hollow hexagon locking screw will positively lock the jet and jet bushes in position. Some degree of lateral clearance is provided between the jet bushes and the bores formed in the main body and the locking screw. In this manner the assembly can be moved laterally until perfect concentricity of the jet and needle is achieved, the screw (15) being slackened for this purpose. This operation is referred to as " centring the jet ", on completion the jet locking nut (15) is finally tightened. See page 19.

In addition to this concentricity adjustment, an axial adjustment of the jet is provided for the purpose of regulating the idling mixture strength.

Since the needle tapers throughout its length, it will be clear that raising or lowering the jet within its bearing will alter the effective aperture of the jet orifice, and hence the rate of fuel discharge. To permit this adjustment the jet is a variably mounted within its bearings and provided with adequate sealing glands.

A compression spring (16) which, at its upper end, serves to compress the small sealing gland (17) and thus prevents any fuel leakage between the jet and the upper jet bearing.

At its lower end this spring abuts against a similar sealing gland, thus preventing leakage of fuel between the jet and the lower jet bearing.

In both locations a brass washer is interposed between the end of the spring and the sealing gland to take the spring thrust. A further sealing gland (19), together with a conical brass washer (20) is provided, to prevent fuel leakage between the jet screw (15) and the main body.

It will be seen from the diagram that the upward movement of the jet is determined by the position of the jet adjusting nut (18) since the enlarged jet head (21) finally abuts against this nut as the jet is moved upwards towards the " weak " or running position.

The position of the nut (18) therefore determines the idling mixture ratio setting of the carburettor for normal running with the engine hot, and is prevented from unintentional rotation by means of the loading spring (22).

The cold running mixture control mechanism comprises the jet lever (23) supported from the main body by the link member (24) and attached by means of a clevis pin to the jet head (21). A tension spring (25) is provided, as shown, to assist in returning the jet-moving mechanism to its normal running position. Connection is made from the outer extremity of the jet lever (23) to a control situated within reach of the driver.

Drillings in the float-chamber attachment bolt (12), the main body of the carburettor, the jet (5) and slots in the upper jet bearing (13) serve to conduct the fuel from the float-chamber to the jet orifice.

It will be seen that the spindle upon which the piston (3) is mounted is hollow, and that it surrounds a small stationary damper piston suspended from the suction chamber cap by means of the rod (26). The hollow

17

interior of the spindle contains a quantity of thin engine oil, and the marked retarding effect upon the movement of the main piston assembly, occasioned by the resistance of the small piston, provides the momentary enrichment desirable when the throttle is abruptly opened. The damper piston is constructed to provide a one-way valve action which gives little resistance to the passage of the oil during the downward movement of the main piston.

An ignition connection (27 in Fig. 8 or 33 in Fig. 9) is provided for use in conjuction with suction-operated ignition advance mechanism, and is fitted to the front carburettor only.

37. THROTTLE AND MIXTURE CONTROL INTERCONNECTION
Fig. 9

A direct connection is provided between the jet movement and the throttle opening. Such an interconnection ensures that the engine will continue to run when the mixture is enriched by lowering the jet, without the additional necessity of maintaining a greater throttle opening than is normally provided by the setting of the slow-running screw (2).

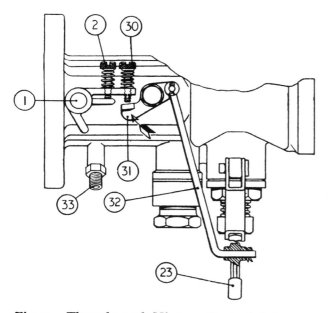

Fig. 9   Throttle and Mixture Control interconnection.

The mechanism involved in this interconnection is shown in Fig. 9. It will be seen that a connecting rod (32) conveys movement from the jet lever (23) to a lever (31) pivoted on the side of the main body casting.

Movement of the jet lever in the direction of enrichment is thus accompanied by an upward movement of the extremity of the lever (31) which, in turn, abuts against the adjustable screw (30) and this opens the throttle to a greater degree than the normal slow-running setting controlled by the slow-running stop screw (2). The screw (30) should be so adjusted that it is just out of contact with the lever (31) when the jet has been raised to its normal running position, and the throttle is shut back to its normal idling condition, as determined by the screw (2).

38. EFFECT OF ALTITUDE AND CLIMATIC EXTREMES ON STANDARD TUNING

The standard tuning employs a jet needle which is broadly suitable for temperate climates at sea level upwards to approximately 3,000 ft. Above this altitude it may be necessary, depending on the additional factors of exteme climatic heat and humidity, to use a weaker tuning than standard.

The factors of altitude, extreme climatic heat, each tend to demand a weaker tuning, and a combination of any of these factors would naturally emphasise this demand. This is a situation which cannot be met by a hard and fast factory recommendation owing to the wide variations in the condition existing and in such cases the owner will need to experiment with alternative weaker needles until one is found to be satisfactory.

If the carburettor is fitted with a spring-loaded suction piston, the necessary weakening may be affected by changing to a weaker type of spring or by its removal.

18

39. CARBURETTOR JET NEEDLES

Two jet needles are available for fitting to the carburettors of the TR2.

(a) FV. For normal motoring.

(b) GC. For high speed motoring and competition driving.

40. TO REMOVE JET NEEDLE

(a) Remove the air-cleaner. See page 10.

(b) Remove the damping piston from the top of the suction chamber.

(c) Withdraw the three suction chamber securing screws and move the carburettor float chamber support arm to one side.

(d) Lift the suction chamber and remove coil spring and washer from piston head.

(e) Remove the piston with jet needle attached from the body of the carburettor and empty away oil in the reservoir.

(f) Loosen screw in base of piston and withdraw jet needle.

41. TO FIT NEEDLE (Fig. 8B)

(a) Ensure that the jet head is loose in the main body of the carburettor by loosening clamp ring.

(b) Ascertain that the jet needle is perfectly straight and position it so that the shoulder is flush with the base of the piston, tighten screw to grip needle. Feed the needle into its recess in the jet head.

NOTE: On no account should the piston with the needle attached be laid down so that it rests on the needle. Failure to observe this point may cause carburation defects due to a bent needle.

(c) Position the washer and the spring on top of the piston and the suction chamber over the piston.

(d) Secure with the three attachment screws with the foremost accommodating the float chamber support arm.

(e) Fill the piston reservoir with thin oil and fit the damper to the suction chamber.

(f) Centralise the jet as described on this page.

(g) Tune the carburettors as described on page 23.

42. CENTRALISATION OF JET (Fig. 8)

(a) Disconnect the throttle linkage to gain access to the jet head (21) and remove damper (26).

(b) Withdraw the jet head (21) and remove adjusting nut (18) and spring (22). Replace nut (18) and screw up to its fullest extent.

(c) Slide the jet head (21) into position until its head rests against the base of the adjusting nut.

(d) The jet locking nut (15) should be slackened to allow the jet head (21) and bearings (13 and 14) assembly to move laterally.

(e) The piston (3) should be raised, access being gained through the air intake and allowing it to fall under its own weight. This should be repeated once or twice and the jet locking nut (15) tightened.

(f) Check the piston by lifting to ascertain that there is complete freedom of movement. If " sticking " is detected operation (d) and (e) will have to be repeated.

(g) Withdraw jet head (21) and adjusting nut (18).

(h) Replace nut (18) with spring (22) and insert the jet head (21).

(i) Check oil reservior and replace damper (26).

(j) Tune the carburettors as described on page 23.

43. TO ASSEMBLE THE CARBURETTOR(S)

Having ensured the cleanliness and the servicability of all component parts, it is suggested that the carburettor(s) are assembled in the following sequence.

The front carburettor differs from that of the rear insomuch that there are certain additions. As and when the additions occur they will be specifically mentioned.

(a) Fit the ignition union to the front carburettor, this utilises the tapped bore which breaks through into the mixture passage.

(b) Position the throttle spindle in the body in such a manner that the spindle protrudes *less* on the left-hand side looking at the air cleaner ends.

19

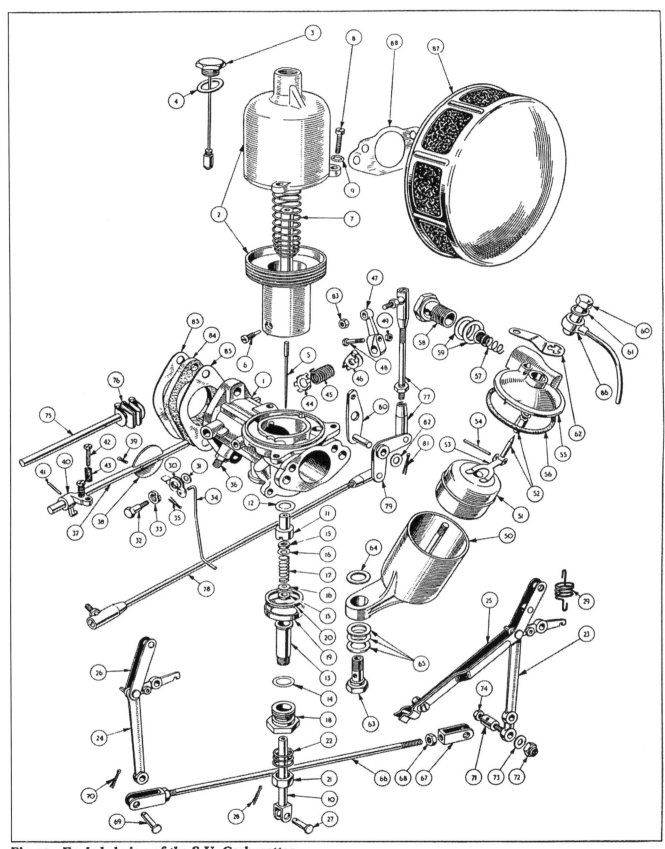

Fig. 10 Exploded view of the S.U. Carburettor.

20

S.U. CARBURETTOR DETAILS (Fig. 10)

Ref. No.	Description	Ref. No.	Description
1	Body assembly.	45	Return spring.
2	Suction chamber and piston assembly	46	End clip.
3	Damper assembly.	47	Throttle lever.
4	Washer.	48	Pinch bolt.
5	Jet needle.	49	Nut for 48.
6	Needle locking screw.	50	Float chamber.
7	Piston spring.	51	Float.
8	Securing screw.	52	Needle and seat assembly.
9	Shake proof washer.	53	Hinged lever.
10	Jet head.	54	Pin for hinged lever.
11	Top half jet bearing.	55	Float chamber cover.
12	Washer.	56	Joint washer.
13	Bottom half jet bearing.	57	Petrol inlet filter.
14	Washer.	58	Banjo bolt
15	Cork gland washer.	59	Fibre washer.
16	Copper gland washer.	60	Cap nut.
17	Spring between gland washers.	61	Aluminium washer.
18	Jet locking nut.	62	Float chamber support arm.
19	Sealing ring.	63	Float chamber attachment bolt.
20	Cork washer.	64	Fibre washer.
21	Jet adjusting nut.	65	Washer.
22	Loading spring.	66	Jet control connecting rod. (Between front and rear jet levers.)
23	Jet lever. (Front carburettor.)	67	Fork end.
24	Jet lever. (Rear carburettor.)	68	Nut on fork end.
25	Jet lever link. (Front carburettor.)	69	Clevis pin.
26	Jet lever link. (Rear carburettor.)	70	Split pin.
27	Clevis pin	71	Choke cable swivel pin.
28	Split pin.	72	Nyloc nut.
29	Jet lever return spring.	73	Plain washer.
30	Rocker lever. (Front carburettor only.)	74	Screw.
31	Washer for 30. ,, ,,	75	Throttle spindle connecting rod.
32	Rocker lever bolt. ,, ,,	76	Folding coupling.
33	Spring washer. ,, ,,	77	Short link rod assembly.
34	Connecting rod.	78	Long link rod assembly.
35	Split pin.	79	Bell crank lever.
36	Ignition connection union. (Front carburettor only.)	80	Pivot lever.
37	Throttle spindle.	81	Split pin.
38	Throttle disc.	82	Plain washer.
39	Throttle disc attachment screws.	83	Nut.
40	Throttle stop. (Front carburettor only.)	84	Insulating packing.
41	Taper pin.	85	Joint washer.
42	Stop adjusting screw.	86	Carburettor splash and overflow pipe.
43	Locking screw spring.	87	Air cleaner.
44	Anchor plate.	88	Air cleaner gasket.

(c) Feed the throttle disc into the slot of the spindle and secure with two countersunk screws. These screws have split shanks which are now opened by the insertion of the screw driver blade.

(d) Position the throttle stop with the two adjusting screws on the shorter end of the throttle spindle of the front carburettor body and secure with the taper pin; to the rear carburettor, fit the throttle stop with the single adjusting screw.

21

(e) Feed the rocker lever bolt through the double coil washer and the rocker lever so that the platform of the lever is on the left viewing the bolt head. This assembly is fitted to the front carburettor with a plain washer between it and the carburettor. Ensure that the rocker lever moves freely.

(f) Fit the throttle spindle return spring anchor plate on the longer end of the spindle and anchor it on the web provided. Follow it with the spring and the end clip then adjust the tension and lock the end clip with the pinch bolt.

(g) To the bottom half of the jet bearing position the copper washer followed by the jet adjusting sealing nut (threaded portion uppermost) spring and secure with the jet adjusting nut. Position the alloy sealing ring, flatter side downwards, and the cork washer over the thread of the jet adjusting nut.

(h) Insert the jet assembly through the jet adjusting nut and bottom half of the jet bearing from below. Position the cork gland washer, the copper gland washer, spring, a second gland washer and cork gland washer on the head of the jet assembly.

(i) Position a copper washer on the shoulders of the upper half jet bearing and, with the shoulder uppermost, balance the top half bearing on the cork gland washer of the jet assembly.

(j) Feed the assembly mentioned in (h) and (i) into the carburettor body and secure with the sealing nut.

(k) Fit the float to the pillar of the float chamber, this is symmetrical and can be fitted either way up.

(l) The needle valve body is secured in the float chamber cover, position valve needle and hinge lever and insert pin. Adjust as described on page 23.

(m) Assemble the splash overflow pipe to the cap of carburettor float chamber with a washer interposed between.

(n) Fit the float chamber cover to the float chamber and attach cap nut as assembled in operation (m). The nut is left loose at this juncture.

(o) Fit the jet needle to the piston assembly and ensure that its lower shoulder is flush with that of the piston.

(p) The piston and jet needle is now fitted to the body assembly so that the brass dowel in the carburettor body locates the longitudinal groove in the piston.

(q) With the smaller diameter of the coil spring downwards, position the spring over the polished stem of the piston.

(r) Fit the suction chamber over the spring and piston stem allowing the spring to position itself outside the suction chamber centre.

(s) The suction chamber is secured to the carburettor body by three screws, these are fitted but left loose at this juncture.

(t) The float chamber is now attached to the carburettor body by the float chamber attachment bolt. Two large bore fibre washers with a brass washer between are positioned between the bolt head and the float chamber and a small bore washer between the float chamber and the carburettor body. With the washers so placed the float chamber is attached to the carburettor body, the attachment bolt is left loose at this juncture.

(u) Looking at the intake end of the carburettor body remove the right-hand suction chamber securing screw (left loose in operation (s)). With a shakeproof washer under its head feed the bolt through the float chamber steady bracket and replace to secure suction chamber. The three screws can now be fully tightened, the cap nut is, however, still left loose. The cap nut of the cover is tightened to secure the splash overflow pipe for tuning purposes when fitted to the car. Attach the jet lever return spring to the position provided betweeen jet assembly and float chamber.

(v) The jet and jet needle are now centralised. See page 19.

(w) The damper assembly is fitted to the suction chamber dry. The oil reservoir is not filled until the carburettors are fitted to the car.

(**x**) Select the jet lever of the front carburettor, identified by having two holes at the extremity of the longer arm. This is attached to the jet assembly by a clevis pin and split pin, position the second end of the lever return spring to the jet lever.

(**y**) Feed the upper end of the tension link through the rocker lever of the front carburettor from behind and the second end through the jet lever. Secure both ends with split pins.

(**z**) Select the front carburettor jet lever link, this is distinguished by the pinch bolt at one end. This is attached to the lug at the rear of the jet assembly and again to the elbow of the jet lever in such a manner that the pinch bolt end of this link points to the rear. Both attachments are made by clevis pins and split pins.

The assembly of the jet lever and jet lever link to the rear carburettor is very similar. Both components are shorter than those fitted to the front carburettor.

44. TO ADJUST THE FUEL LEVEL IN THE FLOAT CHAMBER Fig 11.

The level of the fuel in the float chamber is adjusted by setting the fork lever in the float chamber lid. It is suggested that the following procedure for its adjustment is adopted.

(**a**) Remove the banjo bolt of the fuel connection and collect the two fibre washers and filter.

(**b**) Loosen the screw securing the float chamber support arm to the carburettor body.

(**c**) Withdraw the cap nut from the centre of the float chamber lid and remove washers and splash overflow pipe.

(**d**) Swing the support arm clear to lift the lid of the float chamber and joint washer.

(**e**) The set of the forked lever is correct when, with the lid of the float chamber inverted and the shank of the fork lever resting on the needle of the delivery valve, it is possible to pass a $\frac{7}{16}$″ diameter rod between the inside radius of the forked lever and the flange of the lower face of the cover.

45. CARBURETTOR TUNING Fig. 8.

This should be carried out without the Air Cleaners as it is found they have no effect on balance or performance but their removal considerably facilitates the operation. One clamping bolt of a throttle rod folding coupling should be loosened, the jet connecting rod should be disconnected at one of its fork end assemblies and the choke control cable released.

The rich mixture starting control linkage should also be disconnected by removing one of the clevis pins. This will enable each carburettor to be adjusted independently.

The suction chamber (4) and piston (3) should be removed and the jet needle (6) position checked. The needle shoulder, as shown in the illustration, should be flush with the base of the recess in the piston. The chamber and piston are now replaced.

The oil reservoir should be full and damping affect should be felt when replacing piston when the securing nut is $\frac{1}{4}$″ from the top of the suction chamber.

It is recommended that the adjusting nut (18) is screwed fully home and then slackened back two and a half turns (fifteen flats) as an initial setting.

The throttle adjusting screw (2) on each carburettor should be adjusted until it will just hold a thin piece of paper between the screw and the stop when the throttle is held in the closed position. The throttle butterfly (1) on each carburettor should then be opened by one complete turn of the adjusting screw.

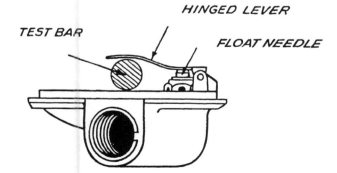

HINGED LEVER

TEST BAR

FLOAT NEEDLE

Fig. 11   **Adjusting the Fuel Level.**

The engine is now ready for starting and, after thoroughly warming up, the speed should be adjusted by turning each throttle adjusting screw an equal amount until the idling speed is approximately 500 R.P.M. The synchronisation of the throttle setting should now be checked by listening to the hiss of each carburettor, either directly or by means of a piece of rubber tubing held near the intake.

The intensity of the noise should be equal and if one carburettor is louder than the other its throttle adjusting screw should be turned back until the intensity of hiss is equal.

After satisfactory setting of the throttle, the mixture should then be adjusted by screwing the jet adjusting nuts up or down on each carburettor until satisfactory running is obtained. The lever tension spring should be connected during this operation. This mixture adjusting may increase the engine idling speed and each throttle adjusting screw must be altered by the same amount in order to reduce speed to 500 R.P.M. and the hiss of each carburettor again compared.

The balance of the mixture strength should be checked by independently lifting the piston of each carburettor no more than $\frac{1}{32}''$. The mixture is correct when this operation causes no change in engine R.P.M. When the engine slows down with this operation it indicates the mixture is too weak and it should be enriched by unscrewing the jet adjusting nut. An increase of engine speed during this operation indicates that the mixture is too rich and, consequently, it should be weakened off by screwing up the jet adjusting nut. The mixture setting should now give a regular and even exhaust beat, it is irregular with a " splashy " type of misfire and a colourless exhaust, the mixture is too weak. A regular or rhythmical type of misfire in the exhaust note, possibly with a blackish exhaust, indicates the mixture is too rich.

The jets of both carburettors should be held against the adjusting nuts before replacing the mixture control linkage, which should be adjusted as necessary, and similarly the throttle should be held tight against their respective idling stops before retightening the folding coupling clamp bolt.

46. CARBURATION DEFECTS

In the case of unsatisfactory behaviour of the engine, before proceeding to a detailed examination of the carburettor, it is advisable to carry out a general condition check of the engine, in respects other than those bearing upon the carburation.

Attention should, in particular, be directed towards the following :—

The ignition system.
Incorrectly adjusted contact breaker gap.
Dirty or pitted contact breaker points, or other ignition defects.
Loss of compression of one or more cylinders.
Incorrect plug gaps.
Oily or dirty plugs.
Sticking valves.
Badly worn inlet valve guides.
Defective fuel pump, or chocked fuel filter.

Leakage at joint between carburettors and induction manifold, or between induction manifold flanges and cylinder head.

If these defects are not present to a degree which is thought accountable for unsatisfactory engine performance, the carburettor should be investigated for the following possible faults.

(a) **Pistons Sticking. Fig. 8.**

The symptoms are stalling and a refusal to run slowly, or lack of power and heavy fuel consumption.

The piston (3) is designed to lift the jet needle (6) by the depression transferred to the top side from the passage facing the butterfly. This depression overcomes the weight of the piston and spring (8). The piston should move freely over its entire range and rest on the bridge pieces (28) when the engine is not running.

This should be checked by gently lifting the piston with a small screwdriver and any tendency for binding generally indicates one of the following faults :—

(i) The damper rod may be bent causing binding and this can be checked by its removal. If the piston is now free the damper rod should be straightened and refitted.

(ii) The piston is meant to be a fine clearance fit at its outer diameter in the suction chamber and a sliding fit in the central bush. The suction chamber should be removed, complete with piston, and the freedom of movement checked after removal of the damper rod. The assembly should be washed clean and very lightly oiled where this slides in the bush and then checked for any tendency of binding. It is permissible to carefully remove, with a hand scraper, any high spots on the outer wall of the suction chamber, but no attempt should be made to increase the clearance by increasing the general bore of the suction chamber or decreasing the diameter of the piston. The fit of the piston in its central bush should be checked under both rotational and sliding movement.

(b) Eccentricity of Jet and Needle Fig. 8.

The jet (14) is a loose fit in its recess and must always be centred by the needle before locking up the clamping ring (15).

(i) The needle should be checked in the piston to see that it is not bent. It will be realised that it does not matter if it is eccentric as the adjustment of the jet allows for this, but a bent needle can never have the correct adjustment. For "Centralisation of Jet", see page 19.

(c) Flooding from Float Chamber or Mouth of Jet. Fig. 8a.

This can be caused by a punctured float (10) or dirt on the needle valve (9) or its seat. These latter items can be readily cleaned after removal of the float chamber lid.

(d) Leakage from Bottom of Jet adjacent to Adjustment Nut.

Leakage in this vicinity is most likely due to defective sealing by the upper and lower sealing gland assemblies

There is no remedy other than removing the whole jet assembly after disconnecting the operating lever and cleaning or replacing the faulty parts. It is very important that all parts are replaced in their correct sequence, as shown in the illustration, and it must be realised that centralisation of the jet and needle and re-tuning will be necessary after this operation.

(e) Dirt in the Carburettor

This should be checked in the normal way by examining and cleaning the float chamber, but it may be necessary if excessive water or dirt is present to strip down and clean all parts of the carburettor with petrol.

(f) Failure of Fuel Supply to Float Chamber

If the engine is found to stop under idling or light running conditions, notwithstanding the fact that a good supply of fuel is present at the float chamber inlet union (observable by momentarily disconnecting this), it is possible that the needle has become stuck to its seating. This possibility arises in the rare cases where some gummy substance is present in the fuel system. The most probable instance of this nature is the polymerised gum which sometimes results from the protracted storage of fuel in the tank. After removal of the float chamber lid and float lever, the needle may be withdrawn, and its point thoroughly cleaned by immersion in alcohol.

Similar treatment should also be applied to the needle seating, which can conviently be cleaned by means of a matchstick dipped in alcohol. Persistent trouble of this nature can only be cured properly by complete mechanical cleansing of the tank and fuel system. If the engine is found to suffer from a serious lack of power which becomes evident at higher speeds and loads, this is probably due to an inadequately sustained fuel supply, and the fuel pump should be investigated for inadequate delivery, and any filters in the system inspected and cleansed.

25

(g) Sticking Jet

Should the jet and its operating mechanism become unduly resistant to the action of lowering and raising by means of the enrichment mechanism, the jet should be lowered to its fullest extent, and the lower part thus exposed should be smeared with petroleum jelly, or similar lubricant. Oil should be applied to the various linkage pins in the mechanism and the jet raised and lowered several times in order to promote the passage of the lubricant upwards between the jet and its surrounding parts.

Service Instruction Manual

SPECIALISED TOOLS

SECTION Q

SPECIALISED TOOLS

SPECIALISED TOOLS
POLICY

Considerable time and care has been taken in the preparation of specialised tools for servicing our Models, as it is realised that efficient servicing is not possible without the correct tools and equipment.

Messrs. V. L. Churchill & Co. Ltd. have designed and are manufacturing on our behalf and this Company has already circulated information concerning these tools, for many have similar applications on the Vanguard, Renown, Mayflower and Eight and Ten H.P. Models.

As the necessity for further tools becomes apparent they will be manufactured, and our agents will receive notice of such items as and when they are introduced.

PARTICULARS OF TOOLS

Brief particulars of approved tools which have been produced are given below. The tool in question should be ordered direct from Messrs. V. L. Churchill & Co. Ltd., Great South West Road, Bedfont, Feltham, Middlesex. Telephone : Feltham (Middx.) 5043. Telegrams : Garaquip, Feltham.

GENERAL
Press and Slave Ring	S 4221

ENGINE
Cylinder Sleeve Retainers	S 138
Sparking Plug Wrench	20SM 99
Connecting Rod Alignment Jig	335
Valve Spring Compressor	S 137
Stud Extractor	450

COOLING SYSTEM
Universal Puller	6312
Water Pump Refacer	S126 and 6300
Water Pump Impeller Remover & Replacer	FTS 127*

CLUTCH
Clutch Assembly Fixture	99A
Clutch Plate Centraliser	20S 72

FRONT SUSPENSION AND STEERING
Front Road Spring Compressor	M 50
Steering Wheel Puller	20SM 3600
Hub Remover for Disc Wheels	M 86
Knock on Wheels	S 132†
Hub Replacer (both types)	S 125
Electronic Wheel Balancer	120
Drop Arm Remover	M 91
Wheel Lock Protractors	121U

* Used in conjunction with S 4221 press.
† Used with S 4221 frame and slave ring.

1

SPECIALISED TOOLS

GEARBOX

Mainshaft Remover	20SM 1
Mainshaft Circlip Installer	20SM 46
Front Oil Seal Protecting Sleeve	20SM 47
Gearbox Extension Remover	20S 63
Constant Pinion Shaft Remover	20SM 66A
Countershaft Needle Roller Retainer Ring Driver	20SM 68
Mainshaft Circlip Remover	20SM 69
Countershaft Assembly Pilot	20SM 76
Countershaft Assembly Needle Roller Retainer	20SM 77
Gearbox Rear Bearing Replacer	20S 78
Gearbox Mainshaft Rear Oil Seal Replacer	20S 87A
Constant Pinion and Mainshaft Bearing Remover and Replacer	S 4615†
Two-way Circlip Pliers	7065
Front Cover Oil Seal Replacer	20SM 73A

REAR AXLE

Half Shaft Bearing Remover	S 4615
Half Shaft Bearing Replacer	M 92
Differential Case Spreader	S 101
Propeller Shaft Flange Wrench	20SM 90
Pinion Bearing Outer Ring Remover	20SM FT 71
Pinion Bearing Outer Ring Replacer	M 70
Pinion Oil Seal Replacer	M 100
Pinion Head Bearing Remover & Replacer	TS 1†
Differential Bearing Remover	S 103★
Differential Bearing Replacer	M 89
Pinion Setting Gauge and Dummy Pinion	M 84
Pinion Bearing Preload Gauge	20SM 98
Rear Hub Extractor (Disc Wheels)	M 86★
Rear Hub Extractor (Knock-on Wheels)	S 132†
Rear Hub Replacer (both type Wheels)	S 125
Rear Hub Oil Seal Replacer	M 29
Backlash Gauges	M 4210

★ Used in conjunction with S 4221 press.
† Used with S 4221 frame and slave ring.

Service Instruction Manual

BRAKES

SECTION R

BRAKES

INDEX

ILLUSTRATIONS

BRAKES

NOTATION FOR Fig. 1.	
Ref. No.	**Ref. No.**
Brake Operation	14 Large shake proof washer
1 Master cylinder to front connection pipe	15 Front to rear connection pipe
2 Two-way connection	16 Flexible hose
3 Banjo bolt	17 Hose locknut
4 Large copper gasket	18 Large shake proof washer
5 Small copper gasket	19 Copper gasket
6 Right to left-hand front connection pipe	20 Three-way connection
7 Front banjo connection	21 Connection attachment bolt
8 Banjo bolt	22 Right-hand brake pipe
9 Large copper gasket	23 Left-hand brake pipe
10 Small copper gasket	24 Rear axle clips
11 Stop light switch	**Clutch Operation**
12 Flexible hose	25 Master cylinder to frame bracket pipe
13 Hose locknut	26 Flexible hose

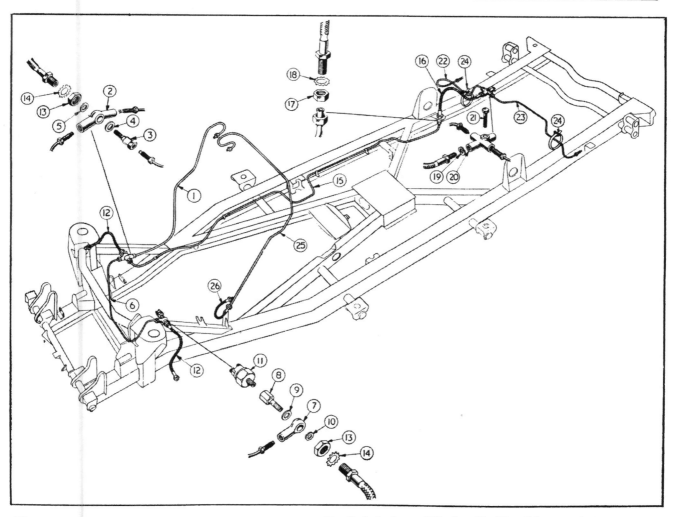

Fig. 1　　　　　　　　　　　**Exploded view of Hydraulic Pipe Lines and Connections.**

1

1. DESCRIPTION

Lockheed Hydraulic Brakes are fitted to all four wheels. Two leading shoe type are used on the front wheels and leading and trailing shoe type on the rear wheels.

A foot pedal operates the brakes hydraulically on all four wheels simultaneously, whilst the handbrake operates the rear brakes only by means of a cable.

The foot pedal is coupled by a push rod to the master cylinder bore in which the hydraulic pressure of the operating fluid is originated. The second bore of the master cylinder is connected to the clutch operating mechanism.

A supply tank, integral with the master cylinder, provides a fluid reservoir for both cylinders, a pipe line consisting of tube, flexible hose and unions connect the master cylinder bore to the wheel cylinders.

The pressure created in the master cylinder, by application of the foot pedal, is transmitted with equal force to all wheel cylinders simultaneously. This moves the piston which in turn forces the brakes shoes outward and in contact with the brake drum. An independent mechanical linkage, actuated by a hand lever, operates the rear brakes by mechanical expanders attached to the rear wheel cylinder and acts as a parking brake. The handbrake is situated in the centre of the car on the right-hand side of the gearbox tunnel. It is operated by pulling the grip rearwards and operating the push button on top by the thumb; when the button is depressed the lever will remain in that rearward position. To release the handbrake it is only necessary to pull the lever rearward sharply and then let it travel forward.

2. ROUTINE MAINTENANCE

Examine the fluid level in the master cylinder periodically and replenish if necessary to keep the level $\frac{1}{2}''$ below the underside of the cover plate.

Do not fill completely. The addition of fluid should only be necessary at infrequent intervals and a considerable fall in fluid level, indicates a leak at some point in the system, which should be traced and rectified immediately.

Ensure that the air vent in the filler cap is not choked, blockage at this point will cause the brakes to drag.

Adjust the brakes when the pedal travels to within $1''$ of the toe board before solid resistance is felt. If it is desired, adjustment may be carried out before the linings have become worn to this extent.

3. BRAKE LINING IDENTIFICATIONS

To afford maximum braking efficiency brake linings of an improved material have been progressively introduced. To enable identification linings are colour marked at their edges.

The following tabulation will give these identification marks and also the Commision number of the car on which they were first used.

BRAKE LINING IDENTIFICATIONS

LINING	IDENTIFICATION	INCORPORATION	REMARKS
DM.7 Front and Rear	3 narrow blue striped markings on lining edges.	FT. TS.1 to TS.3247 R. TS.1 to TS.3219	
DM.8	2 narrow blue and 1 wide blue marking, with aluminium coloured metal impregnation of lining.	FT. TS.3248 to TS.5216 R. TS.3220 to TS.5480	
M.20 Front only. (DM.8 fitted on rear brakes).	5 green stripe markings with bronze coloured metal impregnation	TS.5217 to TS.5480	For 10″ brake only.
M.20 Front and Rear	As above.	TS.5480 and future.	Introduction of 10″ brakes for rear wheels

4. DATA

Front Brakes 10″ × 2¼″

Rear Brakes 9″ × 1¾″ up to Commision No. TS.5481. Rear Brakes 10″ × 2¼″ after TS.5481.

Transverse rear brake cable lengths:

Right-hand 12.97″ ± .06″ 12.47″ ⎱ 10″
Left-hand 26.85″ ± .06″ 26.35″ ⎰ brakes.

These lengths are measured from pin centre of each fork end.

Front brake shoes are interchangeable with one another providing they have the same lining.

Rear brake shoes are interchangeable with one another providing they have the same lining and also interchangeable with front brake shoes of the same diameter and lining type.

5. FRONT BRAKE SHOE ADJUSTMENT

(a) Apply the brakes hard while the car is stationary to position the shoes centrally in the brake drum, then release brake.

(b) Jack up front of car, remove nave plates and road wheels.

(c) Rotate hub until hole provided in brake drum coincides with screwdriver slot in micram adjuster.

(d) Insert screwdriver in slot and turn the adjuster until brake shoes contact the drum, then turn adjuster back one notch.

(e) Repeat operations (c) and (d) with second micram adjuster.

(f) Repeat operations (c), (d) and (e) with second wheel.

(g) Replace wheels and nave plates. Lower car to ground and remove jacks.

(h) Road test car in a quiet thoroughfare.

6. REAR BRAKE SHOE ADJUSTMENT

(a) Chock front wheels and release hand brake. Apply brakes hard to position brake shoes centrally in drums and release.

(b) Jack up rear of car, remove nave plate and road wheels.

(c) Rotate hub until hole provided in the brake drum coincides with screwdriver slot in micram adjuster.

(d) Insert screwdriver in slot and turn the adjuster until brake shoes contact brake drums then turn adjuster cam back one notch.

(e) Repeat operations (c) and (d) with second wheel.

(f) Replace road wheels and nave plates Lower car to ground and remove jack

(g) Road test car in a quiet thoroughfare.

7. HANDBRAKE ADJUSTMENT

Adjustment of the brakes shoes already described automatically readjusts the handbrake mechanism.

The cables are correctly set during assembly and only maladjustment will result from altering the mechanism.

From the compensating linkage to the brake levers mounted on the wheel cylinders are transverse cables which are of a set length when leaving the works. They are however adjustable at their inner ends and should these have been tampered with it is necessary to check the following:

The cable assembled to the right-hand cylinder lever is 12.97″ ± .06″ between centres.

The left-hand is 26.85″ ± .06″, this gives the correct angle of the compensator lever as 17°. Only when a complete overhaul is necessary should the handbrake cables require resetting.

To carry out this operation, the brake shoes should be locked up in the brake drums with the handbrake in the "off" position. Any slackness that is in the cable from compensator to handbrake lever should be removed at the handbrake lever end.

8. TO BLEED HYDRAULIC SYSTEM

Except for periodic inspection of the reservoir in the master cylinder, no attention should be required. If, however, a joint is uncoupled at any time, or air has entered the system the system must be bled in order to expel the air which has been admitted. Air is compressible and its presence in the system will affect the working of the brakes.

The method detailed hereafter is suitable only for the braking system ; the procedure to be adopted when bleeding the clutch is detailed in the " Clutch Section ".

(**a**) Ensure an adequate supply of Lockheed Brake Fluid is in the reservoir of the Master Cylinder Unit and keep the level at least half full throughout the operation. Failure to observe this point may lead to air being drawn into the system and the operation of bleeding will have to be repeated.

(**b**) Clean the bleed nipple on one of the wheel cylinders and fit a piece of rubber tubing over it, allowing the free end of the tube to be submerged in a glass jar partly filled with clean Lockheed Brake Fluid.

(**c**) Unscrew the bleed nipple one full turn. There is only one bleed nipple to each brake.

(**d**) Depress the brake pedal completely and let it return without assistance. Repeat this operation with a slight pause between each depression of the pedal. Observe the fluid being discharged into the glass jar and when all air bubbles cease to appear hold the brake pedal down and securely tighten the bleed nipple. Remove rubber tubing only when nipple is tightened.

NOTE : Check the level of the fluid in the master cylinder frequently and do not allow the level to fall below half full. Seven or eight strokes of the brake pedal will reduce the fluid level from full to half full.

(**e**) Repeat the operation for the remaining three wheels.

(**f**) Top up master cylinder with Lockheed Brake Fluid and road test car.

9. LEAKAGE OF FLUID FROM MASTER CYLINDER

Leakage of fluid from the reservoir of the master cylinder can be explained as follows :

(**a**) Overfilling which allows fluid to be trapped in the filler cap and leak through the breather hole. The fluid level should never be higher than 1″ measured from the top of the filler orifice or ½″ measured from the underside of the cover plate.

(**b**) The breaking up of the filler seal due to foreign matter between it and the rim of the orifice.

(**c**) Leakage has been traced to jets of fluid from one of the cylinder recuperating holes finding its way past a defective filler cap sealing ring or via the breather hole.

The latter condition can be corrected by removing the cover plate and turning it 180° so that the filler cap is no longer directly above the jets.

10. BRAKE AND CLUTCH PEDAL ADJUSTMENT

The pedal adjustment is set when the car is assembled and should not require attention unless the assembly or adjustment has been disturbed.

A minimum clearance of .030″ is necessary between each push rod and the piston which it operates, this free movement can be felt at the pedal pad when it is depressed gently by hand.

The movement at the pedal pad will be magnified owing to the length of the lever and this movement will become between ½″ to ⅝″. Should this free movement not be apparent, first check that the pedals are free on their shaft and not prevented to return by some other fault than insufficient clearance between push rod and piston.

11. ADJUSTING THE BRAKE PEDAL

(**a**) Loosen the jam nut on the shank of the pedal limit stop screw and screw it anti-clockwise approximately ⅛″ away from the master cylinder support bracket.

(**b**) Push the operating push rod end into the master cylinder until it just contacts the piston. Screw up limit stop screw to meet the push rod fork end, but do not allow the rod to be pushed further into the piston. Screw the jam nut so that it makes contact with the master cylinder support bracket.

(**c**) Unscrew the pedal limit stop screw *together with* the jam nut so that a .030″ feeler gauge will pass between nut and support bracket.

(d) Holding the pedal limit stop screw turn the jam nut to the support bracket and tighten.

NOTE: The clutch pedal is set in a similar way but it must be remembered that adjustment at the slave cylinder may also be necessary to obtain the correct free pedal movement.

12. TO REMOVE FRONT LEFT-HAND FLEXIBLE HOSE

(a) Open bonnet and disconnect battery and wires to stop light switch.

(b) Drain the hydraulic system of fluid. Hold hexagon of hose near its bracket.

(c) Withdraw the banjo bolt from the banjo connection. The stop light switch attached to this bolt need not be removed.

(d) Holding the hexagon on the outside of the bracket with a spanner, remove the larger sized locking nut and shake proof washer.

(e) The hose can be withdrawn from its bracket and now removed from the wheel cylinder. Care should be taken to ensure that the entire length of hose is turned whilst it is being removed from the wheel cylinder.

13. TO FIT FRONT LEFT-HAND FLEXIBLE HOSE

Clean all components so that dirt does not enter system.

(a) Secure hose to wheel cylinder.

(b) Thread end of hose through chassis frame bracket and feed on shake proof washer and locknut.

(c) Set hose by holding hexagon with a spanner, tighten locknut to bracket assembly whilst still holding hexagon with spanner.

(d) Fit the larger diameter gasket to the banjo bolt and feed bolt through banjo connection, fit smaller diameter gasket to bolt. Feed bolt into hose end attached to bracket and secure finger tight. It will be seen that there is a gasket between the head of the banjo bolt and the banjo connection and a second gasket between the connection and the thread of the hose protruding through the bracket.

(e) Holding the hexagon of the flexible hose at the outside of the bracket, tighten the banjo bolt.

(f) Screw stop light switch into head of banjo bolt, *still holding* the hexagon of the hose.

(g) Replenish hydraulic reservoir with fresh fluid.

(h) Bleed all brakes as described on page 3.

(i) Check the system for fluid leakage by applying firm pressure to the pedal and inspect the line and connections.

14. TO REMOVE FRONT RIGHT-HAND FLEXIBLE HOSE

(a) Drain hydraulic system.

(b) Holding the banjo bolt of the two-way connection with one spanner remove the Bundy tubing union with a second.

(c) Grip the hexagon of the flexible hose on the outside of the bracket and remove the bolt passing through the centre of the two-way connection.

(d) Still gripping the hexagon of the hose remove locknut and shake proof washer. The flexible hose may now be withdrawn from its bracket.

(e) Remove the flexible hose from the wheel cylinder. Care should be taken to ensure that the entire length of hose is turned whilst it is removed from the wheel cylinder.

15. TO FIT FRONT RIGHT-HAND FLEXIBLE HOSE

Clean all parts and ensure no dirt enters the hydraulic system.

(a) Secure the flexible hose to the wheel cylinder.

(b) Thread end of the hose through chassis frame bracket and feed on shake proof washer and locknut.

(c) Set hose by holding hexagon with a spanner, tighten locknut securely to bracket whilst still holding hexagon with spanner.

(d) Fit the larger diameter gasket to the banjo bolt and feed bolt through two-way connection, fit smaller diameter gasket to bolt and secure bolt to end of hose protruding through chassis bracket.

(e) Hold the hexagon of the flexible hose at the outside of the bracket and tighten the banjo bolt, at the same time ensuring that the two-way connection is not allowed to turn.

(f) Reconnect the Bundy tubing to the head of the connection bolt.

(g) Replenish hydraulic reservoir with fluid.

(h) Bleed all brakes as described on page 3.

(i) Check the system for fluid leakage by applying firm pressure to the pedal and inspect the line and connections.

16. TO REMOVE THE REAR FLEXIBLE HOSE

The hose is first disconnected at its front end adjacent to the right-hand shock absorber bracket.

(a) Drain the hydraulic system of fluid.

(b) Holding the hexagon at the front end of the flexible hose remove the Bundy tubing union nut.

(c) Still holding the hexagon of the hose remove the locknut and shake proof washer. The hose can now be removed from the bracket.

(d) Disconnect hose from three way connection on rear axle. Care should be taken to ensure that the entire length of hose is turned whilst it is removed from the three way connection.

17. TO FIT REAR FLEXIBLE HOSE

Clean all parts thoroughly and ensure that no dirt is allowed to enter the hydraulic system.

(a) Position a gasket on the end of the flexible hose, secure to the three way connection in the rear axle.

(b) Feed foremost end of hose through bracket welded to chassis frame, attach shake proof washer and locknut to end of hose, finger tight.

(c) Holding the hexagon of the hose with a spanner, set it so that the hose is free from any obstructions. Still holding the hexagon secure hose to bracket, with the locknut.

(d) Continuing to hold the hexagon of the hose attach the Bundy tubing and tighten union nut.

(e) Replenish the hydraulic reservoir with fluid.

(f) Bleed all four brakes as described on page 3.

(g) Check the system for fluid leakage by applying firm pressure to the pedal and inspect the line and connections.

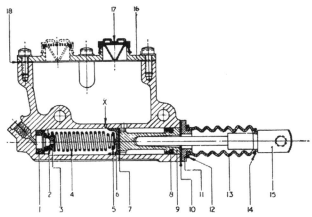

Fig. 2 Sectional view of Brake Master Cylinder. To prevent fluid leakage the cover plate is turned 180° (the dotted outline of the filler cap shows this condition) on later production cars.

NOTATION FOR Fig. 2.

1 Valve seat
2 Valve body
3 Rubber cup
4 Return spring
5 Spring retainer
6 Rubber cup
7 Piston washer
8 Secondary cup
9 Piston
10 Gasket
11 Boot fixing plate
12 Large boot clip
13 Rubber boot
14 Small boot clip
15 Push rod
16 Cover plate
17 Filler cap
18 Gasket
X Port in cylinder bore

18. TWIN BORE MASTER CYLINDER

Description

This unit consists of a body which has two identical bores, one connected to the brakes and the second to the clutch. Each of the bores accommodates a piston having a rubber cup loaded into its head by a return spring; in order that the cup shall not tend to be drawn into the holes of the piston head, a piston washer is interposed between these parts. At the inner end of the bore connected to the brakes, the return spring also loads a valve body, containing a rubber cup, against a valve seat; the purpose of this check valve is to prevent the return to the master cylinder of fluid pumped back into the line whilst bleeding the brake system, thereby ensuring a charge of fresh fluid being delivered at each stroke of the brake pedal and a complete purge of air from the system.

During normal operation, fluid returning under pressure and assisted by the brake shoe pull-off springs, lifts the valve off its seat, thereby permitting fluid to return to the master cylinder and the brake shoes to the " off " position.

There is **no check valve fitted in the bore connected to the clutch,** this precludes the risk of residual line pressure which would tend to engage the clutch, or keep the ball release bearing in contact with the release levers.

The by-pass ports, which break into each bore, ensure that the systems are maintained full of fluid at all times and allow full compensation for expansion and contraction of fluid due to change of temperature.

They also serve to release additional fluid drawn into the cylinder through the small holes in the piston after a brake or clutch application. If this additional fluid is not released to the reservoir, due to the by-pass port being covered by the main cup, as a result of incorrect pedal adjustment, or to the hole being choked by foreign matter, pressure will build up in the systems and the brakes will drag, or the clutch tend to **disengage.**

19. TO REMOVE MASTER CYLINDER

(a) Drain hydraulic system of operating fluid.

(b) Remove the square panel under the dash, which forms the rear wall of the master cylinder pocket from inside the car. Remove also the rubber grommet, from the inside wall of the pocket, to facilitate the withdrawal of the rear master cylinder attachment bolt.

(c) Disconnect the Bundy tubing from the connections at the rear of the master cylinder. Care must be exercised when removing the clutch Bundy tubing; this is connected first to an adapter and then to the cylinder body. It will be necessary to hold the adapter with one spanner, whilst loosening the Bundy tubing nut with a second. The connection for the brake operation is made direct to the master cylinder.

(d) Withdraw the clevis pins from the lever push rod fulcrums by removing the split pins, plain washers and double coil spring washers.

(e) Remove the nuts, lock and plain washers, from the master cylinder attachment bolts and withdraw the bolts, the rearmost one being passed through the aperture in the wall of the pocket into the car.

(f) The master cylinder is now free to be lifted from its support bracket. Empty any fluid that may still be in the reservoir.

20. TO FIT MASTER CYLINDER

(a) Ensure that the connection adapter is secure in the left-hand (clutch) outlet of the master cylinder.

(b) Place the assembly in the master cylinder support bracket, connections to the rear, and secure at the front end, with the attachment bolt and washers, but leave the nut finger tight at this juncture.

(c) The rear attachment bolt is fed in from inside the car, through the aperture in the pocket wall. This bolt passes through two adjustment brackets, one either side of the support bracket. With the washers in place screw on nut finger tight.

(**d**) Connect the Bundy tubing to the master cylinder connections through the aperture at the rear of the master cylinder. The clutch operating pipe is fitted to the adapter on the left and the brake operating pipe, which is on the right, direct to the master cylinder.

(**e**) Attach the piston rod fork ends to the pedals so that the heads of the clevis pins are nearest the centre line of the master cylinder assembly. Secure clevis pins with new split pins after fitting double coil spring and plain washers.

(**f**) Loosen the jam nuts of the adjusting brackets, at both sides of the support bracket, and turn the front nut in a clockwise direction to bring the master cylinder assembly forward to its fullest extent.

(**g**) Secure master cylinder to support bracket by tightening nuts of securing bolts. Lock up jam nuts to the adjusting bracket.

(**h**) Adjust pedal clearance as described on page 4.

(**i**) Replenish fluid reservoir with clean Lockheed Brake Fluid. Bleed brakes as described on page 3. Bleed clutch as described in "Clutch Section" D.

(**j**) Check the system for fluid leaks by applying firm pressure to the foot pedals and inspecting the line and connections for leaks.

(**k**) Replace rubber grommet in wall of master cylinder pocket and the cover at the rear of the pocket.

21. TO DISMANTLE THE MASTER CYLINDER (Fig. 3)

(**a**) Remove the circlip and rubber boot from the master cylinder body and withdraw them together with the push rod fork assembly.

(**b**) Remove the circlip and boot from the fork end assembly.

(**c**) Remove cover plate and joint washer from top of master cylinder body, also remove filler cap.

(**d**) Detach the boot fixing plate and joint washer.

(**e**) Withdraw pistons and washer.

(**f**) By applying low air pressure to the by-pass ports blow out the rubber cups.

(**g**) Tip out the springs and the check valve from the brake operating cylinder.

(**h**) Remove the valve seat from the bottom of the bore.

(**i**) Ease the cup out of the valve body and the secondary cups off the piston.

(**j**) Remove the adapter from the master cylinder body.

22. TO ASSEMBLE THE MASTER CYLINDER (Fig. 3)

Ensure absolute cleanliness during the assembly of these components. Assemble parts with a generous coating of clean Lockheed Brake Fluid.

(**a**) Fit the secondary cups to the pistons so that the lip of the cup faces the head of the piston. Gently work the cup round the groove with the fingers to ensure that it is properly seated.

(**b**) Looking at the open piston bores of the master cylinder, place a valve seal in the bottom of the left-hand (brake operating) bore.

(**c**) Ease the rubber cup into the valve body and fit the body in one end of a return spring, fit a spring retainer on the other end of the spring and insert the assembly, valve leading, into the bore which has the valve seat.

(**d**) Fit the second spring retainer on the second return spring and insert the spring, plain end leading, into the right-hand bore.

(**e**) Insert the main cup, lip leading, into each bore taking care not to damage, or turn back the lip of the cup. Follow with the two piston washers, ensuring that the curved washers are toward the rubber cups.

(**f**) Insert the two pistons, exercising care not to damage the rubber cups.

(**g**) Depress the two pistons, and fit the boot fixing plate, ulitising a new joint washer and securing plate with two screws and shake proof washers.

8

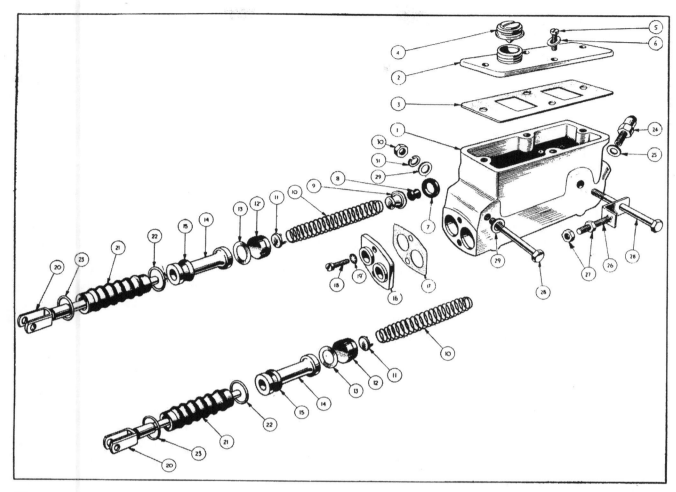

Fig. 3   **Exploded view of Twin Bore Master Cylinder.**

NOTATION FOR Fig. 3	
Ref. No.	Ref. No.
1 Body	17 Gasket between plate and body
2 Cover plate	18 Plate attachment screw
3 Joint washer	19 Shake proof washer
4 Filler cap and baffle	20 Push rod assembly
5 Cover plate attachment screw	21 Push rod boot
6 Shake proof washer	22 Large clip (Boot to fixing plate)
7 Valve seat ⎫	23 Small clip (Boot to push rod)
8 Valve cup ⎬ Brakes only	24 Slave cylinder pipe adapter (clutch)
9 Valve body ⎭	25 Gasket
10 Valve return spring	26 Bracket assembly
11 Spring retainer	27 Jam nut
12 Main cup	28 Master cylinder attachment bolt.
13 Washer between main cup and piston	29 Plain washer. (On front bolt only)
14 Piston	30 Nut
15 Piston secondary cup	31 Lock washers under nuts
16 Boot fixing plate	

(h) Position the cover plate on the body in such a manner that the filler cap is nearer the outlet ports. This will ensure the jets of fluid from the cylinder will impinge upon the plate and so avoid possible leakage through the filler cap. Ensure that the joint washer and filler cap sealing ring are in good order and that the vent hole is clear.

(i) Test the assembly by filling the tank with Lockheed Hydraulic Brake Fluid to within 1″ of the filler orifice top. Then push the piston inward and it should return without any assistance; after a few aplications fluid should be ejected from the outlet connections.

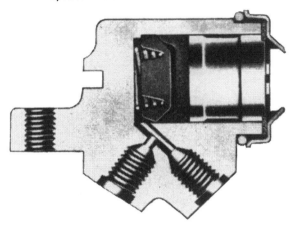

Fig. 4   Sectional view of Front Brake Cylinder.

23. FRONT WHEEL CYLINDERS
Description

The front wheel slave cylinders are mounted rigidly to the back plates inside the brake drums and between the ends of the brake shoes. One cylinder is mounted at the front and the other cylinder at the rear of each brake plate and each cylinder operates one shoe only. They are connected by a bridge pipe.

A single piston in each cylinder acts on the leading tip of its respective shoe, whilst the trailing tip of the shoe finds a floating anchor by utilising the closed end of the actuating cylinder of the other shoe as its abutment.

Between the piston and the leading tip of each shoe is a " Micram " adjuster which is located in a slot in the shoe.

Each front wheel cylinder consists of a body formed with a blind bore to accommodate a piston : a rubber cup, mounted in a cup filler, is loaded upon the piston by a spring which is located in the recess formed in the cup filler.

24. TO REMOVE FRONT WHEEL CYLINDERS

(a) Jack up car, drain off hydraulic fluid, remove nave plate, wheel, and brake drum.

(b) Pull one of the brake shoes against the load of the pull-off springs away from its abutment on the wheel cylinders. Slide the micram mask off the piston cover of the operating piston. On releasing the tension of the pull-off springs the opposite brake shoe will fall away.

(c) Remove the flexible hose as described on page 5.

(d) Unscrew the bridge pipe tube nuts from the wheel cylinders and remove the bridge pipe.

(e) Remove the fixing bolts and lock washers to withdraw wheel cylinders from back plate.

NOTATION FOR Fig. 5

Ref. No.	
1	Front brake plate
2	Wheel cylinder
3	Wheel cylinder body
4	Spring in body
5	Cup filler
6	Cup
7	Piston assembly
8	Rubber seal
9	Wheel cylinder attachment bolt
10	Lock washer
11	Bleed screw
12	Bridge pipe
13	Brake shoe assembly
14	Micram adjuster
15	Micram adjuster mask
16	Brake shoe pull off spring
17	Hub grease catcher
18	Brake drum

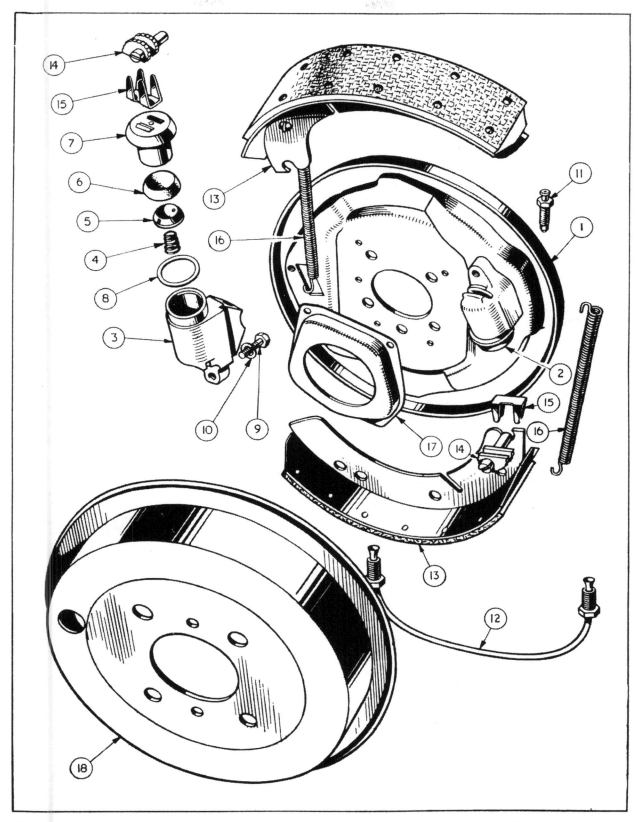

Fig. 5

Exploded view of Front Brake details.

25. TO FIT FRONT WHEEL CYLINDERS

(a) Mount the wheel cylinders on the back plate and secure each with a bolt and lock washer.

(b) Connect bridge pipe to bottom bore of each wheel cylinder, utilising the union nuts trapped on the pipe. Ensure that the pipe is located on its seat before attempting to attach the nut. Tighten nut sufficiently to give and oil and air tight joint.

(c) Attach the flexible hose to the upper bore of the rear cylinder, checking first that the copper gasket is in good order. Fit flexible hose to bracket on the chassis frame as described on page 5.

(d) Fit bleed screw to upper bore of front wheel cylinder.

(e) Fit brake shoes, taking care to locate the " micram " adjusters in the slots in the leading tip of each shoe, with the masks in position.

(f) Fit brake drum and bleed hydraulic system as described on page 3.

(g) Adjust brake as described on page 3.

(h) Check the system for fluid leakage by applying a firm pressure to the pedal and inspecting the pipe line and connections.

(i) Fit road wheel and nave plate. Remove jacks.

26. TO DISMANTLE FRONT WHEEL CYLINDER

(a) Withdraw the piston complete with piston cover from cylinder body.

(b) Apply low air pressure to the flexible hose connection, the rubber cup, the cup filler and spring can readily be removed.

27. TO ASSEMBLE FRONT WHEEL CYLINDER

Ensure absolute cleanliness during the assembly of these components. Assemble parts with a generous coating of clean Lockheed Brake Fluid.

(a) Fit the smaller end of the coil spring over the projection in the cup filler and insert both parts into the cylinder body, with the spring leading.

(b) Follow up with the rubber cup, lip end foremost, taking care not to damage or turn back this lip.

(c) Feed in piston with cover in position.

28. REAR WHEEL CYLINDER

Description

The cylinder, which is fitted in an elongated slot in the rear brake plate, is free to slide in the slot between the tips of the brake shoes which are of the leading and trailing shoe type. The cylinder has a single piston operating on the tip of the leading shoe and this shoe abuts against a fixed anchor block at the bottom of the back plate, the web of the shoe being free to slide in a slot in a block. The trailing shoe is located in a similar manner between the anchor and the closed end of the cylinder and is free to slide and therefore self centring.

The trailing shoes are operated by movement of the reaction of the leading shoe against the brake drum. A " micram " adjuster is located in a slot in the top of the leading shoe.

The wheel cylinder contains a single piston, split in two, the inner piston being hydraulically operated while the outer piston is manually operated by the hand brake lever. A rubber cup mounted in the cup filler is loaded upon the inner piston by a spring. When operated hydraulically, the inner piston abuts against the outer piston leaving the handbrake lever undisturbed, and applies a thrust to the tip of the leading shoe through the dust cover, micram adjuster and mask. When operated manually, an inward movement of the hand brake lever brings the head of the contact lever into contact with the outer piston, thrusting it outwards against the leading shoe without disturbing the inner piston. A rubber boot is fitted to exclude water and foreign matter.

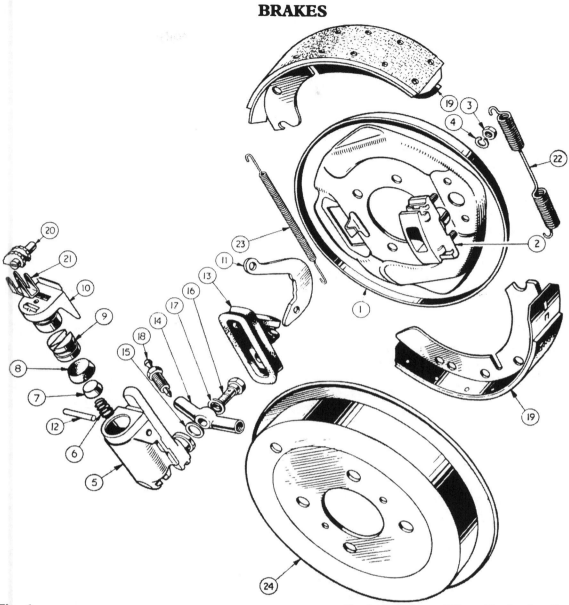

Fig. 6

Exploded view of Rear Brake details.

NOTATION FOR Fig. 6		
Ref. No.		
1	Rear brake plate.	13 Rubber boot.
2	Abutment assembly.	14 Banjo connection.
3	Abutment attachment nut.	15 Small copper gasket.
4	Lock washer.	16 Banjo bolt.
5	Wheel cylinder body.	17 Large copper gasket.
6	Spring in body.	18 Bleed nipple.
7	Cup filler.	19 Brake shoe assembly.
8	Cup.	20 Micram adjuster.
9	Hydraulic piston.	21 Micram adjuster mask.
10	Handbrake piston assembly.	22 Tension spring.
11	Handbrake lever.	23 Brake shoes pull-off spring.
12	Handbrake lever pivot pin.	24 Rear brake drum.

29. TO REMOVE REAR WHEEL CYLINDER

(a) Jack up rear of car. Remove nave plate, road wheel and brake drum. Slacken off micram adjuster.

(b) Drain off hydraulic fluid, disconnect handbrake cables and remove banjo bolt from banjo connection which is situated on the inner side of the brake plate.

(c) Pull the trailing shoe against the load of the pull-off springs and away from its abutment at either end; on releasing tension of the pull-off springs the leading shoe will fall away. Collect the micram adjuster and mask.

(d) Remove the rubber boot and the handbrake piston.

(e) Swing the handbrake lever until the shoulder is clear of the back plate and slide the cylinder casting forward. Pivot the cylinder about its forward end and withdraw its rear end from the slot in the back plate. A rearward movement of the cylinder will now bring its forward end clear of the back plate.

30. TO FIT REAR WHEEL CYLINDERS

(a) Offer up the rear wheel cylinder to the back plate with the handbrake lever to the slot. Engage the forward end of the cylinder in the slot and slide it well forward, taking care to position the lever so that the shoulder clears the back plate. Engage the rear end of the cylinder in the slot and slide it back to hold it in position.

(b) Place the rubber boot over the handbrake lever and ease the boot round the wheel cylinder so that it provides maximum weather protection. Connect handbrake cable to lever, utilising a new split pin for the securing of the clevis pin.

(c) Mount the banjo connection with new copper gaskets on the wheel cylinder and secure with banjo bolt.

(d) Assemble the brake shoes, ensuring that the micram adjuster is in the slot in the leading shoe with the mask in position. Fit the brake drum.

(e) Bleed the hydraulic system as described on page 3. Adjust the brake shoes as described on page 3.

(f) Check the system for fluid leakage by applying firm pressure to the pedal and inspecting the line and connections.

(g) Fit road wheel and nave plate. Remove jacks.

31. TO DISMANTLE REAR WHEEL CYLINDER

(a) Withdraw the piston complete with piston cover from the cylinder body.

(b) Remove the seal from the piston by easing out of its groove.

(c) Drift out the handbrake lever pivot pin to remove handbrake lever.

(d) Apply low air pressure to the inlet connection, the rubber cup, the cup filler and spring can readily be removed.

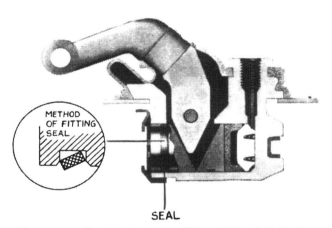

METHOD OF FITTING SEAL

SEAL

Fig. 7   **Sectional view of Rear Wheel Cylinder.**

32. TO ASSEMBLE REAR WHEEL CYLINDER (Fig. 7)

Ensure absolute cleanliness during the assembly of these components. Assemble hydraulic parts with a liberal smear of clean Lockheed Brake Fluid.

(a) Fit the smaller end of the coil spring over the projection in the cup filler and insert both parts into the cylinder body with spring leading.

14

(b) Follow up with the rubber cup, lip end forward, taking care not to damage or turn back this lip.

(c) Insert hydraulic piston into body ensuring that the slot coincides with the lever slot in the cylinder body.

(d) Place the handbrake lever in position and fit pivot pin.

(e) Stretch the handbrake piston rubber seal over the handbrake piston and place with dust cover in cylinder body, ensuring that the hand lever is engaged in the slot of the piston. The seal is to be twisted on its side so that the edge which tends to protrude from the groove enters the bore last.

33. TO REMOVE HYDRAULIC PIPE LINE FROM REAR AXLE

(a) Remove rear flexible hose as described on page 6.

(b) Disconnect the Bundy tubing at the brake plate by withdrawing union nut from banjo connection at each side.

(c) Repeat operation **(b)** at three-way connection.

(d) Remove Bundy tubing from axle by releasing pipe clips at each side.

(e) The three-way connection can be removed after withdrawing bolt and lock washer.

34. TO FIT HYDRAULIC PIPE LINE TO REAR AXLE

The fitting of the Bundy tubing is the reversal of the removal but the following points should be noted:

(a) The olives of the Bundy tubing should be correctly seated before securing the union nut.

(b) The pipe clips should be attached in such a manner that the pipe is in no way squeezed or damaged.

(c) The flexible hose is fitted as described on page 6.

(d) The connections should be inspected for leaks by applying firm pressure to the foot pedal.

35. FITTING REPLACEMENT BRAKE SHOES

(a) Jack up car and remove wheels, brake drums and slacken off all adjustment of micram adjusters.

(b) Remove brake shoes and collect pull-off springs and adjusters.

(c) Fit the replacement shoes and **new** pull-off springs after ascertaining that the brake linings are of the same material (see page 2).

(d) Fit brake drum and adjust brakes as described in page 3.

36. TO REMOVE PEDAL ASSEMBLY

(a) Working under the bonnet, drain both hydraulic systems and remove clevis pins from piston rods of twin master cylinder and disconnect pipe lines from rear of master cylinder.

(b) Remove four nuts and lock washers from front end of master cylinder support bracket adjacent to pedal push rods.

(c) From inside the car withdraw the two bolts and lock washers securing the side flanges of the pedal shaft casing to the bulkhead.

(d) Remove the two bolts and lock washers from the front and rear flange of the shaft casing and remove pedal assembly from bulkhead. The support bracket with master cylinder attached can also be removed from top of bulkhead.

An alternative method of pedal assembly removal is to omit the draining of the hydraulic system and the disconnection of the pipe lines mentioned in operation **(a)**, leaving the master cylinder and support bracket in position.

37. TO FIT PEDAL ASSEMBLY

(a) Working inside the car, secure the pedal assembly to the bulkhead, utilising two bolts and lock washers and the front and rear mounting flanges.

(b) Position the support bracket and master cylinder on the four studs protruding through the bulkhead shelf in such a manner that the clutch and brake piston fork ends engage with the two pedal levers.

NOTATION FOR Fig. 8

1 Pedal shaft cover assembly.
2 Clutch pedal.
3 Brake pedal.
4 Rubber pad for pedals.
5 Pedal pivot bush.
6 Pedal shaft.
7 Supporting bracket for pedal shaft.
8 Lock washer.
9 Bolt securing brackets to shaft.
10 Pedal return spring.
11 Lock washer.
12 Bolt securing pedal assembly to bulkhead.
13 Master cylinder support bracket.
14 Bolt securing pedal assembly and master cylinder support bracket to bulkhead.
15 Lockwasher.
16 Nut securing pedal assembly and master cylinder support bracket to bulkhead.
17 Clevis pin.
18 Double coil spring washer.
19 Plain washer.
20 Split pin.
21 Jam nut.
22 Pedal limit stop.

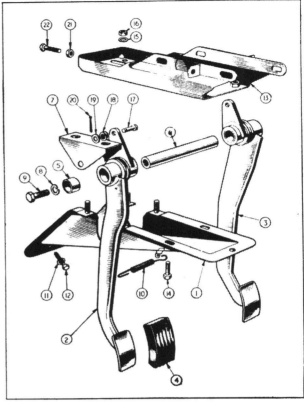

Fig. 8   Exploded view of Pedal Assembly. (R.H.S. shown.)

Attach bracket to bulkhead utilising four nuts and plain washers, these nuts are left loose at this juncture.

(c) Inside the car the pedal assembly is further secured to the bulkhead by two bolts and lock washers, these bolts are fully tightened.

(d) Under the bonnet, tighten the four nuts mentioned in operation (b). Connect the two pipe lines to their appropriate outlet ports and attach pedal levers to master cylinder fork end assemblies, utilising clevis pins.

(e) Adjust pedal clearances as described on page 4.

(f) Replenish reservoir with Lockheed Hydraulic Fluid.

(g) Bleed and adjust clutch as described in Clutch Section "D".

(h) Bleed brakes as described on page 3.

(i) Adjust brake shoes as described on page 3.

38. TO DISMANTLE PEDAL ASSEMBLY

(a) Suitably identify the pedals relative to their positions.

(b) Release the tension of the return spring by withdrawing end from the anchoring tab. The spring can now be removed from the pedal.

(c) Withdraw the two bolts and lock washers from pedal shaft support brackets and remove these brackets.

(d) Drift out pedal shaft.

(e) Lift out pedal assemblies from pedal shaft cover.

39. TO ASSEMBLE PEDALS

During assembly note the marked components and return them to their original positions.

(a) Fit the pedals to the shaft cover assembly in such a manner that the wall of the cover pressing is accommodated in the recess in the revolving collar on each pedal.

(b) Feed the pedal shaft through the pivots.

(c) Position the support brackets on the shanks of the welded bolts and allow the cut away side to drop into the recess of the revolving collar. Slight pressure may be necessary to bed this bracket.

(d) Secure the bracket to the pedal shaft, utilising two bolts and lock washers, one each side.

(e) Hook the return springs in the shaft of the pedals and anchor the other to the welded tab.

40. TO REMOVE HANDBRAKE LEVER

(a) Chock the wheels, jack up the car and release handbrake.

(b) Remove the bakelite handle grip and tape the thread for protection.

(c) Withdraw the three self tapping screws securing the draught excluder plate to floor. Remove plate and draw draught excluder up the handbrake lever.

(d) Working under the car, withdraw the clevis pin from the front fork end of the handbrake cable after first removing split pin and washer.

(e) Release the tabs of the locking plate and withdraw two bolts securing attachment plate.

(f) Remove the nyloc nut, locking pivot bolt to chassis frame.

(g) Withdraw pivot bolt. The handbrake lever can be drawn downward through the floor.

41. TO FIT HANDBRAKE LEVER

(a) Feed the pivot bolt through, first the lever assembly and then the mounting plate.

(b) Working beneath the car feed the lever through the floor assembly and attach lever to chassis by the pivot bolt which is left loose at this juncture.

(c) Utilising two bolts and a locking plate secure the lever mounting plate to the chassis frame and lock bolts with tabs of locking plate.

(d) Tighten the pivot bolt leaving the lever freedom of movement and attach the locking nut to the pivot bolt from inside the cruciform. When tightening this nut the head of the pivot bolt must be held to ensure the freedom of movement of the lever.

(e) Attach the fork end of the cable to the brake lever and secure with clevis pin, split pin and plain washer.

(f) Working inside the car, feed the draught excluder on to the lever and secure to floor with plate and three self tapping screws.

(g) The tape can now be removed from the thread and the bakelite grip screwed into position.

(h) Lower the car and remove chocks from the wheels. No readjustment of the handbrake should be necessary as the lengths of the cables have not been altered.

42. TO DISMANTLE HANDBRAKE ASSEMBLY

(a) Remove the bakelite grip and protect thread with tape.

(b) Detach the attachment plate from the ratchet by removing the bolt and nyloc nut.

(c) Remove the split pin and plain washer from the clevis pin, applying pressure to the press button at the top of the hand, withdraw clevis pin. This will allow the ratchet to become disengaged from the pawl and enable it to be withdrawn.

(d) Releasing the pressure on the button and allow it to protrude through the lever casing under the influence of the spring. Remove button from push rod, followed by the spring and plain washer.

(e) The push rod and pawl can now be withdrawn from the lower end of the lever and the pawl removed from the push rod.

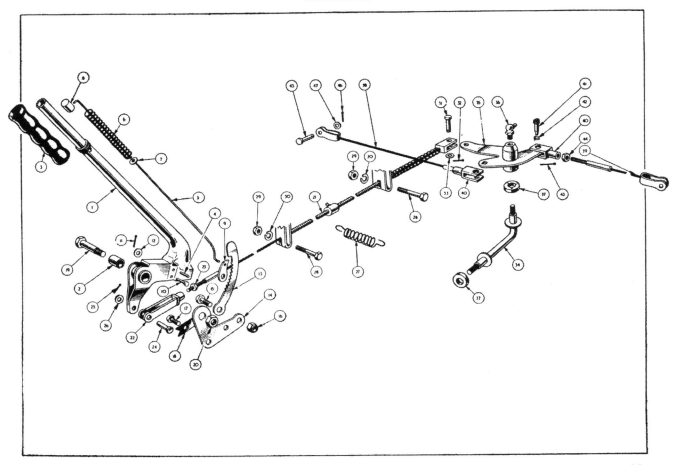

Fig. 9   **Exploded view of Hand Brake Assembly.**

NOTATION FOR Fig. 9	
Ref. No.	**Ref. No.**
1 Lever assembly.	25 Split pin.
2 Lever pivot bush.	26 Plain washer.
3 Handbrake lever grip.	27 Anti-rattle spring.
4 Pawl stop mills pin.	28 Bolt.
5 Pawl release push rod.	29 Nut.
6 Pawl release spring.	30 Lock washer.
7 Plain washer between spring and lever.	31 Clevis pin.
8 Push rod button.	32 Split pin.
9 Pawl.	33 Plain washer.
10 Clevis pin, pawl to lever.	34 Compensator bar assembly.
11 Split pin.	35 Compensator lever assembly.
12 Plain washer between split pin and lever.	36 Grease nipple.
13 Ratchet.	37 Felt seal.
14 Attachment plate.	38 R.H. cable assembly 12.97″ long. 12.47″ ⎫ 10″
15 Set screw. Ratchet to attachment plate.	39 L.H. cable assembly 26.85″ long. 26.35″ ⎭ brakes.
16 Nyloc nut.	40 Fork end.
17 Set screw. Ratchet to attachment plate.	41 Swivel pin.
18 Tab washer on setscrews.	42 Anti-rattle spring.
19 Pivot bolt.	43 Split pin.
20 Nyloc nut.	44 Jam nut.
21 Cable assembly (handbrake to compensating lever).	45 Clevis pin.
22 Fork end.	46 Split pin.
23 Jam nut.	47 Plain washer.
24 Clevis pin.	

43. TO ASSEMBLE HANDBRAKE ASSEMBLY

(a) Feed the push rod into the lever from below so that its shape corresponds with that of the handle.

(b) Attach the pawl to the push rod so that it points rearward.

(c) Allow the push rod to protrude through the upper portion of the handle and feed on a plain washer and coil spring, followed by the button. Apply pressure to the button to compress spring.

(d) Hold the pressure on the button and feed the ratchet, teeth facing forward, into the lower portion of the casing, ensuring that it is positioned well inside the lever. Manipulate the pawl until its fulcrum hole is aligned with the hole in the lever and insert the clevis pin; pressure on the button can now be released. Secure clevis pin with plain washer and split pin.

(e) Secure the attachment plate to the ratchet, utilising a bolt and nyloc nut. Tighten the nut sufficiently to allow the attachment plate to swing on the ratchet. Failure to observe this instruction will result in imperfect handbrake operation.

(f) The tape protecting the thread can now be removed and the grip fitted.

44. TO REMOVE HANDBRAKE CABLES

(a) Let off the handbrake, lock the rear brakes on by turning the micram adjuster.

(b) Withdraw the split pins and clevis pins at each end of the handbrake cable assembly.

(c) Release the tension of the spring securing the brake cable to the gearbox tunnel. Withdraw the two bolts from the cable abutment brackets and remove cable assembly.

(d) Withdraw the split pins and clevis pins attaching the transverse cables to the levers on the brake backing plate.

(e) Remove the split pins and clevis pins at their inner ends, taking care to collect the anti-rattle springs. Remove cables from car.

(f) The compensator assembly can be removed from the axle by turning lever and bar assemblies independently in an anti-clockwise direction.

45. TO FIT HANDBRAKE CABLES

The fitting is the reversal of the removal but the following points should be noted :—

(a) The transverse cables should be of the correct length. R.H. 12.97"±.06" L.H. 26.85"±.06". These measurements for 10" brakes are 12.47" and 26.35" respectively.

(b) All cables and fulcrums should be thoroughly greased before fitting.

(c) The bar assembly is attached to the axle with a new felt seal and then turned back one turn. This instruction also applies to the lever assembly when fitted to the bar assembly.

(d) The handbrake is adjusted as described on page 3.

Service Instruction Manual

EXHAUST SYSTEM

SECTION S

EXHAUST SYSTEM

INDEX

ILLUSTRATIONS

EXHAUST SYSTEM

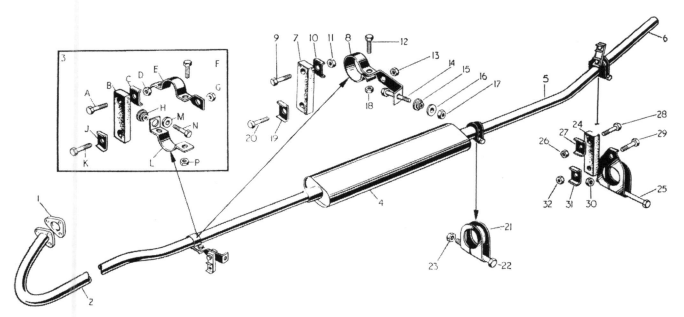

Fig. 1

Exploded view of Exhaust System.

NOTATION FOR Fig. 1.	
Ref. No.	**Ref. No.**
1 Exhaust flange joint.	10 Clamp plate.
2 Front exhaust pipe.	11 Attachment nut.
3 Prior to Commission No. TS.4310 only :	12 Clamp bolt
A Attachment bolt to chassis.	13 Attachment nut.
B Flexible mounting strip.	14 Attachment bolt to chassis frame.
C Clamp plate.	15 Rubber and steel grommet.
D Attachment nut.	16 Rubber and steel grommet.
E Exhaust pipe clip (Upper half).	17 Attachment nut.
F Clamping bolt.	18 Clamp nut.
G Attachment nut.	19 Clamp plate.
H Rubber and metal grommet.	20 Attachment bolt (Support bracket).
J Clamp plate.	21 Pipe clip.
K Attachment bolt.	22 Pinch bolt.
L Exhaust pipe clip (Lower half).	23 Nut for pinch bolt.
M Rubber washer.	24 Flexible mounting strip.
N Attachment bolt (Lower half clip).	25 Pinch bolt.
P Nut for clamp bolt.	26 Attachment nut.
4 Silencer.	27 Clamp plate.
5 Tail pipe assembly.	28 Attachment bolt to chassis.
6 Tail pipe extension.	29 Pipe clip attachment bolt.
7 Flexible mounting strip.	30 Nut for pinch bolt.
8 Exhaust pipe support bracket.	31 Clamp plate.
9 Attachment bolt to chassis.	32 Clip attachment nut.

EXHAUST SYSTEM

1. DESCRIPTION

The manifolds are attached to one another by studs in the aluminium alloy induction manifold and lugs moulded in the cast iron exhaust manifold. There is no " hot spot " for easy starting.

The exhaust system is situated on the right-hand side of the engine and passes down to the rear of the car through the centre of the cruciform to a position adjacent to the left-hand chassis member. The front exhaust pipe is attached to the engine by a flange and is flexibly mounted to the chassis frame at a point forward of the cruciform centre. This attachment also secures the pipe to the outside of the silencer. Cars with Commission No. TS.4310 and before has this clip in two halves as shown in Fig. 1.

Two types of silencers have been used in production, the former 18″ silencer being changed for a 24″ type at Commission No. TS.2532. A modified tail pipe incorporating a 12″ silencer can be fitted, at the owner's discretion, to the shorter type silencer, if the exhaust note is considered too loud (see Fig. 2). This modified tail pipe fits into the main silencer and is attached with the existing clip.

At the rear the tail pipe is attached to the chassis by a flexible mounting strip and the clip secures the chromium plated extension piece inside the tail pipe.

2. MAINTENANCE

The exhaust system should be inspected periodically to ensure its correct function. Attention should be paid to the gaskets at the cylinder head, carburettor and front exhaust pipe flanges to ascertain their condition. If signs of " blowing " are detected then gasket must be replaced as soon as possible. Manifold gaskets should be replaced as a pair and no gasket should ever be used twice.

The flexible mounting strips should be inspected and replaced if any deterioration is apparent.

The position of the silencer assembly in relation to the cruciform centre should always be such that during any vibrationary period the exhaust system cannot come into contact with the cruciform.

3. TO REMOVE AND DISMANTLE EXHAUST SYSTEM

(a) Working from the **rear of the car** loosen the bolt of the **rear pipe clip** attachment and withdraw exhaust pipe extension.

(b) Withdraw the **lower bolt** securing pipe clip attachment to flexible mounting strip and collect nut and lock washer.

(c) Loosen the **pinch bolt** of the pipe clip attachment at the rear of the silencer and withdraw tail pipe assembly.

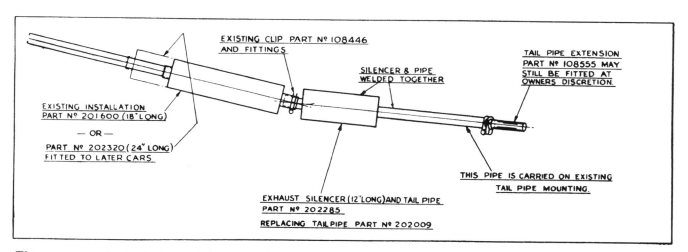

Fig. 2   **Method of supplementary silencing provided by 18″ Silencer, fitted prior to Com. No. T.S.2532.**

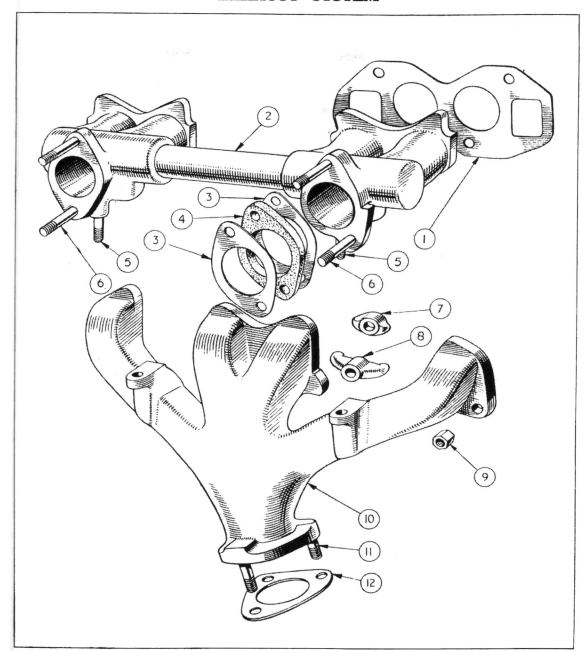

Fig. 3.   **Manifold details.**

NOTATION FOR Fig. 3.	
Ref. No.	Ref. No.
1 Manifold gasket.	7 Small manifold clamp.
2 Inlet manifold.	8 Large manifold clamp.
3 Joint washer.	9 Manifold securing nut.
4 Insulating washer.	10 Exhaust manifold.
5 Exhaust manifold attachment stud.	11 Exhaust pipe attachment stud.
6 Carburettor attachment stud.	12 Flange joint washer.

(d) Loosen the pinch bolt (or bolts) forward of the cruciform centre and withdraw the silencer rearward.

(e) Remove the lower bolt attaching flexible mounting strip to chassis frame. Remove the nut and bolt, collecting the rubber grommet and rubber washer securing the bracket to the chassis frame.

(f) The front exhaust is detached from the exhaust manifold by the removal of three nuts with spring washers. After the joint is broken the front exhaust pipe is moved clear of the car.

4. TO FIT EXHAUST SYSTEM

The fitting of the system is the reversal of the removal but the following points should be noted:

(a) It is suggested that work is started at the front as each component fits into the one in front.

(b) Each mounting should be left loose and finally tightened when the position of the silencer is set. The front tube of the silencer assembly which passes through the cruciform centre will need setting to avoid the possibility of it vibrating against the cruciform centre. The mountings can be tightened progressively from front to rear.

(c) If the tail pipe incorporating the small silencer is being fitted it is attached in a similar manner to the pipe and uses the existing clips. Fig. 2.

5. TO REMOVE MANIFOLDS

(a) Remove the carburettors as described in the "Fuel" Section P.

(b) Disconnect the exhaust pipe at the flange by removing the three nuts and spring washers.

(c) Remove the eight nuts, spring washers and six clamps. Both manifolds together with the gaskets can be removed from the combustion head.

(d) The manifolds can be separated by removing the two nuts and spring washers situated below the carburettor mounting flanges.

6. TO FIT MANIFOLDS

The fitting is the reversal of the removal but the following points should be noted :—

(a) New gaskets should be used and so ensure gas tight joints.

(b) The manifolds should be attached to the cylinder head before finally tightening the inter-connection nuts.

(c) The carburettors must be synchronised before the car is ready for the road.

SERVICE INSTRUCTION MANUAL

SUPPLEMENT

TR3 MODELS

Issued by

SERVICE DIVISION, THE STANDARD MOTOR CO. (1959) LTD.

COVENTRY, ENGLAND

TR3

FOREWORD

Certain modifications have been incorporated in the TR3 models and have been made with a view to enhancing its appearance and performance.

Whilst the general specification of the TR3 agrees very largely with that for the earlier models, there are certain differences which are set out in this Supplement.

The nature of the modifications made to the later model cannot be easily incorporated in the earlier models, nor is it the intention of The Standard Motor Company (1959) Limited that they should.

Service Instruction Manual Supplement

TR3 MODELS

GENERAL DATA

SECTION A

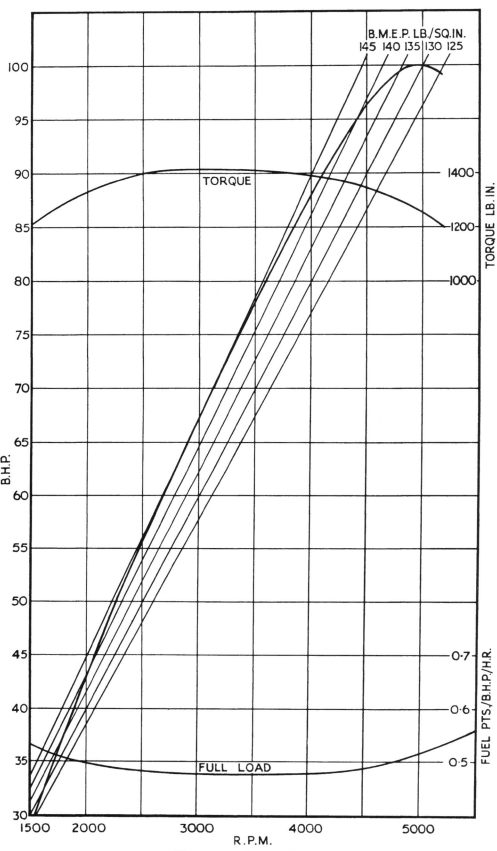

Fig. 1   Power Curve.

GENERAL DATA

GENERAL DATA

The information given in this section should be studied in conjunction with that given in the appropriate pages of the main Manual.

CAMSHAFT BEARINGS

Vandervell shell bearings are fitted to the 2nd, 3rd and rear journals.

CARBURETTORS

Two S.U. Type H6 carburettors are fitted. The early TR3 cars were fitted with carburettors having "TD" needles, but this needle was changed to type "TE" early in normal production and was, at Engine No. TS.10037E, superseded by type "SM". Where replacement needles are required for carburettors fitted with the early needles, both needles should be replaced by the "SM" needle.

PERFORMANCE DATA

95 B.H.P. at 4,800 R.P.M.

TRANSMISSION

Ratios

	O/D Top	Top	O/D 3rd	3rd
Gearbox	.82	1.00	1.08	1.325
Overall	3.03	3.7	4.02	4.9

	O/D 2nd	2nd	1st	Rev.
Gearbox	1.64	2.00	3.38	4.28
Overall	6.07	7.4	12.5	15.8

Engine Speed at

	10 m.p.h.	10 km.p.h.
O/D Top	410 R.P.M.	245 R.P.M.
Top	500 „	310 „
O/D 3rd	540 „	340 „
3rd	660 „	410 „
O/D 2nd	820 „	510 „
2nd	1,000 „	620 „
1st	1,680 „	1,050 „
Rev.	2,130 „	1,325 „

1

Service Instruction Manual Supplement

TR3 MODELS

ENGINE

SECTION B

ENGINE

1. CYLINDER BLOCK

Vandervell replaceable shell bearings have been introduced for the 2nd, 3rd and rear camshaft journals. These are manufactured to very fine limits, and whilst certain fitting precautions must be observed, line boring of the assembled bearings is unnecessary. Removal of the rear bearing will necessitate the removal of the sealing disc behind it, which, in turn, will require the removal of the gearbox, clutch and flywheel. The tool illustrated in Fig. 1 is designed to assist in the removal and replacement of the bearings.

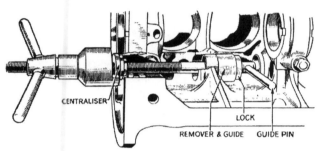

CENTRALISER

LOCK

REMOVER & GUIDE   GUIDE PIN

Fig. 1. Fitting Intermediate Camshaft Bearing using Churchill Multipurpose Tool No. 32 with Adaptors S.32-1.

(a) Camshaft Bearings

To remove, proceed as follows :—

(i) Using a suitable tool, drift the sealing disc out of the rear camshaft bearing housing.

(ii) Unscrew and remove the three shouldered setscrews and plain washers which retain the bearings in position.

(iii) Assemble the extracting tool and adaptors into the cylinder block as illustrated, and withdraw each bearing in turn.

(b) To Fit New Bearings

See Fig. 1 and observe the following: The oil feed holes must be correctly aligned and when drawing the bearings into position all possible precautions should be taken to ensure that these **do not turn** and so misalign the holes. Ensure also that the locating hole in each bearing is centrally disposed in the tapped hole which accommodates the locating screw. Failure to observe this instruction may result in the bearing becoming distorted when the locating screw is tightened.

Fit a plain steel washer of $\frac{1}{16}$" thickness (1.588 mm.) between the head of each locating screw and the cylinder block. Refit or replace the camshaft sealing disc if necessary.

2. ALUMINIUM PEDESTALS FOR ROCKER SHAFT

New rocker pedestal brackets of aluminium alloy were incorporated in normal production at Engine No. TS.12564E. The new metal, by reason of its higher degree of expansion when hot, enables the same rocker clearances to be used for exhaust valves as were previously applied only to the inlets. This reduction in the exhaust valve clearances has the advantage of reducing " tappet " noise when the engine is cold without any sacrifice of performance. Where it is desired to fit the new pedestal brackets, these should be fitted as a complete set, the part numbers being as follows :—

Aluminium Pedestal Bracket (Plain)—
3 off—Part No. 112546

Aluminium Pedestal Bracket (Drilled)—
1 off—Part No. 112545

3. PISTONS

From Engine No. TS.9731E, the pistons are fitted with :—

1 Plain ring.
1 Taper ring.
1 Oil scraper ring.

4. COMBUSTION HEAD

To further improve performance, " High Port " type combustion heads where incorporated in production at Engine No. TS.9350E. In countries where high octane fuel is unobtainable, the compression ratio may be lowered to 7.5/1 by the use of a compression plate, Part No. 200906. This plate must be used in conjunction with a steel " Corrojoint " gasket, Part No. 202775 in addition to the normal gasket.

When using this low compression plate it will also be necessary to use Champion L.10 sparking plugs gapped to 0.025" and special push rods, Part No. 114048.

1

5. **ENGINE OIL FILTER**

In order to give the maximum protection to the engine when subjected to high speed or rally conditions, a new filter of the " full flow " type has been introduced on the TR3 models. This type of filter ensures that all the oil in circulation passes through the filtration system.

The " full flow " type of filter was introduced into normal manufacture at Engine No. TS.12650E., part numbers affected by this change being as follows :—
Oil filter assembly, Part No. 301994, is replaced by Part No. 203271.

The replacement Element, Part No. 101963, remains the same for both types of filter.

The oil pressure on the " full flow " type of filter remains at 70 lbs. per sq. in. with an oil temperature of 70°C. at an engine speed of 2,000 r.p.m.

The new filter assembly can be fitted if desired to an engine prior to TS.12650E.

6. **SUMP**

A special cast aluminium sump, Part No. 301318, and tray, Part No. 201984, are available as optional extras.

Fig. 2 Oil cleaner " full-flow " type.

Service Instruction Manual Supplement

TR3 MODELS

CLUTCH

SECTION D

CLUTCH

CLUTCH DRIVEN PLATE ASSEMBLY

An improved clutch driven plate incorporating a Belleville washer friction centre was fitted after Engine No. TS.7830E. (TR2).

The new driven plate can be recognised by four small tongues (or tabs) protruding through the spring retaining plates adjacent to the longer side of the splined hub and by the colour of the six cushioning springs, white and light green.

HYDRAULIC OPERATING MECHANISM

This is described under " Girling Brakes and Hydraulic Clutch " in the " Brake " supplement.

1

Service Instruction Manual Supplement

TR3 MODELS

REAR AXLE

SECTION F

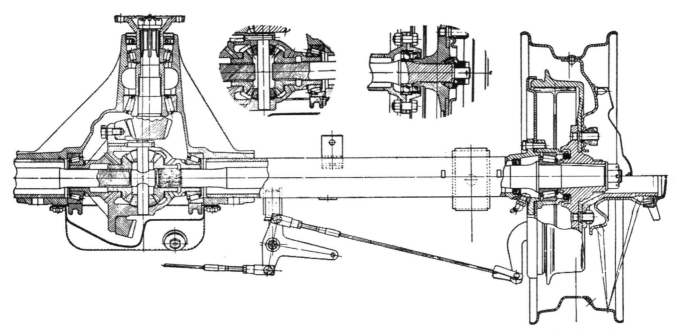

Fig. 1   Rear axle section (inserts indicate axle arrangement for cars up to Commission No. T.S. 1300).

REAR AXLE

1. GENERAL

A new rear axle assembly, Part No. 302177, bearing the Serial No. 13511, was introduced at Commission No. 13046 and fitted on all subsequent cars.

The major differences incorporated in the new axle include new half shaft and hub assemblies, a thrust button mounted on the differential cross-pin and adjustable taper roller hub bearings, as shown in Fig. 1. The sectioned insert views indicate the axle arrangement for cars prior to this change.

Fig. 2   Rear hub lubricator.

2. LUBRICATION OF REAR HUB BEARINGS

The rear hub bearings are each lubricated by a grease nipple located behind the brake backing plate and facing downwards, as shown in Fig. 2.   The nipples should receive a small but regular supply of grease, as specified on pages 9-12 (Section " A "). Six strokes of the hand grease gun every 6,000 miles (10,000 km.) will normally be sufficient, as it is inadvisable to overload with grease.

3. AXLE SHAFT, WHEEL BEARINGS AND OIL SEALS

(a) To Dismantle
The procedure is as follows :—

   (i) Jack up the rear of car, remove road wheel, unscrew two securing setscrews and detach brake drum from hub.

Fig. 3   Extraction of rear hub.

   (ii) Withdraw split pin and remove castellated nut from end of half shaft and remove rear hub with extractor, as shown in Fig. 3.

   (iii) Remove six setscrews securing the brake backing plate and bearing housing to the axle sleeve outer flange, then detach the bearing housing complete with the bearing outer ring.

Fig. 4 Extracting outer ring of hub bearing from housing.

Note.—Removal of a half shaft does not normally require detachment of the brake backing plate, but if its removal is necessary, then the

3

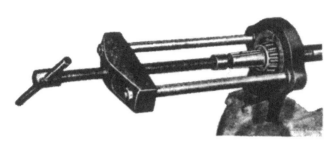

Fig. 5   Removing hub bearing inner ring.

Fig. 6 Fitting hub bearing inner ring to axle shaft.

Fig. 7   Fitting oil seal into bearing housing.

brake fluid pipe and the hand-brake attachments must first be disconnected and the backing plate subsequently removed.

(iv) Extract the bearing outer ring from the housing, as shown in Fig. 4, after first tapping out the oil seal which should be renewed during re-assembly.

(v) Withdraw the axle shaft and inner bearing ring. After first removing the driving key, the bearing inner ring is then removed by using the extractor, as shown in Fig. 5.

(b) **Inspection**
Inspect bearing for looseness and roughness; the axle shaft for cracks and worn splines; the hub for loose wheel studs and worn keyway. Replace all parts which are excessively worn or defective in any way.

Note.—When inspecting rear axle hub bearings, apply as much load as possible by hand, as this enables noise and roughness to be more readily detected.

(c) **To Re-assemble**
Continue as follows :—

(i) Using a special tool, drive the hub bearing inner ring on to the axle shaft, as shown in Fig. 6, and refit key.

Fig. 8   Fitting oil seal into axle sleeve.

(ii) Draw the bearing outer ring into the housing by using the same tool as shown in Fig. 4, and install a new oil seal (Fig. 7).

(iii) Exercising care to avoid damage to the fabric face of the seal, thread the assembled bearing housing on to the shaft and refit hub, plain washer and castellated nut, tightening this to a torque of 125-145 lbs. ft. (17.29-19.71 kg. metres) and securing it with a split pin.

(**iv**) Examine the inner oil seal and, if a replacement is necessary, proceed as shown in Fig. 8. Oil seal renewal is recommended in all cases of axle overhaul.

(**v**) Replace the original shim pack over the spigoted portion of the axle sleeve, followed by the brake backing plate.

(**vi**) Again exercising care in the case of the inner oil seal, thread the assembled axle shaft through the seal and into the axle casing. After locating the shaft splines in those of the sun wheel, secure the bearing housing by inserting and tightening six setscrews with lock-plates.

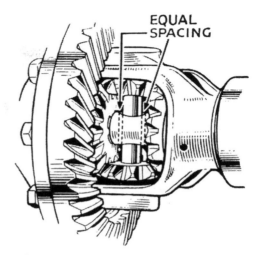

Fig. 10 **Showing position of differential cross-pin in relation to thrust block.**

towards and away from the axle casing. The dial indicator will then record the axle shaft end-float.

Adjustment is effected by adding to, or subtracting from the shim pack interposed between the axle sleeve flange and the brake backing plate, thus increasing or decreasing respectively the axle shaft end-float.

Important.—In addition to the existence of the specified end-float, it is important that the thrust block which separates the inner extremities of the two axle shafts, should have a clearance on the cross-pin, as shown in Fig. 10. To ensure centralization of the thrust block with the cross-pin, the shim packs behind both backing plates will be approximately of equal thickness.

(**vii**) Replace brake drum, road wheel and, before removing the lifting jack, it is essential to grease the hub bearing.

Fig. 9   **Checking axle shaft end float.**

Axle Shaft End-Float
The specified axle shaft end-float is 0.004″-0.006″ (0.102-0.152 mm.).

This can be checked by mounting a dial indicator on the backing plate, as shown in Fig. 9, then moving the hub

4. **DIFFERENTIAL AND PINION ASSEMBLIES**
Except for the addition of a thrust block (item 15, Fig. 13), the crown wheel and pinion assemblies remain the same as fitted to the previous axle. Therefore, instructions for the servicing and adjustment of these assemblies are unaltered.

5. HIGH SPEED AND COMPETITION WORK

(a) Rear Axle Assembly—ratio 4.1/1

A rear axle of the above ratio is available for high speed and competition work but is **only suitable for cars fitted with Overdrive.** The installation and servicing procedure is the same as for standard ratio axles.

Crown wheel (41 teeth)   Part No. 202579

Pinion (10 teeth)   Part No. 202580

Complete axle assembly (for wire wheels)   Part No. 505179

(for disc wheels)   Part No. 503930

(b) Speedometer

The following special ratio speedometers are necessary when using 4.1/1 axles :—

Speedo — Kilo.   Part No. 113632

Speedo — Mile   Part No. 113631

(c) Centre Lock Adaptors (Wire Wheel)

These splined hub extensions are attached to the hubs by shorter studs than normally used for disc wheels. Figs 11. and 12 show the extensions being fitted and the existing studs sawn-off flush with the outside of the wheel nuts.

Hub Extension
(L H) Part No 202447
(R.H.) Part No. 202446

Knock-off Wheel Nut
(L.H.) Part No. 107949
(R.H.) Part No. 107948

Fig. 11 Fitting splined hub extension to normal hub.

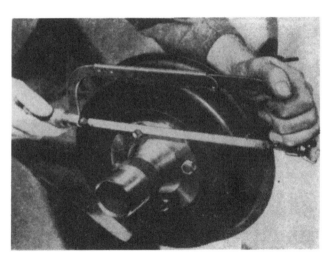

Fig. 12 Reducing length of studs to enable wire wheels to be fitted.

Ref. No.	Description	Ref. No.	Description	Ref. No.	Description
	NOTATION FOR EXPLODED ARRANGEMENT OF REAR AXLE (Fig. 13)				
1	Axle casing assembly.	18	Crown wheel securing bolt.	35	Fibre washer for (34).
2	Bearing cap setscrew.	19	Plain washer for (18).	36	Axle half shaft.
3	Spring washer.	20	Three hole lockplate for (18).	37	Rear hub bearing.
4	Axle case breather.	21	Two hole lockplate for (18).	38	Hub bearing housing.
5	Fibre washer.	22	Pinion head bearing.	39	Oil seal for hub bearing housing.
6	Drain plug.	23	Adjusting shims for (22).	40	Adjusting shims for hub bearing.
7	Differential bearing.	24	Bearing spacer.	41	Lockplate.
8	Adjusting shims for (7).	25	Pinion tail bearing.	42	Setscrew for securing housing.
9	Differential casing.	26	Adjusting shims for (25).	43	Hub.
10	Differential sun gear.	27	Pinion shaft oil seal.	44	Road wheel attachment stud.
11	Thrust washer for (10).	28	Pinion driving flange.	45	Hub driving key.
12	Differential planet gear.	29	Driving flange securing nut.	46	Hub securing nut.
13	Thrust washer for (12).	30	Plain washer for (29).	47	Plain washer for (46).
14	Cross pin.	31	Split pin for (29).	48	Split pin for (46).
15	Thrust block.	32	Rear cover.	49	Cover plate securing screw.
16	Lock pin for securing (14).	33	Joint washer for (32).	50	Spring washer for (49).
17	Crown wheel and pinion.	34	Oil filler plug.	51	Axle tube oil seal.

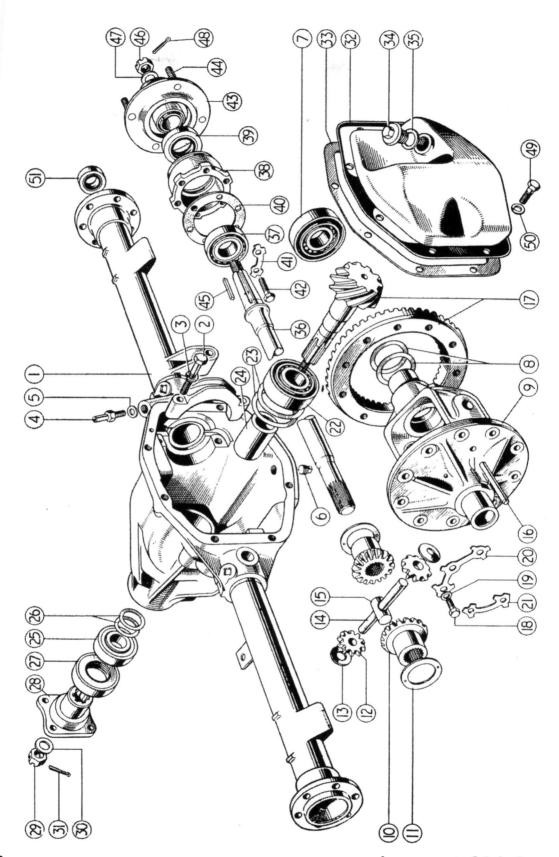

Fig. 13

7

Arrangement of Axle Components.

371

Service Instruction Manual Supplement

TR3 MODELS

FRONT SUSPENSION

AND

STEERING

SECTION G

FRONT SUSPENSION AND STEERING

NYLON BEARINGS

LOWER INNER WISHBONE ATTACH-MENT

These bearings supersede the rubber bushes and were introduced into production at car Commission No. TS.9121.

They are as follows :—

(a) Nylon bearing. 4 off. Pressed into each wishbone arm.

(b) Steel bush. 4 off. Fitted to fulcrum pin.

(c) Sealing rings. 8 off. Fitted to outside edge of nylon washers.

(d) Nylon washers. 8 off. Fitted each side of wishbone arms.

When these bearings are being fitted the following instructions must be observed :—

(i) Fit the rubber sealing rings to the nylon washers.

(ii) Press the nylon bushes into the inner ends of the wishbone arms.

The following four instructions supersede the operations x and xii on page 12 in the main manual Front Suspension Section " G ".

(i) Smear the fulcrum pin situated on the upper face of the chassis frame with grease and feed on the steel bushes.

(ii) Smear the outside of the bushes with grease and feed on a pair of nylon washers complete with sealing rings.

(iii) Feed the wishbone arms on to the steel bushes on the inner fulcrum pin and on the shackle pins of the vertical link simultaneously; followed by a second pair of nylon washers and sealing rings.

(iv) Fit the triangular support plates and secure with nyloc nuts but leave finger tight at this juncture. Refer to the main section of the manual as mentioned previously for the wishbone arm outer attachments.

Service Instruction Manual Supplement

TR3 MODELS

BODY

SECTION N

BODY

1. BODY SPECIFICATION

Provision is made for the installation of an occasional bench seat in the luggage compartment immediately behind the driver and passenger seats.

2. REVEAL MOULDING AND GRILLE

A chromium plated moulding is fitted to the front rim of the air intake with a grille mounted immediately behind.

3. STAINLESS STEEL WING BEADING

The stainless steel wing beadings are positioned between the front and rear wings and the body of the car.

4. PASSENGER SEAT

A folding squab seat is now fitted to allow easier access to the luggage space behind or to the occasional seat if the car is so fitted.

5. OCCASIONAL REAR SEAT

These seats are an optional extra on the TR3 models. Provision is made for the installation of this seat in each car.

6. TO REMOVE REVEAL MOULDING AND GRILLE

(a) Withdraw the self-tapping screw from each end of two horizontal grille bars.

(b) Ease the upper portion of the grille into the air intake and withdraw the assembly when it is inclined approximately 30°.

(c) Slide the moulding joint plates to one side to expose the joint in the mouldings.

(d) Remove the nuts and lock washers from the stud plates securing the mouldings to the air intake and withdraw the two half mouldings. Access to these nuts entail working behind the front cowling.

(e) The stud plates can be withdrawn by sliding them to the end of each half moulding.

7. TO REFIT REVEAL MOULDING AND GRILLE

(a) Slide the joint plates on to the ends of any one moulding.

(b) Position five stud plates in the upper half of the beading and six in the lower half at intervals to align with the holes in the air intake periphery.

(c) Attach the two halves of the reveal mouldings to the air intake with nuts and lock washers.

(d) Slide the joint plates from one moulding to the other and position in such a manner that the joint is covered.

(e) Feed the grille in, top side first, and settle the extremities of the vertical struts adjacent to the reveal moulding already installed.

(f) Secure the grille by four self-tapping screws one at each end of two of the horizontal bars.

8. TO REMOVE OR FIT WING BEADING

This is effected by removing or fitting the front and rear wings as described on page 4 of the Body Section in the main portion of the Manual.

9. TO REMOVE PASSENGER SEAT SQUAB

(a) Remove the cushion from the seat pan.

(b) Remove the two domed nuts at the base of the seat squab.

(c) Spring the squab from the seat pan.

10. TO FIT PASSENGER SEAT SQUAB

(a) Position the seat squab on the seat pan studs and attach with the dome nuts.

(b) Fit the back of the seat cushion under the spring clip at the rear of the seat pan and settle cushion into position.

11. TO FIT OCCASIONAL SEATS

  (a) Slide the driver and passenger seats forward to their fullest extent.

  (b) Lift up the carpet at the rear of the two seats and remove the two bolts and washers so exposed.

  (c) Make two small holes in the carpet to align with the tappings in the floor assembly.

  (d) Withdraw the two chrome headed bolts and washers from the trim at the rear of the passenger compartment.

  (e) Position the occasional seat behind the driver and passenger seats and secure with **four** bolts removed during the previous operations.

12. TO REMOVE OCCASIONAL SEAT

  (a) Withdraw the four attachment bolts and plain washers.

  (b) Remove the seat from the rear of the passenger compartment.

  (c) Return the bolts and plain washers to their tappings for safe keeping.

2

TR2 & 3 "HARD TOP" INSTALLATION

Description

The "Hard Top" is of pressed steel construction, incorporating channel sections which are spot welded to the main panel. These channels considerably stiffen the assembly and also accommodate the front and rear mounting brackets. The sides of the main panel are folded to form a "U" section which further strengthens the construction and also serves as a means of securing the drip channels and draught sealing rubbers. The "Hard Top" is supplied completely trimmed less the rubber sections and the rear window light, all of which are included in the kit for fitting at a final stage.

The "Hard Top" Kit No. 900771 (less side screens) contains the following items :—

Detail No.	Description	No. off.
800840	Drip channel R.H.	1
800839	Drip channel L.H.	1
603328	Seal rubber screen top	1
603116A	Seal rubber backlight	1
603116B	Seal rubber backlight filler	1
602269	Seal strip waist	1
603089	Seal rubber drip channel	2
553132	Backlight (Perspex)	1
553742	Hard top (Metal)	1
	Headlining (fitted)	1
602299	Tapping plate	5
WQ0305	Spring washer	10
502406	Screw (drip channel attachment)	24
602295	Fixing screw (screen bracket)	3
602938	Protection plate (side cappings)	2
500488	Round headed screw (protection plate)	10
603189	Bracket (windscreen fitting)	3
602380	Fix washer (windscreen fitting)	3
602939	Protection plate (rear cappings)	3
602943	Washer	5
602326	Bridge piece	5
602327	Dome headed screw	5
501434	Screw (bracket to screen)	6
TR.6503	Round headed screw (tapping plate)	10

Detail No.	Description	No. off.
500229	Round headed PK screw (drip channel)	10
502233	Countersunk screw (screen rubber)	12
WN.0705	Shake proof washers (bracket to screen)	6
CD13515	Cup washers	22

When required for TR2 cars, sliding side lights must be ordered separately.

Caution

The fitting of "Hard Tops" to early cars may present some difficulty due to the inconsistencies of body dimensions, a feature not uncommon with hand made bodies. Present manufacturing methods make use of more elaborate assembly jigs which result in the maintenance of close body tolerances. A further cause of a badly fitting "Hard Top" may be the result of bent screen pillars which have become displaced by heavy drivers or passengers pulling on the windscreen to remove themselves from the car. When fitting, the "Hard Top" must be initially positioned from the screen, it should be appreciated, therefore, that any misplacement of the screen itself will move the "Hard Top" out of position at the rear of the car. No difficulty should be experienced with cars in normal condition after Commision Number TS.6824.

I. CAR PREPARATION

To prepare the car for the installation of a "Hard Top", prepare as follows :—

(a) Starting at the screen rail, remove the hood by lifting the fasteners from around the edge of the body.

(b) Release the hood webbing at the rear by removing the two flat headed screws and the two hood fastener screws. (Fig. 1.)

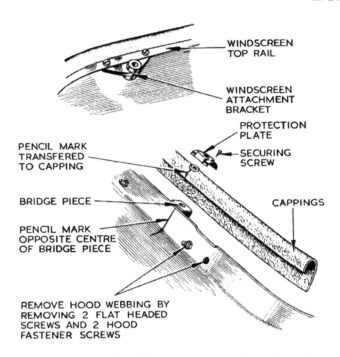

Fig. 1   **Marking and fitting Rear Cappings.**

(c) Unscrew and remove the four countersunk screws securing the hood frame to the body, and then lift the frame out of the body.

(d) Detach the five cappings from the elbow rail after removing the P.K. securing screws. Remove the two wood blocks from the elbow rail. (Fig. 2.)

(e) Remove the millboard from the front of the petrol tank after removing the P.K. securing screws. (Fig. 2.)

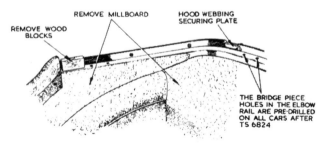

Fig. 2   **Showing Cappings removed.**

NOTE—The P.K. screws securing the bottom of the millboard can be removed after lifting the rear of the carpet.

2. HARD TOP PREPARATION

(a) Loosely assemble the three windscreen attachment brackets on the "Hard Top" front rail.

(b) If not already fitted, insert the three shorter angle brackets through slots in the stiffener rail at the rear of the "Hard Top", and a longer angle bracket at each side. Using short flat headed screws and lockwashers, secure the brackets in position. (Fig. 4.)

NOTE—It will be necessary to neatly cut the trim fabric to allow entry of the brackets into the slots in the stiffener channel.

3. WINDSCREEN ATTACHMENT BRACKETS—TO FIT

CAUTION—To guard against the possibility of damage to paintwork, masking tape should be applied to that part of the body which will be in contact with the hard top during fitting operations.

(a) Position the hard top on the car and feed the assembled brackets under the windscreen top rail.

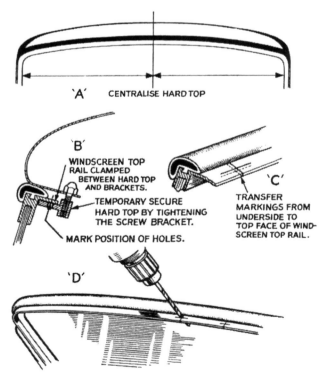

Fig. 3a, b, c and d   **Positioning "Hard Top" and drilling Screen Rail.**

Centralise the "Hard Top" over the windscreen and temporarily secure by tightening the three attachment brackets. (Fig 3A.)

(b) Mark the position of the attachment bracket holes on the underside of the windscreen top rail (Fig 3B). Slacken off the brackets then remove the "Hard Top" from the car.

(c) Mark the top side of the screen exactly in line with the markings previously made on the underside (Fig. 3C). Using a No. 11 drill, carefully drill six holes from the above screen and $\frac{3}{16}''$ from the edge. (Fig. 3D.)

(d) Remove the windscreen attachment brackets from the "Hard Top" and finally secure to the underside of the windscreen top rail by six chromium plated screws and lock washers. (Fig. 1.)

4. BRIDGE PIECES—TO FIT

(a) Loosely secure the five bridge pieces to the angle brackets previously fitted in the rear stiffener rail. (Fig. 4.)

(b) Reposition the "Hard Top" to the car and secure to the three windscreen attachment brackets. The bridge pieces will now be resting on the elbow rail channel. (Fig. 4.)

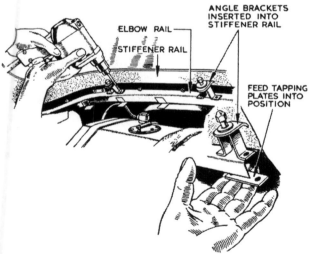

ELBOW RAIL
STIFFENER RAIL
ANGLE BRACKETS INSERTED INTO STIFFENER RAIL
FEED TAPPING PLATES INTO POSITION

Fig. 4 Drilling the Elbow Rail and installing Bridge Pieces.

NOTE : **Drilling should only be necessary on Cars prior to TS.6824.**

(c) Mark the position of the bridge pieces on the elbow rail and identify them to these positions. Release the "Hard Top" at the windscreen and remove from the car.

(d) Using a No. 11 drill, drill ten holes through the markings on the elbow rail.

(e) Remove the bridge pieces from the "Hard Top" and secure to the elbow rail channel, using flat headed screws which screw into tapping plates fed into position under the channel. (Fig. 4.)

5. REAR CAPPINGS—TO FIT

(a) Opposite to the centre of each bridge piece, scribe a line with a pencil on the body protection tape. Attach the cappings to the body, loosely securing with the P.K. screws. (Fig. 1.)

(b) Over the cappings, extend the markings previously scribed on the body. Remove the cappings from the car. Scribe the inside of the cappings exactly in line with the marks on the outside. (Fig. 1.)

(c) Using a $\frac{3}{8}''$ drill, drill the cappings at the positions marked on the insides and ensure that when drilled, the holes are aligned with those in the bridge pieces.

(d) Attach the millboard to the front of the petrol tank and secure with P.K. screws. Refit the cappings over the bridge pieces and secure.

(e) Select the three narrow protection caps and position these on the rear cappings, aligning the centre holes with the threaded centres of the bridge pieces. Drill the cappings through the protection caps and secure with P.K. self tapping screws. The two larger caps are fitted in a similar manner to the side elbow rails. (Fig. 1.)

(f) Fit four countersunk screws and chromium washers in the holes previously used to accommodate the hood bracket screws. Remove the protecting tape from the body of the car.

5

BODY

6. DRIP CHANNELS—TO FIT
(See Figs. 5 and 6)

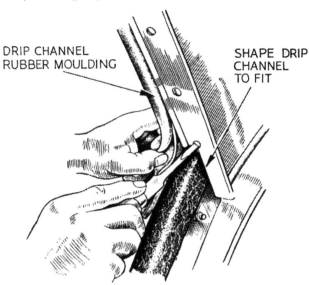

DRIP CHANNEL RUBBER MOULDING

SHAPE DRIP CHANNEL TO FIT

Fig. 5   **Fitting lower part of Drip Channel.**

After correctly shaping the ends of the drip channels, position the channels and draught rubbers " C " on the " Hard Top " as illustrated and secure with the screws " A " and " B ". (Fig. 6.)

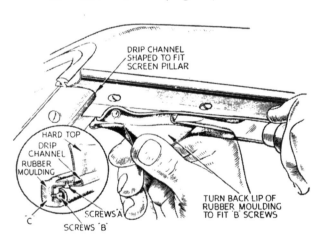

DRIP CHANNEL SHAPED TO FIT SCREEN PILLAR

HARD TOP DRIP CHANNEL RUBBER MOULDING

TURN BACK LIP OF RUBBER MOULDING TO FIT 'B' SCREWS

C   SCREWS 'A'   SCREWS 'B'

Fig. 6   Fitting Drip Channel and Draught Rubber.

7. SEALING RUBBERS—TO FIT

Using " Seelastik ", secure the rubber mould " D " (Fig. 7) to the rear lower edge, and the rubber section " E " (Fig. 8) to the front top edge of the " Hard Top ".

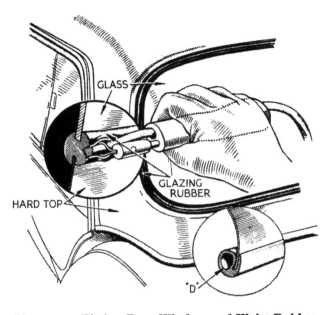

GLASS

GLAZING RUBBER

HARD TOP

'D'

Fig. 7   **Fitting Rear Window and Waist Rubber.**

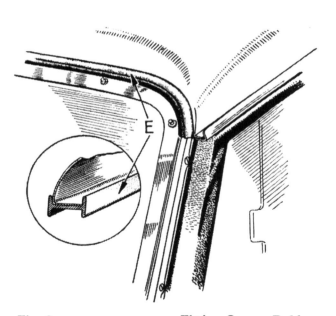

E

Fig. 8   **Fitting Screen Rubber.**

8. HARD TOP—TO FIT

Re-position the " Hard Top " on the body and after loosely assembling the attachment bolts, progressively tighten them until the " Top " is finally secured.

9. REAR WINDOW LIGHT—TO FIT

(See Fig. 7)

Fit the rubber moulding around the glass with the filler section positioned towards the rear of the car. Offer the glass with the rubber attached, to the aperture in the "Hard Top" and with the help of an assistant manipulate the inner rubber lip into position. Using a special tool (see Fig. 7) finally secure the glass by feeding the filler strip into position. With the aid of a "Seelastik gun", complete the installation by forcing Sealastik compound between the "Hard Top" and the outer lip of the glazing rubber to effect water sealing.

Service Instruction Manual Supplement

TR3 MODELS

FUEL SYSTEM

SECTION P

FUEL SYSTEM

1. PETROL TANK

The petrol tank has been modified slightly to accommodate the occasional seat, its capacity is thereby reduced to 12 gallons.

2. FLEXIBLE FUEL PIPES

A flexible fuel pipe connects the twin carburettors and is integral with a short feed line which is connected to the Bundy tubing at a point adjacent to the thermostat housing.

3. CARBURETTORS

S.U. H6 type carburettors are fitted to this engine. This carburettor has a four-point mounting but is similar in other respects to the H4 used on the TR2, is identical in operation and requires the same maintenance.

Carburettors fitted to early cars were equipped with " TD " needles, while with later cars " TE " needles were used, this needle in turn was superseded by type " SM " at Engine No. TS.10037E. When needles are changed for any reason a pair of type " SM " should be fitted.

4. AIR CLEANERS

The air cleaners are similar to those fitted to the TR2 apart from the off-set mountings.

5. INLET MANIFOLD

This has been modified to accommodate the four-point fixing H6 carburettor, and manifolds fitted to engines after TS.9350E have a larger bore to align with the enlarged throat area of the high port combustion head.

6. TO REMOVE FLEXIBLE FUEL HOSE ASSEMBLY

(a) Hold the hexagon of the flexible hose assembly and disconnect the union nut of the rigid pipe adjacent to the thermostat housing.

(b) Withdraw the banjo bolt from one carburettor, collecting the gauze filter and retaining spring.

(c) Repeat operation (b) with the second carburettor.

7. TO FIT FLEXIBLE FUEL HOSE ASSEMBLY

(a) Position the filter assembly in the rear carburettor float chamber, spring first.

(b) Feed a fibre washer on to the banjo bolt, followed by the banjo connection and a second fibre washer, and then attach to the rear carburettor and leave finger tight at this juncture.

(c) Repeat operation (a) and (b) with the front carburettor.

(d) Holding the hexagon of the flexible hose with a spanner, attach the union nut of the rigid supply pipe and secure to give a petrol tight joint.

(e) Adjust the position of the banjo connections on the float chambers of the twin carburettors so as to avoid any strain, and tighten banjo bolts to give a petrol tight joint.

(f) Start the engine and observe the fuel pipes for leaks.

8. CARBURETTOR DETAILS

The instructions given for the H4 carburettor as fitted to the TR2 apply to the H6 type apart from the four-point mountings. The jet needles at present used in normal manufacture are of the " SM " type, although with early releases of the TR3 model the " TD " or " TE " needle was fitted.

The " TD " or " TE " needles in both carburettors should be replaced by type " SM " if damage or wear justifies the exchange in either unit.

9. AIR CLEANERS

The air cleaners have off-set mounting and must be positioned on the carburettor air intake in such a manner that the off-set is rearward.

10. INLET MANIFOLD

The inlet manifold is removed and fitted as those fitted to the TR2 engine.

Service Instruction Manual Supplement

TR3 MODELS

BRAKES

SECTION R

BRAKES

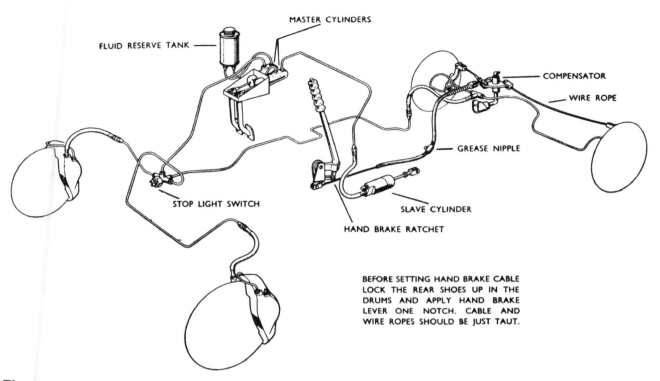

MASTER CYLINDERS

FLUID RESERVE TANK

COMPENSATOR

WIRE ROPE

GREASE NIPPLE

STOP LIGHT SWITCH

SLAVE CYLINDER

HAND BRAKE RATCHET

BEFORE SETTING HAND BRAKE CABLE
LOCK THE REAR SHOES UP IN THE
DRUMS AND APPLY HAND BRAKE
LEVER ONE NOTCH. CABLE AND
WIRE ROPES SHOULD BE JUST TAUT.

Fig. 1   **Brake and Clutch layout**

GIRLING BRAKES AND HYDRAULIC CLUTCH
(From Chassis No. TS.13101)

1. DESCRIPTION

The brakes on the front wheels are the Girling Disc Brakes and on the rear are Girling HL.3 Drum Brakes. All four wheels are hydraulically operated by foot pedal operation, directly coupled to a CV master cylinder in which the hydraulic pressure is originated. A supply tank which provides fluid reserve for both brake and clutch systems is installed to allow for fluid replenishment.

An independent mechanical linkage (see Fig. 6), actuated by a hand lever control, operates the rear brakes by levers attached to the wheel cylinder bodies, thus acting as a hand or parking brake.

2. FRONT BRAKES (Fig. 1)

The front brakes are the 11″ dia. Girling Disc Brakes, which are extremely simple in construction, consisting of the 11″ disc

Fig. 2   **Front Disc Brakes Assembly.**

which is made from high quality cast iron and cast iron calipers mounted to a support bracket.

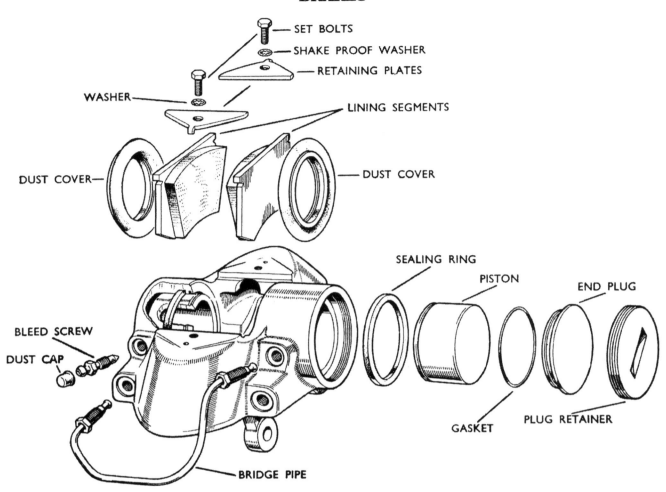

SET BOLTS
SHAKE PROOF WASHER
RETAINING PLATES
WASHER
LINING SEGMENTS
DUST COVER
DUST COVER
SEALING RING
PISTON
END PLUG
BLEED SCREW
DUST CAP
GASKET
PLUG RETAINER
BRIDGE PIPE

Fig. 2   Exploded arrangement of Disc Brake Caliper Assembly.

Due to the simplicity of these disc brakes the only normal servicing which will be carried out by the owner or garage will be the replacement of worn lining segments, seals and boots of the hydraulic caliper.

(a) Lining Segment Replacement (Fig. 2)

Jack up the front of car and remove road wheels. On the top of the caliper body are two setscrews which secure the segment retaining plates. The release of these will enable the retaining plates to be raised out of engagement with the casting and swung through an arc of 180°. The segments are then fully exposed and can be lifted out of the caliper.

Under no circumstances should attempts be made to reline worn segments and these must be replaced by new parts.

In order to fit new segments the pistons in the caliper bore should be pushed to the bottom, and the new segments placed into position. When the segments are positioned correctly, the retaining plates should be replaced in their original position and the setscrews tightened down.

The replacement of segments is then complete and bleeding is unnecessary, but the foot pedal should be pumped until a solid resistance is felt.

Jack down the front of the car and road test.

(b) Caliper Cylinder Maintenance To Replace the Rubber Seals

In order to replace the rubber " O " rings or seals it is necessary to remove the caliper assembly from the vehicle. The brake segments should be removed in the manner described above.

Instead of pushing the pistons to the bottom of the bore withdraw them from the caliper body, taking great care not to damage the bores. The sealing rings may then be removed by inserting a blunt tool under the seals and prising out, taking care not to damage the locating grooves. Examine the bores and pistons carefully for any signs of abrasion or "scuffing." No attempt should be made to remove the end plug retainer, as this is screwed in tightly by mechanical means.

It is important that in cleaning the components no petrol, paraffin, trichlorethylene or mineral fluid of any kind should be used. Clean with methylated spirits and allow to vaporise, leaving the component clean and dry.

After cleaning and examining, lubricate the working surfaces of the bores and piston with clean genuine Girling Crimson Brake and Clutch Fluid.

(c) Assembling

Fit new rubber seals into the grooves of caliper cylinder bore. Locate the rubber dust cover with the projecting lip into the groove provided which is the outer one of the cylinder bore.

Insert the piston, closed end first, into the bore, taking great care not to damage the polished surface. Push the piston right home and then engage the outer lip of the rubber boot into the groove of piston.

The replacement of the lining segments as described under the heading " Segment Replacement " will retain the pistons in position.

Refit the caliper assembly to the support bracket by means of the two securing bolts ensuring that the disc passes between the two lining segments.

Re-connect the pressure hose and bleed the brake, as described under " Bleeding the System."

2. DISCS

To ensure that the brake functions at maximum efficiency a check should be made to see that the disc runs truly between the segments. The maximum run-out permissible on the disc is .004″.

(For instructions regarding wheel bearing settings refer to page 7, Section " G," in the main part of this manual.) If excessive run-out is present this will cause the knocking back of the pistons which will possibly cause judder.

If it is found that the discs have been damaged in any way, which is extremely unlikely, it will be necessary to remove the discs from the car in order for them to be " trued " up. Under no circumstances should more than .060″ be removed, with the finish to be 32 micro ins. maximum measured circumferentially and 50 micro ins. measured radially.

3. REAR BRAKES (Figs. 3 and 4)

From the illustration it will be seen that they are of the drum type with a wheel cylinder and adjuster affixed to a backplate supporting the two shoes which are held in position by two return springs. The shoes, which are hydraulically operated by the Girling single acting wheel cylinder (incorporating lever handbrake mechanism), are not fixed but are allowed to slide and centralize. Lining wear is adjusted by a Girling wedge type mechanical adjuster common to both shoes. At the cylnder end, the leading shoe is located in a slot in the

Fig. 3   **Rear Drum Brake Assembly.**

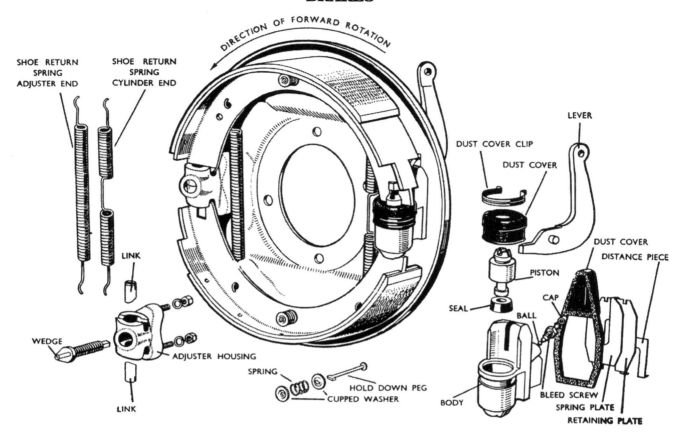

SHOE RETURN SPRING ADJUSTER END · SHOE RETURN SPRING CYLINDER END · DIRECTION OF FORWARD ROTATION · LEVER · DUST COVER CLIP · DUST COVER · LINK · DUST COVER · DISTANCE PIECE · PISTON · CAP · WEDGE · SEAL · BALL · ADJUSTER HOUSING · SPRING · HOLD DOWN PEG · CUPPED WASHER · BODY · BLEED SCREW · LINK · SPRING PLATE · RETAINING PLATE

Fig. 4   **Details of Rear Brake Assembly.**

piston, while the trailing shoe rests in a slot formed in the cylinder body. At the adjuster end the shoe ends rest in slots in the adjuster links. The shoes are supported by platforms formed in the backplate, these being held in position by two hold-down springs fitted on each shoe with a peg passing through a hole in the backplate.

The adjuster consists of an alloy housing with studs, which is spigoted and secured firmly to the inside of the backplate by nuts and spring washers.

The housing carries two opposed steel links, the outer end slotted to take the shoes, and the inclined inner faces bearing on inclined faces of the hardened steel wedge (the axis of which is at right angles to the links).

The wedge has a finely threaded spindle with a square end which projects on the outside of the backplate. By rotating the wedge in a clockwise direction the links are forced apart and the fulcrum of the brake shoe expanded.

A piston and seal moves in the highly

finished bore of a light alloy die cast wheel cylinder body, whilst a slot, machined in the opposite end of the body, serves to carry the trailing shoe. The cylinder, incorporating a bleed screw with rubber cap, is attached to the back plate by spring clips which allow it to slide laterally. The handbrake lever pivots on, and projects at right angles through the back plate.

When the brake is applied, the piston under the influence of the hydraulic pressure moves the leading shoe and the body reacts by sliding on the backplate to operate the trailing shoe.

The handbrake lever is pivoted in the cylinder body and when operated, the lever tip expands the leading shoe and the pivot moves the cylinder body and with it the trailing shoe.

(a) Dismantling

If it is found necessary to remove a rear wheel cylinder, the following procedure should be followed :—

(i) Jack up the vehicle, remove the wheels, and disconnect the rod from handbrake lever.

(ii) Remove the brake drum and shoes. Disconnect the pressure pipe union from the cylinder, and remove the rubber dust cover from rear of backplate.

(iii) By using a screwdriver, prise the retaining plate and spring plate apart, then tap the retaining plate from beneath the neck of the wheel cylinder.

(iv) Withdraw the handbrake lever from between the backplate and wheel cylinder.

(v) Remove the spring plate and distance piece, and finally the wheel cylinder from the backplate.

(b) Refitting the Rear Wheel Cylinder

Mount the wheel cylinder on to the backplate with the neck through the large slot. Replace the distance piece between cylinder neck and backplate, with the open end away from handbrake lever location. The two cranked lips must also be away from the backplate.

Insert the spring plate between the distance piece and backplate, also with open end away from handbrake lever location and the two cranked lips away from the backplate.

Replace handbrake lever. Locate the retaining plate between the distance piece and spring plate (open end towards the handbrake lever), tap into position until the two cranked tips of the spring plate locate in the retaining plate.

Fit the rubber dust cover. Attach the pressure pipe union to the cylinder and connection to the handbrake lever. Replace the shoes, brake drum, and bleed the system. Finally re-fit wheels.

(c) Fitting Replacement Shoes

(i) Jack up the car and remove road wheels and brake drums.

(ii) Remove the holding down springs by turning the washer under the peg head. Lift one of the shoes out of the slots in the adjuster link and wheel cylinder piston. Both shoes complete with springs can then be removed. Place a rubber band round the wheel cylinder to keep piston in place.

(iii) Clean down the backplate, check wheel cylinders for leaks and freedom of motion.

(iv) Check adjusters for easy working and turn back (anti-clockwise) to full "off" position. Lubricate where necessary with Girling White Brake Grease.

(v) Smear the shoe platforms and the operating and abutment ends of the new shoes with Girling White Brake Grease.

(vi) Fit the two new shoe return springs to the new shoes (with the shorter spring at the adjuster end) from shoe to shoe and between shoe web and backplate. Locate one shoe in the adjuster link and wheel cylinder piston slots, then prise over the opposite shoe into its relative position. Remove rubber band. Insert the hold down peg through hole in backplate, and replace spring and cupped washers smeared with Girling White Brake Grease.

(vii) Make sure drums are cleaned and free from grease, etc., then refit.

(viii) Adjust brakes.

(ix) Refit road wheels and jack down.

Note.—The first shoe has the lining positioned towards the heel of the shoe and the second shoe towards the toe or operating end in both L.H. and R.H. brake assemblies.

Several hard applications of the pedal should be made to ensure all the parts are working satisfactorily and the shoes bedding to the drums, then the brakes should be tested in a quiet road before normal running is resumed.

Handbrake Setting—refer to Fig. 1.

5

4. RUNNING ADJUSTMENTS

The front disc brakes are entirely self-adjusting. The rear brakes are adjusted for lining wear at the brakes themselves, and on no account should any alteration be made to the hand brake cable for this purpose (Fig. 1).

One common adjuster is provided for each brake assembly. Adjustment of both rear wheels is identical.

Release the handbrake and jack up the car. Turn the square end of the adjuster on the outside of each rear brake backplate in a clockwise direction until a resistance is felt, then slacken back two clicks, when the drum should rotate freely.

Immediately after fitting replacement shoes it is advisable to slacken one further click to allow for possible lining expansion, reverting to normal adjustment afterwards.

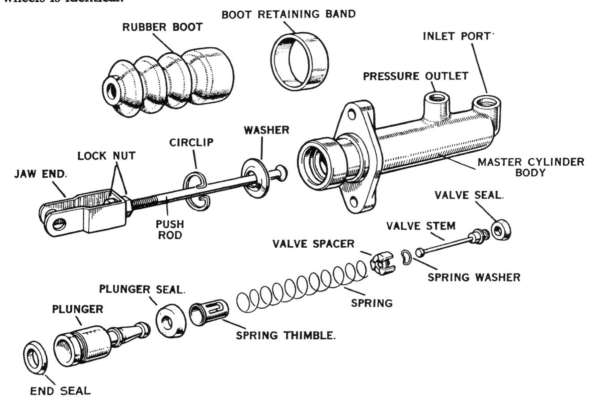

Fig. 5

C.V. Girling Brake and Clutch Master Cylinder.

CLUTCH HYDRAULIC OPERATING MECHANISM

5. HYDRAULIC CLUTCH OPERATION

A slave cylinder mounted on the side of the clutch housing is mechanically connected to the clutch operating mechanism. This assembly, by reason of its hydraulic connection, is actuated by a Girling C.V. master cylinder to which the suspended clutch pedal is coupled.

When pressure on the clutch pedal is applied, the piston of the master cylinder displaces the fluid in the cylinder which in turn moves the piston of the slave cylinder, pushing against the lever of the clutch thrust race.

(a) The CV Master Cylinder (For Brake and Clutch, Fig. 5)

This is the Girling CV Type, which consists of an alloy body with a polished finished bore. The inner assembly is made up of the push rod, dished washer, circlip, plunger and seal, plunger seal, spring thimble, plunger return spring, valve spacer, spring washer, valve stem and valve seal. The open end of the cylinder is protected by a rubber dust cover.

(b) Dismantling

Disconnect the pressure and feed pipe unions from the cylinder and remove the securing bolts and clevis pin from jaw end. Pull back the rubber dust cover and remove the circlip with a pair of long nosed pliers. The push rod and dished washer can then be removed. When the push rod has been removed the plunger, with seal attached, will then be exposed. Remove the plunger assembly complete. The assembly can then be separated by lifting the thimble leaf over the shouldered end of the plunger. Ease the pressure seal off the plunger and remove back seal. Depress the plunger return spring allowing the valve stem to slide through the elongated hole of the thimble, thus releasing tension of spring.

Remove thimble, spring and valve complete. Detach the valve spacer, taking care of the spacer spring washer which is located under the valve head. Remove the seal from the valve head.. Examine all parts, especially the seal, for wear or distortion, and replace with new parts where necessary.

(c) Assembling

Replace the valve seal so that the flat side is correctly seated on the valve head. The spring washer should then be located with dome side against the underside of the valve head, and held in position by the valve spacer, the legs of which face towards the valve seal. Replace the plunger return spring centrally on the spacer, insert the thimble into the spring and depress until the valve stem engages through the elongated hole of the thimble, making sure the stem is correctly located in the centre of the thimble. Check that the spring is still central on the spacer. Refit new plunger seal on to the plunger with flat of seal seated against the face of plunger, and a new back seal with lip of seal facing plunger seal. Insert the reduced end of plunger into the thimble until the thimble leaf engages under the shoulder of the plunger. Press home the thimble leaf.

Smear the assembly well with Girling brake and clutch fluid, and insert the assembly into the bore of the cylinder, valve end first, easing the plunger seal lips in the bore. Replace the push rod with the dished side of washer under the spherical head into the cylinder, followed by the circlip which engages into groove machined in the cylinder body.

Replace the rubber dust cover, refit the cylinder to the chassis and bleed the system.

6. THE CLUTCH SLAVE CYLINDER (Fig. 6)

The slave cylinder is of simple construction, consisting of alloy body, piston with seal, piston stop, spring and bleed screw, the open end of the cylinder being protected by a rubber dust cover. The cylinder is mounted to the clutch housing by a flange and two bolts.

(a) Dismantling

Remove the rubber dust cap from the bleed nipple, attach a bleed tube, open the bleed screw threequarters of a turn and pump the clutch pedal until all the fluid has been drained. Unscrew the pressure pipe union and remove the bolts from the flange. The cylinder can then be removed.

Remove the rubber cover and piston stop, then, by using an air line, blow out the piston and seal.

The spring will also be removed. Examine all parts, especially the seal, and replace if worn or damaged.

(b) Assembling

Place the seal on to the stem of the piston, with the back of the seal against the piston, replace the spring with small end on stem, smear well with Girling Crimson Brake and Clutch Fluid, and insert into cylinder. Replace the piston stop and stretch rubber dust cover over cylinder. Mount the cylinder in steel clip, making sure the push rod enters the hole in the rubber boot. Secure the cylinder by the two bolts, and screw in the pipe union.

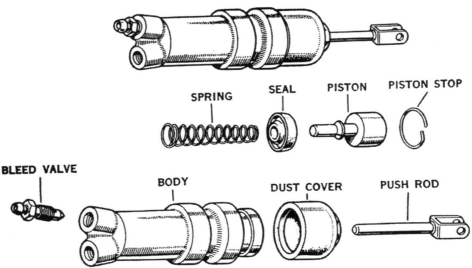

SPRING · SEAL · PISTON · PISTON STOP

BLEED VALVE · BODY · DUST COVER · PUSH ROD

Fig. 6 · Clutch Slave Cylinder.

(c) Bleeding

Remove the bleed screw dust cap, open the bleed screw approximately three-quarters turn and attach a tube, immersing the open end into a clean receptacle containing a little Girling Crimson Brake and Clutch Fluid. Fill the master cylinder reservoir with genuine Girling Crimson Brake and Clutch Fluid, and by using slow full strokes pump the pedal until the fluid entering the container is free from air bubbles. On a down stroke of the pedal, nip up the bleed screw, remove the bleed tube and replace the dust cap. After bleeding, top up the reservoir to its correct level of approximately three-quarters full.

7. GENERAL MAINTENANCE

(a) Replenishment of Hydraulic Fluid for both Brake and Clutch Systems

Inspect the reservoir at regular intervals and maintain at about three-quarters full by the addition of Girling Crimson Brake and Clutch Fluid.

Great care should be exercised when adding brake fluid to prevent dirt or foreign matter entering the system.

Important.—Serious consequences may result from the use of incorrect fluids, and on no account should any but Girling Crimson Brake and Clutch Fluid be used. This fluid has been specially prepared and is unaffected by high temperatures or freezing.

Never top up the system with any other fluid.

(b) Bleeding the Hydraulic System

Bleeding is necessary any time a portion of the hydraulic system has been disconnected, or if the level of the brake fluid has been allowed to fan so low that air has entered the master cylinder.

With all the hydraulic connections secure and the reservoir topped up with fluid, remove the rubber cap from the L.H. rear bleed nipple and fit the bleed tube over the bleed nipple, immersing the free end of the tube in a clean jar containing a little Girling Brake and Clutch Fluid.

Unscrew the bleed nipple about three-quarters of a turn and then operate the brake pedal with slow, full strokes until the fluid entering the jar is completely free of air bubbles.

Then during a down stroke of the brake pedal, tighten the bleed screw sufficiently to seat, remove bleed tube and replace the bleed nipple dust cap. **Under no circumstances must excessive force be used when tightening the bleed screw.**

This process must now be repeated for each bleed screw at each of the three remaining brakes finishing at the wheel nearest the master cylinder. Always keep a careful check on the reservoir during bleeding, since it is most important that a full level is maintained. Should air reach the master cylinder from the reservoir, the whole operation of bleeding must be repeated.

After bleeding, top up the reservoir to its correct level of approximately three-quarters full.

Never use fluid that has just been bled from a brake system for topping up the reservoir, since this fluid may be to some extent aerated.

Great cleanliness is essential when dealing with any part of the hydraulic system, and especially so where the brake fluid is concerned. Dirty fluid must never be added to the system.

GENERAL ADVICE ON HYDRAULIC COMPONENTS

The following precautions should be studied carefully and observed punctiliously by all concerned.

Essential Precautions

Always Exercise extreme cleanliness when dealing with any part of the hydraulic system.

Never Handle rubber seals or internal hydraulics parts with greasy hands or greasy rags.

Always Use Girling Crimson Brake and Clutch Fluid from sealed quart tins.

Never Use fluid from a container that has been cleaned with petrol, paraffin or trichlorethylene.

Never Put dirty fluid into the reservoir, nor that which has been bled from the system.

Always Use clean Girling Brake and Clutch Fluid or alcohol for cleaning internal parts of hydraulic system.

Never Allow petrol, paraffin or trichlorethylene to contact these parts.

Always Examine all seals carefully when overhauling hydraulics cylinders and replace with genuine Girling spares, any which show the least sign of wear or damage.

Always Take care not to scratch the highly finished surfaces of cylinder bores and pistons.

Always Use WAKEFIELD / GIRLING Rubber Grease No. 3 (Red) for packing rubber boots, dust covers and lubricating parts likely to contact any rubber components.

Never Use Girling White Brake Grease or other grease for this purpose.

Always Replace all seals, hoses and gaskets with new ones if it is suspected that incorrect fluids have been used or the system contaminated with mineral oil or grease. Drain off the fluid, thoroughly wash all metal parts and flush out all pipes, etc., with alcohol or clean Girling Crimson Brake and Clutch Fluid

Never Use anything else for this purpose.

Always Use a particular container (reserved for this purpose) for bleeding the system, and always maintain in a clean condition.

Never Use a receptacle which has been cleaned with petrol, paraffin or trichlorethylene.

Always Remember that your safety and the safety of others may depend on the observance of these precautions at all times.

Service Instruction Manual Supplement

TR3 MODELS

EXHAUST SYSTEM

SECTION S

EXHAUST SYSTEM

EXHAUST SYSTEM

The exhaust system is unchanged from the TR2 apart from the new manifold gasket fitted to the enlarged port combustion head after engine number TS.9350E.

Brooklands Books Ltd., P.O. Box 146, Cobham, Surrey KT11 1LG, England
Phone: (44) 1932 865051
E-mail: sales@brooklands-books.com   www.brooklands-books.com

ISBN 9780948207693   Part No. 502602   Ref: T122WH   2T5

OFFICIAL TECHNICAL BOOKS

Brooklands Technical Books has been formed to supply owners, restorers and professional repairers with official factory literature.

Workshop Manuals

TR2 & TR3	502602	9780948207693
TR4 & TR4A	510322	9780948207952
TR5, TR250 & TR6 (Glove Box Autobooks Man.)		9781855201835
TR5-PI Supplement	545053	9781869826024
TR250 Supplement	545047	9781783181759
TR6 inc. TC & PI	545277/E2	9781869826130
TR7	AKM3079B	9781855202726
TR7	Autobooks Manual	9781783181506
TR8	AKM3981A	9781783180615
Spitfire Mk 1, 2 & 3 & Herald 1200, 12/50, 13/60 & Vitesse 6	511243	9780946489992
Herald 948, 1200, 12/50, 13/60 Autobooks Man.		9781783181513
Spitfire Mk 4	545254H	9781869826758
Spitfire 1500	AKM4329	9781869826666
Spitfire Mk 3, 4, 1500(Glove Box Autobooks Man.)		9781855201248
2000 & 2500	AKM3974	9781869826086
GT6 Mk 1, 2, 3 & Vitesse 2 Litre	512947	9780907073901
GT6 Mk 2, GT6+ & Mk 3 & Vitesse 2 Litre - Mk 2 1969-1973	Autobooks Manual	9781783181322
Stag	AKM3966	9781855200135
Stag	Autobooks Manual	9781783181490
Dolomite Sprint	AKM3629	9781855202825

Parts Catalogues

TR2 & TR3	501653	9780907073994
TR4	510978	9780907073949
TR4A	514837	9780907073956
TR250 US	516914	9781869826819
TR6 Sports Car 1969-1973	517785A	9780948207426
TR6 1974-1976	RTC9093A	9780907073932
TR7 (1975-1978)	RTC9814CA	9781855207943
TR7 1979+	RTC9828CC	9781870642231
TR7 & TR8	RTC9020B	9781870642651
Herald 13/60	517056	9781869826154
Vitesse 2 Litre Mk 2	517786	9781869826147
Stag	519579	9781870642996
GT6 Mk 1 and Mk 2 /GT6+	515754/2	9781783180448
GT6 Mk 3	520949/A	9780948207938
Spitfire Mk 3	516282	9781870642873
Spitfire Mk 4 & Spitfire 1500 1973-1974	RTC 9008A	9781869826659
Spitfire 1500 1975-1980	RTC9819CB	9781870642187
Dolomite Range 1976 on	RTC9822CB	9781855202764

Owners Handbooks

Triumph Competition Preparation Manual TR250, TR5 and TR6		9781783180011
TR4	510326	9780948207662
TR4A	512916	9780948207679
TR5 PI	545034/2	9781855208544
TR250 (US)	545033	9780948207273
TR6	545078/1	9780948207402
TR6-PI	545078/2	9781855201750
TR6 (US 73)	545111/73	9781855204348
TR6 (US 75)	545111/75	9780948207150
TR7	AKM4332	9781870642736
TR8 (US)	AKM4779	9781855202832
Stag	545105	9781855206830
Spitfire Mk 3	545017	9780948207181
Spitfire Mk 4	545220	9781870642439
Spitfire Mk 4 (US)	545189	9781855207967
Spitfire 1500	RTC9221	9781870642453
Spitfire Competition Preparation Manual		9781870642606
GT6	512944	9781855201583
GT6 Mk 2 & GT6+	545057	9781855201422
GT6 Mk 3	545186	9780946489848
GT6, GT6+ & 2000 Competition Preparation Manual		9781855200678
2000, 2500 TC and 2500S	AKM3617/2	9781855202788
Herald 1200 12/50	512893/6	9781855200616
Herald 13/60	545037	9781855201415
Vitesse 2 Litre	545006	9781855200746
Vitesse Mk 2	545070/2	9781855200418
Vitesse 6	511236/5	9781855207974

Carburetters

SU Carburetters Tuning Tips & Techniques	9781855202559
Solex Carburetters Tuning Tips & Techniques	9781855209770
Weber Carburettors Tuning Tips and Techniques	9781855207592

Truimph - Road Test Books

Triumph Herald 1959-1971	9781855200517
Triumph Vitesse 1962-1971	9781855200500
Triumph 2000 / 2.5 / 2500 1963-1977	9780946489237
Triumph GT6 Gold Portfolio 1966-1974	9781855202443
Triumph TR6 Road Test Portfolio	9781855209268
Triumph Spitfire Road Test Portfolio	9781855209534
Triumph Stag Road Test Portfolio	9781855208933

From Triumph specialists, Amazon or all good motoring bookshops.

Brooklands Books Ltd., P.O. Box 146, Cobham, Surrey, KT11 1LG, England, UK
Phone: +44 (0) 1932 865051   info@brooklands-books.com
www.brooklands-books.com

www.brooklandsbooks.com